Information Systems for Business

An Experiential Approach

Edition 4.0

France Bélanger
Virginia Tech

Craig Van Slyke
Louisiana Tech University

Robert E. Crossler
Washington State University

Prospect
Press

Founded in 2014, Prospect Press serves the academic discipline of Information Systems by publishing essential and innovative textbooks across the curriculum including introductory, emerging, and upper-level courses. Prospect Press offers reasonable prices by selling directly to students. Prospect Press provides tight relationships among authors, publisher, and adopters that many larger publishers are unable to offer. Based in Burlington, Vermont, Prospect Press distributes titles worldwide. We welcome new authors to send proposals or inquiries to Beth.golub@prospectpressvt.com.

Editor: Beth Lang Golub
Production Management: Peter Holm, Sterling Hill Productions
Cover Design: Annie Clark

eTextbook:
Edition 4.0
ISBN: 978-1-943153-87-9
Available from Redshelf and VitalSource

Printed Paperback:
Edition 4.0
ISBN: 978-1-943153-88-6
Available from Redshelf

For more information, visit https://www.prospectpressvt.com/textbooks/belanger-information -systems-for-business-4-0

To all the students who have inspired us
FB, CVS, and REC

To my family and friends for their relentless
support, and to Pierre, for a life worth living
FB

To Tracy, my long-suffering and loving wife, and to each
and every one of the bears for all the joy their hijinks bring
CVS

To Crystal and my children, who bring such joy and
purpose to my life. Thanks for encouraging me through
the process of creating this edition of the text.
REC

About the Authors

Dr. France Bélanger, *Virginia Tech* France Bélanger is the R. B. Pamplin Professor and Tom and Daisy Byrd Senior Faculty Fellow at Virginia Tech. Dr. Bélanger has been teaching information systems (IS) for more than 25 years in the United States, as well as in Canada, Ethiopia, Portugal, and New Zealand. She has published several books and articles on IS education-related topics, including a highly cited book on distance learning. She has won several international awards, including the 2008 IEEE Education Society Research Award. Dr. Bélanger is a proponent of learner-centered approaches to IS education and has developed a number of tools and materials to facilitate such learning. She has taught online and face-to-face classes for undergraduate, graduate, professional, and executive students, in addition to PhD students. She led numerous workshops related to education, learner-centered teaching, and mentoring.

Dr. Craig Van Slyke, *Louisiana Tech University* Craig Van Slyke is the Mike McCallister Eminent Scholar Chair in Computer Information Systems at Louisiana Tech University. Prior to joining Louisiana Tech, he was dean and professor of computer information systems at the W. A. Franke College of Business at Northern Arizona University. Dr. Van Slyke has more than 20 years of university teaching experience, primarily in the area of information systems. He has published extensively on information systems education and has coauthored four IS textbooks. Dr. Van Slyke has held leadership positions in Southern Association for Information Systems and Association for Information Systems Special Interest Group on Education. In addition, he has applied the principles of learner-centered education to a variety of courses, including the Introduction to IS course for undergraduates and the Core IT course for MBA students.

Dr. Robert E. Crossler, *Washington State University* Robert E. Crossler is the Philip L. Kays Distinguished Associate Professor in Information Systems and the Management, Information Systems, and Entrepreneurship Department chair in the Carson College of Business at Washington State University. Dr. Crossler has more than 15 years of university teaching experience, primarily in information systems. He has published on information systems education and facilitated innovative teaching workshops. In addition, he has applied the principles of learner-centered education to a variety of courses, including the Computer Concepts course for freshmen, the Introduction to IS course for juniors and seniors, the Core IT course for MBA students, database courses at the undergraduate and master's levels, and information security courses at the undergraduate level.

Contents

Preface   **xv**

Mission/Purpose   xv
Target Audiences and Approach   xv
Key Features of the Book   xvii
Supplemental Materials   xviii
Changes in Edition 4.0   xix
Substantive Detailed Changes   xix
Overview of the Content and Organization   xxi
Acknowledgments   xxiv

SECTION 1 · INTRODUCTION   **1**

Chapter 1 | The Value of Information   **1**

Learning Objectives   1
Chapter Outline   1
Information and Emerging Technologies   2
Focusing Story: How Information Cost Me (a Lot of) Money   2
Learning Activity 1.1: The Rapid Growth of Information   *3*
Learning Activity 1.2: A Day in My Technology-Free Life   *4*
Stats Box 1.1: Top IoT Application Areas   5
Learning Activity 1.3: Autonomous Vehicles   *7*
Data, Information, Knowledge, and Wisdom   9
Learning Activity 1.4: How I Use Information   *11*
Information Literacy   11
Uses of Information   12
Learning Activity 1.5: How Businesses Use Information   *12*
Business Example Box 1.1: Promising Information Careers   14
Information and Your Career   14
Chapter Summary   16
Review Questions   16
Reflection Questions   16
Additional Learning Activities   17
References   18
Glossary   18

Chapter 2 | Introduction to Information Systems   **19**

Learning Objectives   19
Chapter Outline   19

Why All Business Professionals Need to Be Information
Systems Managers   19
Focusing Story: Registering for Classes in 1981   20
Learning Activity 2.1: Identifying My Information Systems   *21*
Overview of Systems   21
Learning Activity 2.2: Components of a System   *22*
Learning Activity 2.3: Components of a Business Information System   *23*
Foundations of Information Systems   23
How Information Systems Help Us Deal with Information   27
Stats Box 2.1: Growth of Applications in Apple's App Store   28
Business Example Box 2.1: How Uber Provides On-Demand Service
with Information Systems   29
Learning Activity 2.4: Information Systems and Change   *30*
How Information Systems Facilitate Organizational Change   30
Business Example Box 2.2: AgTech—Making Farming More
Efficient and Sustainable   32
Common Information Systems   33
Chapter Summary   35
Review Questions   35
Reflection Questions   36
Additional Learning Activities   36
References   37
Glossary   37

Chapter 3 | Evaluating Information   **39**

Learning Objectives   39
Chapter Outline   39
Being a Smart Information Consumer   39
Focusing Story: The Personal Consequences of Misinformation   40
Learning Activity 3.1: How Good Is This Information?   *41*
Stats Box 3.1: The Cost of Poor Information Quality   42
Business Example Box 3.1: Business Impacts of Information Quality   43
Information Overload and the Need to Evaluate Information   43
Learning Activity 3.2: My Online Life and Information Overload   *44*
Dealing with Information Overload   45
Information Quality   45
Learning Activity 3.3: Dimensions of Information Quality   *46*
Business Example Box 3.2: Information Quality as an Ethical Issue   48
Evaluating Information   48
Learning Activity 3.4: Evaluating Information Sources   *49*
Chapter Summary   52
Review Questions   52
Reflection Questions   53

Additional Learning Activities   53
References   55
Glossary   55

Chapter 4 | Gaining Strategic Value from Information   57

Learning Objectives   57
Chapter Outline   57
Strategic Information Systems   57
Focusing Story: Health Care on My Smartphone   58
Strategic Planning Process   59
    Learning Activity 4.1: Trends in Top Management Concerns *61*
Frameworks for Strategic Information Systems   63
SWOT Analyses in Practice   64
Evaluating Strategic Initiatives   69
    Learning Activity 4.2: What Does This Company Need for Success? *71*
Stats Box 4.1: CSFs for Improving Business in Australia   72
Hypercompetition: Sustainability of Competitive Advantage   72
    Learning Activity 4.3: Competitive Advantage at TRIPBAM *73*
    Learning Activity 4.4: Disrupting an Established Market! *74*
Chapter Summary   74
Review Questions   75
Reflection Questions   76
Additional Learning Activities   76
References   77
Glossary   78

SECTION 2 · MANAGING INFORMATION   81

Chapter 5 | Storing and Organizing Information   81

Learning Objectives   81
Chapter Outline   81
Focusing Story: The Database behind Facebook   82
Stats Box 5.1: Digital Data   82
    Learning Activity 5.1: Data for an Amazon Order   *83*
Overview of Relational Databases   83
Databases versus Spreadsheets: When to Use a DBMS   84
    Learning Activity 5.2: Connecting Data Elements   *85*
Database Diagrams   88
iTunes Match   90
Online Databases   91
    Learning Activity 5.3: Finding Business Databases Online *91*
Big Data   92

Learning Activity 5.4: Fitness Trackers   *93*
Big Data at Facebook   94
Chapter Summary   96
Review Questions   97
Reflection Questions   97
Additional Learning Activities   97
Glossary   98

Chapter 6 | Analyzing Information for Business Decision-Making   **101**
Learning Objectives   101
Chapter Outline   101
Focusing Story: How a Spreadsheet Helped Me Make a Hard Decision   102
The Importance of Good Decision-Making Skills   102
Learning Activity 6.1: Information and Decision-Making   *103*
Learning Activity 6.2: What Kind of Information Do I Need?   *103*
Using Information for Decision-Making   104
Types of Decisions   104
Business Example Box 6.1: Why Do Managers Make Bad Decisions?   105
A Decision-Making Process   106
Learning Activity 6.3: Requirements, Goals, and Criteria   *109*
Information Retrieval and Analysis Tools   111
Learning Activity 6.4: "What If" I Get a 75% on the Final?   *113*
Chapter Summary   116
Stats Box 6.1: The Cost of Spreadsheet Errors   117
Review Questions   118
Reflection Questions   119
Additional Learning Activities   119
References   121
Glossary   121

Chapter 7 | Transmitting Information   **123**
Learning Objectives   123
Chapter Outline   123
Introduction and Definitions   123
Focusing Story: The Rise of Mobile Wi-Fi: Hotspots   124
Learning Activity 7.1: How Does That Car Drive Itself?   *125*
Types of Networks   126
The Internet   129
Learning Activity 7.2: Broadband in Developed Countries   *129*
Learning Activity 7.3: Who Is Using the Internet?   *130*
Business Example Box 7.1: Internet Application of the Day:
Distance Learning   133
Learning Activity 7.4: A Future Internet? Internet2 and Business   *134*

Networking Architectures   135
   Learning Activity 7.5: Architecture and Principles   *141*
Communicating Information in Modern Organizations   142
The Power of Wikinomics (Crowdsourcing)   143
Stats Box 7.2: Diffusion of Innovations Timeline   145
   Learning Activity 7.6: Web 3.0 versus the Internet of Things   *146*
Chapter Summary   146
Review Questions   147
Reflection Questions   148
Additional Learning Activities   148
References   149
Glossary   150

Chapter 8 | Securing Information   **153**

Learning Objectives   153
Chapter Outline   153
Introduction and Definitions   153
Focusing Story: My Mac Is More Secured than Your Windows-
   Based PC!   154
   Learning Activity 8.1: How Protected Is Your Computer?   *157*
Information Security Threats   158
   Learning Activity 8.2: Detecting Phishing   *160*
   Learning Activity 8.3: What Do Hackers Break into the Most?   *162*
Business Example Box 8.1: Even Big Technology Companies Can
   Be Hacked   163
Business Example Box 8.2: Costs of Data Breaches   167
Security Technologies and Solutions   168
   Learning Activity 8.4: How Strong Is Your Password?   *170*
   Learning Activity 8.5: Biometrics on the PC   *173*
Stats Box 8.1: Security Policy Violations   178
   Learning Activity 8.6: Breaking the Encryption   *180*
Business Example Box 8.3: Expert Advice on Protecting Your Home
   Wireless Network   181
   Learning Activity 8.7: IoT Security   *182*
   Learning Activity 8.8: Where Is the Security?   *183*
Chapter Summary   183
Review Questions   185
Reflection Questions   185
Additional Learning Activities   186
References   187
Glossary   188

Chapter 9 | Protecting the Confidentiality and Privacy of Information   **193**

Learning Objectives   193
Chapter Outline   193
Focusing Story: Perceptions of Anonymity   194
Introduction and Definitions   195
Information Privacy Threats   195
Data Collection   195
   Learning Activity 9.1: Privacy and Smartphones   *196*
   Learning Activity 9.2: Concern for Information Privacy   *197*
   Learning Activity 9.3: Privacy Pizza   *198*
Consequences of Privacy Violations   198
Stats Box 9.1: Perceptions of Monitoring   199
Technologies and Solutions for Information Privacy   200
Stats Box 9.2: Fraud Type from Identity Theft   201
Stats Box 9.3: The Black Market for Stolen Identities   202
Anonymous Browsing   203
   Learning Activity 9.4: Surveillance Societies around the World   *204*
   Learning Activity 9.5: Privacy Policy Creation   *205*
Business Example Box 9.1: Privacy and Facebook   207
Government Information Privacy Regulations   208
Business Example Box 9.2: GDPR in Practice   209
Mobile Information Privacy   210
   Learning Activity 9.6: Why Your Adviser Cannot Talk to Your Parents   *210*
   Learning Activity 9.7: PAPA, Privacy Policies, and FERPA   *211*
IoT and Privacy   211
Privacy and Ethics   212
Ethical Decisions   212
   Learning Activity 9.8: Making Ethical Decisions   *213*
Relationship between Security and Privacy   215
Chapter Summary   215
Review Questions   216
Reflection Questions   216
Additional Learning Activities   216
References   218
Glossary   219

SECTION 3 · USING INFORMATION   **221**

Chapter 10 | Developing Information Systems   **221**

Learning Objectives   221
Chapter Outline   221
Focusing Story: The $6 Billion Software Bug   222

Time, Cost, and Quality   223
Software Development Methodologies   223
   Learning Activity 10.1: Determining Requirements   *224*
Traditional Systems Development Life Cycle   224
   Learning Activity 10.2: Advantages and Disadvantages of the SDLC   *229*
Alternative Methodologies   229
Stats Box 10.1: Information Technology Failures   230
   Learning Activity 10.3: Developing by Modeling   *231*
   Learning Activity 10.4: Comparing the Methods   *234*
Build or Buy Decision   234
   Learning Activity 10.5: App Developers Needed!   *235*
Using Open Source Software in Business   237
   Learning Activity 10.6: How Open Source Software Impacts Build
     versus Buy   *237*
Business Example Box 10.1: Open Android versus Closed iOS   238
Outsourcing Information Systems   239
   Learning Activity 10.7: What to Outsource   *239*
Business Example Box 10.2: Reversing the Outsourcing Decision   241
Chapter Summary   243
Review Questions   244
Reflection Questions   245
Additional Learning Activities   246
References   247
Glossary   247

Chapter 11 | Information-Based Business Processes   **249**

Learning Objectives   249
Chapter Outline   249
What Is a Process?   249
Focusing Story: Improving Processes Is for Everyone!   250
   Learning Activity 11.1: How Many Steps in This Process?   *251*
Process Modeling   252
   Learning Activity 11.2: Model This!   *252*
Technology and Processes   254
Process Improvement   256
Business Example Box 11.1: Banking Processes Improved via Blockchain   258
   Learning Activity 11.3: Informal Business Processes   *258*
   Learning Activity 11.4: Redesign This!   *258*
Business Example Box 11.2: Quicken Loans Rocket Mortgage:
   Moving the Mortgage Process Online   260
Chapter Summary   261
Review Questions   261
Reflection Questions   262

Additional Learning Activities   262
References   263
Glossary   263

Chapter 12 | Enterprise Information Systems   265

Learning Objectives   265
Chapter Outline   265
Focusing Story: Supply Chain Innovations at Walmart   266
Enterprise Systems   267
  Learning Activity 12.1: Finding the Components of an Enterprise System 268
Enterprise Resource Planning (ERP)   270
  Learning Activity 12.2: The Online Beer Game   273
Stats Box 12.1: Top ERP Vendors   274
Supply Chain Management Systems   274
Business Example Box 12.1: ERP Is Not Always a Horror Story:
  Cisco Implementation on Time and Budget   275
Stats Box 12.2: Supply Chain's Impact on Company Performance   277
  Learning Activity 12.3: Blockchain and Supply Chain Management 278
Customer Relationship Management Systems   278
  Learning Activity 12.4: Customer Relationship Management   280
Stats Box 12.3: CRM Vendors   281
  Learning Activity 12.5: Self-Servicing   282
Business Example Box 12.2: DUFL Gains Competitive Advantage
  through Customer Service   283
Customer-Managed Interactions (CMI)   284
Warehouse Automation and Robotics   284
Chapter Summary   284
Review Questions   285
Reflection Questions   286
Additional Learning Activities   286
References   287
Further Readings   288
Glossary   288

Chapter 13 | Information for Electronic Business   291

Learning Objectives   291
Chapter Outline   291
Introduction to E-Business   291
Focusing Story: The iPod and the Music Industry   292
  Learning Activity 13.1: Why Is E-Business Important?   293
Stats Box 13.1: Retail E-Commerce in the United States   295
E-Business Models   296
Business Example Box 13.1: Adapting Its Business Model: Amazon.com 297

E-Business Enablers 298
E-Business Impacts 301
Design for E-Business 302
Learning Activity 13.2: Why Would I Trust Them or Buy from Them? 303
Business-to-Business (B2B) 303
Learning Activity 13.3: B2C versus B2B 304
Search Engine Optimization 306
Learning Activity 13.4: Where Is My Web Page? 307
Trends in E-Business 309
Stats Box 13.2: The Global Nature of E-Business 310
Chapter Summary 310
Review Questions 311
Reflection Questions 312
Additional Learning Activities 313
References 313
Glossary 314

Chapter 14 | Information and Knowledge for Business Decision-Making   317

Learning Objectives 317
Chapter Outline 317
Focusing Story: Managing Knowledge by Texting 318
Knowledge Management 318
Learning Activity 14.1: How You Manage Knowledge 319
Why Managing Knowledge Is Important 319
Main Processes for Knowledge Management 321
Business Example Box 14.1: The Issue of Trust in Knowledge
  Management 325
Learning Activity 14.2: Wikis for Managing Knowledge 326
Knowledge Management Technologies 326
Decision Support Systems and Collaboration Systems 328
Learning Activity 14.3: Using a Decision Support System 329
Learning Activity 14.4: Slack for Student Team Projects 331
Business Analytics 332
Business Example Box 14.2: Explaining Profitability Problems
  through Visualizations 336
Learning Activity 14.5: Effective Visualizations 337
Artificial Intelligence 341
Business Example Box 14.3: Applications of AI 343
Business Example Box 14.4: Agricultural Robots 345
Learning Activity 14.6: Are AI and Robots Good or Bad for Society? 346
Chapter Summary 346
Review Questions 347
Reflection Questions 348

Additional Learning Activities   348
References   349
Glossary   350

SECTION 4 · APPENDICES

Note that the following appendices are included in the e-textbook only. The appendices are also posted on the title's website at https://www.prospectpressvt .com/textbooks/belanger-information-systems-for-business-4-0/. (Scroll to the red horizontal menu bar, then click on "Student Resources.") Users of the printed textbook can freely access the appendices at the website.

Appendix A: Computer Hardware
Appendix B: Computer Software
Appendix C: Access Fundamentals
Appendix D: Advanced Access
Appendix E: Advanced Database Concepts
Appendix F: Excel Fundamentals
Appendix G: Advanced Excel
Appendix H: Networking Basics
Appendix I: Security and Privacy
Appendix J: Funding Information Technology
Appendix K: Managing IS Projects

Preface

Mission/Purpose

An interesting paradox often exists in introductory information systems courses. Information systems touch almost every aspect of students' lives, yet students are often detached and uninterested in the introductory course. Often they do not see the point of taking such a class, unless they are information systems majors. This book is meant to generate an interest in information systems for everyone, by taking a learner-centered approach and providing experiential learning activities and reflection questions throughout the book. In addition to being effective in terms of learning outcomes, learner-centered education is also more satisfying for the learner. The focus of a learner-centered class is on a learner-teacher partnership with shared responsibility for learning. The method not only introduces important concepts but also deepens the learning through active application of the concepts and reflection upon what has been learned. The learner-centered approach fits well with the characteristics of millennial-generation students who are team oriented, value continuous learning, and seek frequent feedback.

The purpose of the book is to engage students while helping them become intelligent consumers of information. The book puts information center stage to allow the relevance of information systems to be put into context, thus increasing student interest. Our goal is to help students understand that they will be using information throughout their personal and professional lives. Information systems not only produce information but also help us make better use of information. By focusing on information rather than systems, students are grounded in the end goal and are better able to understand why knowledge of information systems is important. This translates into greater engagement; students are more engaged when they understand the relevance of what they are learning.

Target Audiences and Approach

The book is primarily directed at students who are taking the Introduction to IS course as juniors or seniors, or even first-time non-IS master's students. Students who are not IS/IT majors often fail to see the value in the course, especially if the course is traditionally structured and focused on conceptual knowledge. Our approach to this text is based on our experience in reaching these students. Allow us to describe three experiences that have helped shape our approach.

I (Bélanger) was asked to take over and improve our Introduction to Information Systems master's course targeted at master of accountancy (and a few MBA) students. For years, the students had been complaining about the course, which is a requirement for completion of their degree. When I first taught the class, I quickly realized that the students did not want to be there and did not care about the content. I did bring in cases, as I had used the case approach in most of my master's classes. I quickly realized that the case discussion went well but that the remainder of the "lecture" material was pretty much lost on them. In redesigning the class for the second semester, I decided to include "practice" slides for every subtopic, resulting in shorter lectures and "practice" elements throughout. The most obvious impact was that students really wanted to know why certain things happened when they acted on some technology-related issue, be it developing a privacy policy, testing their computer's security, or finding information about an emerging technology. The resulting discussions were animated, and the students came to class, even though I made attendance nonmandatory! The most blatant example of the impact of practice on student interest is the second exercise of the semester (Learning Activity 1.2), in which students have to spend five or six hours without using any information technology and then blog about it. I am always surprised every time to see the depth of their realization about the importance of IT in their lives . . . and in their future careers. It starts the semester on a very different foot from what I had experienced with the traditional lecture (and case) approach.

I (Van Slyke) was asked to teach a junior-level Introduction to IS course in a large-section format (200+ students per section). The first semester was somewhat of a disaster. It was the first time I had ever received evaluations below the mean. For the next semester, two changes were made. One change was to add a third exam. The more meaningful change was to have an activity in every class meeting. Instead of a 75-minute lecture, each meeting had an activity approximately halfway through the class. The activities typically lasted 5 or 10 minutes. After the activity, I led a discussion based on the activity. For example, during the class on supply chain management, students were asked to get into groups of two or three and come up with three ways in which supply chain management (SCM) systems and enterprise resource planning (ERP) systems are similar and three ways in which they differ. During the activity, my fellow instructor and I (the course was team taught, although I was primarily responsible for the lectures) would wander around the auditorium answering student questions. We knew we had a winner when we saw many students flipping back through their notes on ERP, which had been covered a few weeks earlier. After the activity, I would ask students to share their findings. The findings were insightful and showed a solid grasp of the basic concepts of *both* SCM and ERP systems. The results of the more active, learner-centered class sessions were remarkable. Not only did students perform better on the exams, but our evaluations went up almost a full point on a five-point scale as well. A nice side benefit was that class sessions were much more enjoyable for the students and the instructors.

Finally, I (Crossler) had spent the first few years of my career teaching a freshman-level introduction to information systems course. I followed an approach I like to refer to as "death by PowerPoint." After a few years of teaching this course, I realized that I was bored by this course as well and could only imagine what my students felt. It was at this point that I learned about experiential learning and started thinking about how to make the study of information systems much more engaging. Almost all undergraduate students seem to like technology, so why couldn't they like learning about how it is used in business? As someone who spent a lot of time researching in information security and privacy and knowing that students enjoyed their smartphones, I started creating ways for students to discover the security and privacy issues that were present on a device they used every day. Once these were uncovered, learning about the threats and opportunities to protect themselves became an activity in which they were much more engaged. When the opportunity to become an author on the second edition of this textbook, I jumped at the opportunity as the approach to the textbook was consistent with how I thought information systems should be taught, through engagement. As a result, teaching the introductory courses with this textbook livened up the classroom and helped more students to see enjoyment in learning about this exciting and fast-changing field.

These experiences led us to embrace a learner-centered, active approach in all our classes (both graduate and undergraduate). Such an approach seems to be especially effective for today's students. They seem to appreciate being an active participant, rather than a passive vessel, in the learning process.

Key Features of the Book

The book is designed to support an active, learner-centered approach, with the guiding model of "praxis," or putting learning into practice. Students learn best when they actively apply their learning rather than simply learn to parrot facts and concepts. The book facilitates a learning cycle of "learn-do-reflect," which closely follows the ideas of experiential learning.

To ensure continued relevance, the book will be updated on a frequent basis, since information technologies and the use of information in business will continue to evolve. An important feature of the book is that it is concise but complemented with substantial additional (optional) material in the Appendices. The conciseness of the book helps students who today are used to receiving information in small chunks; they live in a world of Facebook wall postings, tweets, and text messages. Forty- and fifty-page chapters, filled with details, are incongruent with their world. As a result, we offer tightly focused, concise chapters that provide the necessary information without being bogged down in unnecessary detail. Students will actually fill in many of the details themselves through the active learning exercises and online materials. Our goal is to help students understand how to utilize information systems rather than just regurgitate facts. As such, the chapters are not meant to give every detail on every topic but rather provide broad guidelines for students to explore the topics of interest.

The activities and content of the text embrace a constructivist perspective to learning. We try to connect the new information to what most students will already know, either from general life experience or from material covered earlier in the course. For example, the SCM/ERP activity described above connected the new SCM concepts to the ERP concepts covered earlier in the course.

In the text, we implement the concepts of experiential learning in the following ways:

- *Concise, focused chapters:* Textbooks often try to cover all possible aspects of a topic. While this is appropriate for many courses, in the introductory course, we believe it is better to focus on key information. Because of this, our chapters are concise and focused.
- *Focusing stories:* Each chapter opens with a focusing story accompanied by focusing questions that ask students to reflect on the story. This approach helps students ground the content in a concrete example. It also demonstrates the relevance of the material, which increases student engagement. The topics are specifically chosen to be relevant and understandable to most students.
- *Designed for active learning:* For each major topic, we provide activities that are designed to help students deepen their learning through application.
- *Reinforcing end-of-chapter materials:* Each chapter provides additional learning activities and reflection and review questions that help students synthesize the various concepts in the chapter. There are three types of reinforcing elements. Review questions are straightforward questions designed to help students check their declarative knowledge of the material. Reflection questions require deeper thinking. Many of these questions require synthesis of multiple concepts. Additional learning activities are active-learning exercises that reinforce the material in the chapter through application and synthesis.
- *Detailed instructor's manual:* An important element of this book is the instructor's manual, which provides guidance to instructors on how to use the book in various settings and on how to apply the learning-centered approach to the material provided in the chapter. The instructor's manual provides detailed teaching notes and suggestions regarding the duration of each section and learning activity.
- *Online resources for active learning:* Links to tutorial videos are available on the book's website. (Click to https://www.prospectpressvt.com/text books/belanger-information-systems-for-business-4-0, scroll to the red horizontal menu bar, then click on "Student Resources.")
- *Cases:* Recommended cases from sources such as the Ivey Business School or the Harvard Business School are provided for some of the chapters so that instructors can also use this book in a case-based class.

Supplemental Materials

In addition to the instructor's manual, supplemental materials include test banks and slide decks. These resources are specifically designed to support the active-learning approach. In addition, there are test questions that are constructed to assess deeper levels of learning beyond simple declarative knowledge. The instructor's material is available at the book's website. (Click to https://www.prospectpressvt.com/textbooks /belanger-information-systems-for-business-4-0, scroll to the red horizontal menu bar, then click on "Instructor Resources." If you do not have a password, you will be able to request one here. Prospect Press qualifies all requests to confirm they are from instructors.)

Changes in Edition 4.0

We are excited to provide this new edition, which has been created to provide up-to-date, relevant material in this ever-changing world of information systems.

Substantive detailed changes (beyond text editing) are identified below. Overall, we updated all statistics and links in every chapter, changed numerous business example boxes to include more current and relevant examples, and updated or changed most of the focusing stories and stats example boxes. We also created new focusing stories for some of the chapters and appendices. Several major topics were added, including mobile privacy, two-factor authentication, blockchain, ransomware, and bitcoins. Some outdated material was removed. We also updated the hands-on software activities and examples to the most recent versions of the relevant software. Finally, we created several new learning activities to provide instructors with more opportunities to apply the learner-centered approach in their classes.

As of December 2020, all URLs have been checked and updated as necessary.

Substantive Detailed Changes

- *Chapter 1:* A new section on emerging technologies was added, as was Learning Activity 1.3, which concerns autonomous vehicles.
- *Chapter 2:* Learning Activity 2.4 (Information Systems and Change) was updated to use the example of universities' responses to COVID-19. The business example about Coopers Brewery (Business Example Box 2.2) was replaced with a more current discussion about agricultural technology (AgTech).
- *Chapter 3:* The focusing story (Biased Information in a Trusted Source) was changed to a more current story pertaining to misinformation from a high-ranking government official. Business Example Box 3.1, was changed from résumé fraud to the business impacts of information quality.
- *Chapter 4:* Learning activities and examples throughout the chapter were updated to timely and student-relatable contexts, statistics were updated, new examples were added, clarity was added to the priority matrix and additional learning activities were updated by removing one older activity and adding two new cases.
- *Chapter 5:* The "Storing Data" section was updated to include a discussion of data lakes. The "Retrieving and Disseminating Data" section was updated to include data analytics.

- *Chapter 6:* The focusing story was changed to be about making a hard decision related to quitting a job and returning to graduate school.
- *Chapter 7:* The chapter was completely revised with Internet of Things concepts weaved throughout. Terminology was updated to include hotspots, 5G broadband, and IPv6 and remove old networks like Personal Area Networks. Statistics were all updated. Several examples were modified or replaced with content related to distance learning and remote work as a result of the 2020 pandemic. Business process as a service was added. Outdated learning activities were also removed or updated.
- *Chapter 8:* The chapter was completely revised to include discussions of new devices. IoT has also been weaved in to various discussions throughout the chapter. Definitions were added for clarity. All learning activities were revised to include current tools and concepts. All statistics were revised. Several business example boxes were replaced with new examples. Additional learning activities were updated with cases on cyberespionage, activities with BYOD, and on IoT security.
- *Chapter 9:* The focusing story was updated with several new examples. IoT was included in several examples throughout the chapter. All statistics were updated. Tables were updated with recent tools. GDPR and CCPA were added to the privacy law discussion. A new section on IoT privacy was added. Accordingly, some of the reflection questions were updated. Two additional learning activities were updated or changed to include CCPA, GDPR, and privacy outside of the USA and Europe.
- *Chapter 10:* The focusing story was updated with 2018–2020 examples. Statistics were all updated. A new learning activity (10.3) was added, which uses Mendix for modeling. The spiral model was removed and only a brief mentioned was retained. A new learning activity was added on salaries and need for application developers. Learning activities were renumbered accordingly. The text was updated with discussions of open source web servers and with more recent examples throughout. Additional Learning Activity 10A6 was updated to focus on IoT development and a new activity (10.7) was added, which focuses on outsourcing software development.
- *Chapter 11:* The business process improvement section was updated to provide more clarity. Business Example Box 11.1 was added to show how the banking process is improved using block chain.
- *Chapter 12:* The focusing story was updated to include new information regarding how Wal-Mart manages their supply chain. Learning Activity 12.3, which discusses blockchain and supply chain management, was added.
- *Chapter 13:* The e-business section of this chapter was substantially rewritten to better explain the integrated nature of business and consumer e-business. This resulted in the removal of the Mobile Business and Global E-Business sections.

• *Chapter 14:* The discussion of executive information systems was changed to a discussion of dashboards. Learning Activity 14.4, which relates to the use of Slack for student teams, was added. The Business Intelligence section was changed to Business Analytics, and sections on business analytics methods and information visualization were added. A section on artificial intelligence (AI) was added. This section includes Business Box 14.3, which provides examples of business-related AI applications, and also includes Business Box 14.4, which discusses agricultural robots. Learning Activity 14.6, which concerns the societal impacts of AI and robotics, was also added.

Overview of the Content and Organization

The book is organized around three main sections of content. Section I provides introductory material, Section II focuses on topics related to managing information, and Section III discusses various uses of information. Section IV comprises a set of appendices that provide hands-on coverage of Microsoft Office tools and other topics that may be used to support and extend the main chapters.

Each chapter starts with a focusing story and includes learning activities for each major topic. The chapters all end with a summary of the chapter, review questions, reflection questions, additional learning activities, references, further readings (if applicable), and a glossary.

A brief summary of the chapters follows.

Section I: Introduction

Chapter 1: The Value of Information

Chapter 1 sets the stage for the rest of the book by discussing information and its role in business. The focus here is on understanding what information is and how it is used both personally and by businesses. We classify information use into three areas: communication, process support, and decision-making. This provides the framework that is used throughout the book. We also discuss the wisdom hierarchy (data, information, knowledge, wisdom) that provides further grounding for students.

Chapter 2: Introduction to Information Systems

This chapter gives a high-level view of information systems. A brief overview of key systems concepts is also provided. This overview gives students the foundation knowledge necessary to better understand information systems. In addition, the chapter provides an overview of information and information systems as strategic resources. The chapter also builds a case for the importance of understanding information systems even if one is not an information systems professional.

Chapter 3: Evaluating Information

With the tremendous amount of information available today, it is especially import-ant to be able to evaluate information. This chapter provides an overview of informa-tion quality and presents a methodology for evaluating information. Also, it dis-cusses how to deal with information overload, information quality, and source credibility.

Chapter 4: Gaining Strategic Value from Information

The material in this chapter expands the concepts introduced in the first three chap-ters to explore the strategic value of information. The chapter provides an overview of the key frameworks and concepts related to the role of information and informa-tion systems in the creation and maintenance of competitive advantage. The chapter includes popular frameworks that can be used to identify strategic initiatives and others that can be used to evaluate such initiatives. It also includes a discussion of whether competitive advantage can be sustained.

Section II: Managing Information

Chapter 5: Storing and Organizing Information

Databases are the core of information systems. Many business students have consid-erable experience with using information systems, but few have any understanding of the databases that sit at the core of these systems. This chapter serves two pur-poses. First, it introduces and provides a high-level overview of relational databases, database management systems, and data warehousing. The intent is to provide foun-dational knowledge that is useful for users rather than the in-depth knowledge nec-essary for IS professionals. Second, the chapter discusses how to retrieve information from databases and online resources. (Note that more detail regarding databases and database management systems is available in Section IV: Appendices.)

Chapter 6: Analyzing Information for Business Decision-Making

Making good decisions requires information, but that is not enough. You must also be able to analyze information. In this chapter, we want to help you understand the basics of business-oriented decisions and the process by which they are made. In Chapter 14, you will learn about systems specifically designed to support decision-making. Here the focus is on understanding more about the decision-making process and tools that can help you retrieve and analyze information as part of that process.

Chapter 7: Transmitting Information

This chapter covers basic networking concepts, including the Internet, Web, and Internet of Things. To demonstrate the relevance of the topics covered, the chapter opens with a discussion of cloud computing. The chapter concludes with consider-ations and tools for collaboration, which have become very important to business, with its increasingly globally distributed workforce.

Chapter 8: Securing Information

This chapter provides an overview of information security concepts and issues and the technologies used to handle information security. This includes discussions of personal information security, security threats, security technologies, and approaches.

Chapter 9: Protecting the Confidentiality and Privacy of Information

This chapter gives an overview of information privacy and the technologies and techniques used to help protect information privacy. The ethics of privacy are also discussed. Issues are discussed from both the individual and organizational perspectives.

Section III: Using Information

Chapter 10: Developing Information Systems

In this chapter, we provide a high-level view of the information systems development process. We focus on the perspective of an end user who may be involved in the development process. We achieve this by helping students understand the goals, processes, and roles involved with systems development.

Chapter 11: Information-Based Business Processes

This chapter discusses how information supports business processes. The concept of a process is introduced, along with some familiar examples. Emphasis is placed on how information and its flow are critical to the successful completion of processes. In addition, the idea of redesigning processes to make use of information technology is discussed.

Chapter 12: Enterprise Information Systems

This chapter includes a general overview of enterprise information systems. ERP, supply chain management systems, and customer relationship management (CRM) systems are discussed. The chapter makes the point that virtually all business professionals interact with one or more of these systems. Examples of how each type of system is applied are also included.

Chapter 13: Information for Electronic Business

Because of its importance to individuals and businesses, we devote an entire chapter to e-business. Common e-business models are discussed. A brief discussion of e-government is also included.

Chapter 14: Information and Knowledge for Business Decision-Making

Making good decisions requires good information and good decision-making processes, skills, and tools. This chapter ties together various elements of the book by showing how information is central to the decision-making process. The chapter provides students with an understanding of what is necessary for good decision making and introduces a decision-making methodology. Because ultimately an organization's only sustainable competitive advantage lies in how its employees apply knowledge to business problems, in this chapter we discuss the concept of knowledge and various approaches to knowledge management.

Acknowledgments

Writing a book is always a major undertaking, and we are fortunate to have received the help of many individuals. We wish to thank Beth Lang Golub at Prospect Press and the production team at Sterling Hill Productions, led by Peter Holm. Also, thank you to the following individuals who responded to the User Survey to provide feedback for Edition 4.0:

Mariano Álvarez Diente, IE University
Steve Clements, Eastern Oregon University
Mike Dohan, Lakehead University
Judith Gebauer, University of North Carolina Wilmington
C. Matt Graham, University of Maine, Maine Business School
Bob Gregory, Bellevue University
Guillermo de Haro, IE University
Divakaran Liginlal, Carnegie Mellon University
Radney Redd, Florida Southern College
George Schell, University of North Carolina Wilmington
Efthimios Tambouris, University of Macedonia
Paul Witman, Cal Lutheran University

We also are grateful to the following individuals for their insightful suggestions and ideas for prior editions of this text:

Linda Jo Calloway, Pace University
Mandy Yan Dang, Northern Arizona University
Richard Egan, New Jersey Institute of Technology
C. Matt Graham, University of Maine
Bob Gregory, Bellevue University
Brian Hall, Champlain College
Beverly Kahn, Suffolk University
Mike Pangburn, University of Oregon
Dan Phelps, Carnegie Mellon University Qatar
Joseph Pugh, Immaculata University
David Reavis, Texas A&M Texarkana
Radney Redd, Florida Southern College
Ali Vedadi, Mississippi State University
Gavin Yulei Zhang, Northern Arizona University
Paul Witman, Cal Lutheran University

The Value of Information

Learning Objectives

By reading and completing the activities in this chapter, you will be able to:

- Describe several important emerging technologies, including 5G, the Internet of Things, and artificial intelligence
- Compare and contrast data, information, and knowledge
- Explain the concepts of connectedness and usefulness as they relate to information
- Name the skills required for information literacy
- Discuss the importance of information literacy in a business career
- Name and describe the three main uses of information in businesses
- Compare and contrast information systems careers and information analysis careers

Chapter Outline

Information and Emerging Technologies
Focusing Story: How Information Cost Me (a Lot of) Money
    Learning Activity 1.1: The Rapid Growth of Information
    Learning Activity 1.2: A Day in My Technology-Free Life
    Learning Activity 1.3: Autonomous Vehicles
Data, Information, Knowledge, and Wisdom
    Learning Activity 1.4: How I Use Information
Information Literacy
Uses of Information
    Learning Activity 1.5: How Businesses Use Information
Information and Your Career

This book is about information systems. Although it seems obvious, it is worth noting that there are two parts to "information systems": information and systems. All too often, people tend to get caught up in the "systems" part (the technology) and do not pay enough attention to the "information" part. But the information is really the critical part. If the information is not valuable, there is not much need for the system. Because of this, we begin the book by focusing on the value of information.

However, it is also important to understand that knowing about technology can help you better deal with information. Information technology helps us organize, locate, and use information. So while we start the book by focusing on information, much of the book helps you understand the technologies and systems that can help you deal with information.

Information and Emerging Technologies

The world of technology changes rapidly, bringing equally rapid changes in how we live and work. The homes and workplaces of thirty years ago would seem strange today. You would not see people checking email and social media on their mobile phones. There would be no digital assistants ordering pizza at the sound of your voice. When driving to an unfamiliar place, you would consult a map—GPS was

FOCUSING STORY

How Information Cost Me (a Lot of) Money

Here is a personal story that illustrates the value of good information or, more accurately, the consequences of *not* having the right information. After running my own business for several years, I decided that it was time to try something different. One of my clients wanted to expand their use of computers to a new department and needed someone with my capabilities. Since they were one of my long-term clients, I knew the company and its employees quite well. The company and its needs were a good match for me, so I discussed the possibilities with the president of the company. He was very enthusiastic about my coming to work for him. At the end of our discussion, he asked me my salary requirements. I knew what salary I needed to meet my living expenses and I gave him a figure. He immediately agreed; I instantly knew that I made a mistake, one that would cost me quite a bit of money. My mistake was in not gathering information on the market salaries for that particular type of job. Had I been armed with this simple bit of information, I could have increased my salary by at least 20%. That is quite a bit of money to leave on the table! The thing to remember about this little tale is that better information usually results in better outcomes.

While this might seem like a simple example, it illustrates how being "information deficient" can have serious consequences. The decisions that business professionals face every day often have significant impacts. Making good decisions requires having good information. With the rapid pace of today's business world, being successful in a business career requires being able to understand what information is required and being able to efficiently and effectively gather, evaluate, and analyze that information. Because of this, the theme of this book is information. We will help you understand information and the systems businesses use to deal with the vast quantity of information they generate and use. Our goal is for you to build the knowledge and skills you will need to be successful in your career.

Focusing Questions

1. What were the consequences of not having competitive salary information?
2. Suppose I did have information on competitive salaries, but that information was inaccurate. What are the possible consequences of having poor-quality information?

reserved largely for military and government applications. If you wanted to listen to music while driving, you would pop in a cassette tape (if you were lucky) or listen to whatever the radio station disc jockey wanted you to hear.

In the workplace, few desks would have computers, and the computers you did see would seem bulky and primitive by today's standards. Video conferencing would be unknown to most workers. Employees who needed to stay in touch while away from the office relied on pagers (commonly called beepers) that, at best, included only the phone number of the paging party. Although mobile phones existed, they were rare and expensive.

Technology has brought rapid, seemingly constant change to our lives, and we can expect even more changes in the next few decades. The technologies behind these changes have one thing in common—they deal with information. As abilities to gather, store, process, transmit, and output information have improved, clever people have found new ways to apply these improved abilities in ways that bring irrevocable changes to the world. In the remainder of this section, we discuss some emerging technologies that are likely to change our lives even more. Throughout these discussions, keep in mind that all of these technologies are about using information in interesting ways.

If you ask futurists or information systems experts what emerging technologies will have the greatest influence on the future, you will receive many different answers. Despite the disagreements, we think that most lists will include three technologies—fifth-generation cellular networking (**5G**), the **Internet of Things** (IoT), and **artificial intelligence** (AI). These emerging technologies are interesting primarily because of the applications they enable, especially when we combine all three. (We discuss each of them in more detail in later chapters.)

5G, which is currently being implemented in many areas of the world, offers much faster data transmission speeds, wider coverage areas, and better response times than fourth-generation cellular (4G). In other words, 5G is considerably better at handling information than prior generations of cellular networking. (In the next section, you will learn about how data and information differ. For now, we use the terms interchangeably.) 5G's enhanced data-handling capabilities offer many possibilities for new technology applications that depend on the ability to send and

---

LEARNING ACTIVITY 1.1

The Rapid Growth of Information

Did you know that the amount of data and information around us is constantly increasing? Think about information that you have access to that you did not have five or even two years ago. As you do, watch the video at the following link, then consider the questions below: https://www.youtube.com/watch?v=u06BXgWbGvA

1. How does the rate of change in information affect us?
2. Do you think the data in this video is current? Why?
3. What does this mean to you as you finish school and embark on a career?

LEARNING ACTIVITY 1.2

A Day in My Technology-Free Life

Most of us interact with information technology (IT) throughout the day, although we may not always be aware of it. Some uses of IT are relatively obvious, such as using email or surfing the Web. Others are not readily apparent, such as texting a friend or using a debit card. Your assignment is to go one half day (six hours when you are awake) without using any information technology and then write a short paper reflecting on the experience. Here are a few of the things you *will not* be able to use:

- Facebook, email, or the Web (Do not even turn on your computer!)
- Text messaging
- Debit or credit cards (unless the merchant uses an old-fashioned imprint machine)
- Smartphone applications (apps)
- And many, many others

You need to time this so that you will not have any issues with completing assignments from this or other classes. Also, if there is an emergency that requires your use of IT, disregard this assignment and deal with the emergency. It may also be a good idea to inform those with whom you communicate regularly, such as family and close friends, that you are going without information technology. This may prevent them from thinking it is an emergency when you are not answering text messages.

At the end of the half day, prepare a one- to two-page reflection paper that addresses the following questions:

1. What did you learn from your experience?
2. What technology was the most difficult to live without?
3. What kinds of tasks were the most difficult to complete without technology?
4. Were there any benefits of being "tech free"?

receive information rapidly. These capabilities are especially useful when combined with the Internet of Things.

The Internet of Things refers to an almost limitless network of connected sensors that are embedded in physical objects ranging from refrigerators to airplanes. These sensors collect data from physical objects and the surrounding environment, process the data in limited ways, and then transmit the data to computers through the Internet. These computers process the data further and send data back to the sensors, which then can send commands to other processors embedded in the physical objects. For example, traffic sensors can send data to a traffic control system that sends data to smart traffic signals, adjusting the timing of signals to improve traffic flow.

Many interesting IoT-based applications already exist. You have probably heard about "smart homes" that include IoT-enabled devices such as smart climate control and lighting systems, surveillance systems, video doorbells, smart refrigerators, and smart locks, among many others. If you own a smart home, you can adjust its thermostat when leaving school to ensure that your home is comfortable when you arrive there. If someone approaches your front door, your video doorbell will alert you and

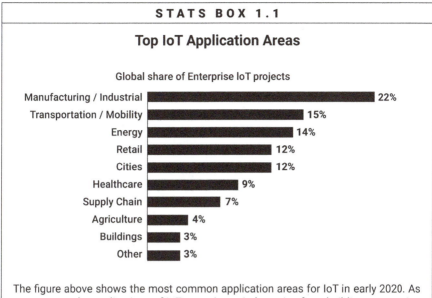

STATS BOX 1.1

Top IoT Application Areas

Global share of Enterprise IoT projects

Manufacturing / Industrial	22%
Transportation / Mobility	15%
Energy	14%
Retail	12%
Cities	12%
Healthcare	9%
Supply Chain	7%
Agriculture	4%
Buildings	3%
Other	3%

The figure above shows the most common application areas for IoT in early 2020. As you can see, the applications of IoT are quite varied, ranging from building automation and maintenance to transportation. Keep in mind, however, that this list is dynamic and will likely change quite a bit as new application areas for IoT develop.

Source: https://iot-analytics.com/top-10-iot-applications-in-2020/

allow you to take appropriate action. You can even check the contents of your refrigerator to see if you need to pick up milk on the way home. Did you forget to lock your door when you left home? No problem—just lock it remotely from your mobile phone.

Artificial intelligence (AI) is used to describe a family of technologies that approximate human cognitive abilities, such as reasoning, planning, learning, problem-solving, perception, and language processing. It is a bit ironic to include AI in a discussion of emerging technologies—the roots of AI go back to Alan Turing's work in the 1930s—but AI and the applications it enables continue to evolve rapidly. In addition, we include AI in this discussion because it will be a key ingredient in future technology applications.

The term AI has been applied to many different tools and techniques, such as expert systems, machine learning, natural language processing, and intelligent agents, among others. (Note that there is quite a bit of disagreement about whether some of these are truly AI. At this point, we are not concerned with such debates.) Although AI has matched and even exceeded human intelligence in specific areas, such as playing games like chess and Go, AI cannot currently match human intelligence in tasks that require broad general knowledge, or that include significant ambiguity.

Possibilities for the future become very interesting when we consider how these three technologies can work together. Information technology applications are built around the ability to gather, communicate, and process information. These three capabilities align nicely with the main functions of IoT (gather), 5G (communicate), and AI (process). The fusion of these technologies has been referred to as "intelligent connectivity" (https://iot-analytics.com/how-5g-ai-and-iot-enable-intelligent-connectivity/). As these technologies continue to improve, expect to see many exciting new and sometimes revolutionary applications of information systems.

The critical point to remember here, and throughout this book, is that information has been, and continues to be, the driving element behind advances in how we apply information technologies. In short, it is all about information. The purpose of IoT is to gather information, the point of 5G is to communicate information (data) faster and more effectively, and the point of AI is to use information more effectively. If you want to understand the possibilities as new technologies emerge, think about how they will help organizations and individuals deal with and use information. In the rest of this section, we examine a few interesting applications that will grow in importance as IoT, 5G, and AI continue to develop.

IoT, combined with 5G and AI, will make smart cities more common. The term "smart city" is really a philosophy rather than a single application. The idea behind smart cities is to use IoT to collect data, which is then used to better manage resources and to more effectively provide services. Different cities will deploy different specific applications and technologies depending on their citizens' needs.

For example, cities that suffer from serious traffic congestion might pursue the smart traffic control system described earlier. Smart cities use technologies such as IoT-enabled sensors to control traffic by sensing when areas become congested and adjusting traffic lights to ease the traffic jams and route traffic to less congested routes. AI can be used to determine the optimal timing of traffic signals based on traffic density, weather, time of day, and other factors. These systems will be even more effective with the increasing use of autonomous vehicles, which are also enabled by IoT and will improve as 5G rolls out. These same networks may also improve traffic safety. For example, a traffic control system with IoT sensors might detect a pedestrian jaywalking (crossing a road at a location other than a crosswalk) and quickly send the data to an AI system that also would receive a signal indicating an approaching autonomous vehicle. The AI system might use a 5G network to send a command to have the autonomous vehicle apply emergency braking.

Other smart city applications include environmental monitoring, energy management, intelligent transportation systems, and management of public spaces. Even refuse collection is being integrated into smart city networks. For example, sensors can detect how full trash containers are so that workers can be dispatched to containers that are becoming full, but not to those that are still empty. There are dozens of other smart city applications, and more will emerge as the underlying technologies become more common and cities learn how to better take advantage of these and other technologies.

LEARNING ACTIVITY 1.3

Autonomous Vehicles

For decades, self-driving vehicles were the domain of science fiction, but that sci-fi dream is on the verge of becoming a reality. Review the SAE International Levels of Driving Automation website (https://www.sae.org/news/2019/01/sae-updates-j3016 -automated-driving-graphic) and study the different levels of driving automation. Take a few minutes and think about the possibilities and problems related to autonomous vehicles. What level do you think is ideal? What problems do you see for autonomous vehicles at the higher levels? Would you use a ride-share autonomous vehicle? Why or why not?

Autonomous vehicles are another exciting application of IoT, 5G, and AI. Although fully autonomous vehicles are currently rare, many newer cars have partial automation. In these vehicles, the human driver is responsible for monitoring and reacting to the driving environment, but automated systems alert the driver to potential hazards, such as a stopped vehicle ahead. In some cases, the automated system will take over certain systems. For example, an automatic braking system can apply a vehicle's brakes when an obstacle is detected. (Note: Not all of these systems use IoT, and many are self-contained and thus do not require 5G connectivity.)

According to SAE International (formerly known as the Society of Automotive Engineers), there are six levels of automated driving systems (Shuttleworth 2019). These range from no automation (Level 0) to full automation (Level 5). Level 2 systems provide some automation of systems such as steering, braking, and acceleration, but the driver must constantly monitor the driving environment and be ready to take control when necessary. Many automakers currently provide some Level 2 support through "active safety features," such as automated braking and lane-keeping systems and adaptive cruise control. Level 3 autonomous vehicles are becoming more common. These vehicles are fully autonomous under certain conditions but still require driver attention.

Autonomous vehicles work because of the convergence of sensor technologies (including IoT), communication networks, and AI. Sensors include video cameras, limited range radar systems, and position and proximity sensors. These sensors send and receive data through various networks, some of which are external to the vehicle. For example, proximity sensors could interact with traffic control systems through 5G networks to automatically slow a vehicle approaching a traffic light that is about to change from green to red. Sensors in other vehicles could also communicate with each other to coordinate actions to allow smooth traffic flow and avoid collisions. AI systems will use the information from IoT sensors and other information sources to take evasive action to avoid potential collisions. As you might imagine, the details behind all of this are extremely complicated and well beyond our scope. The important thing to remember is that autonomous vehicles depend on the various systems' abilities to gather, communicate, and use information, as is the case with any system.

Virtual reality (VR) and the related areas of augmented reality (AR) and mixed reality represent another emerging technology area that will increase in importance and capabilities thanks to 5G, IoT, and AI. VR, augmented reality, and mixed reality are all similar in that they involve using digital technologies to represent or add to physical reality. Virtual reality is a fully immersive experience in which the technology makes you sense that you are in a different environment. The reality is virtual in that it approximates a reality using three-dimensional, computer-generated images to represent an environment. The VR system allows you to explore and interact with the computer-generated environment. This lets users do things like play a game in a simulated environment, walk around a virtual museum, or practice complicated surgery.

Augmented reality is a bit different. AR systems overlay information on top of elements of the real world. Games such as Pokémon Go and Real Strike are perhaps the best known applications of augmented reality. But AR has more serious applications as well. For example, AR can give surgeons a sort of x-ray vision by sensing the positions of surgical tools and superimposing the positions on a display of a patient's CT scan. AR systems can help maintenance technicians by overlaying important information such as part numbers, diagnostic and repair procedures, and safety warnings on top of the technician's view of the machine being repaired. IoT sensors can make this informational overlay even more useful by adding information such as the current temperature of a part or other diagnostic information.

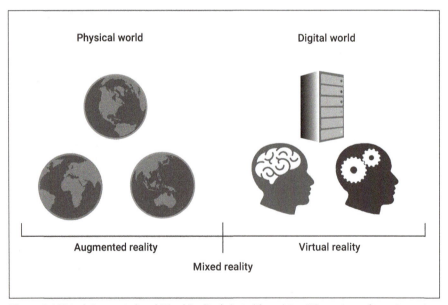

Figure 1.1 Virtual, Augmented, and Mixed Reality (adapted from: https://docs.microsoft.com/en-us/windows/mixed-reality/images/mixedrealityspectrum.png)

Mixed reality lies somewhere between virtual and augmented reality. It takes AR a step further by combining elements of the real and virtual worlds. This allows users to interact with both physical and virtual objects. The ability for physical and virtual objects to coexist opens the possibilities for applications in many areas, including surgery, training, retailing, real estate, and entertainment, among others. Mixed reality can even help remote teams work together in collaborative virtual environments. Eventually, it will seem like they are working face-to-face.

Figure 1.1 shows the virtuality continuum, which illustrates the relationships among AR, VR, and mixed reality. As you can see from the figure, AR starts with the physical world and layers virtual elements onto it, while VR exists in the digital world. Mixed reality blends the two worlds into a single environment. As the technologies that enable VR, AR, and mixed reality continue to improve, the possibilities for these areas to transform many aspects of our lives will be almost endless.

As we close this section on emerging technologies, keep two things in mind. First, we have only scratched the surface of these exciting technologies. In fact, you could take entire courses on each of the technologies and applications we have discussed. So, if you feel like you do not fully understand them, do not worry. We will provide additional details about IoT throughout the book. We will also discuss 5G in Chapter 7, and AI in Chapter 14. Second, remember that all these applications and all emerging information technologies to come depend on information— information is the lifeblood of information systems and information technologies. In the next section, we dig more deeply into information and the related concepts of data, knowledge, and wisdom.

Data, Information, Knowledge, and Wisdom

Consider the following question. Is 62 good or bad? Of course, this is a meaningless question, but why is this so? The number 62 is an item of **data** (or datum), but it lacks any real meaning. Let us combine 62 with some other bits of data. Maybe 62 is the score you received on an exam. Perhaps it is the score you need to receive on your final exam to retain an A in a class. When we combine the original data with other related data, we now have **information** that helps us interpret whether 62 is good or bad. Once we have the meaningful information, we need to interpret it to determine what actions should be taken. (In our example, you might decide to spend more time studying for the final exam in a class for which you need a much higher score.) When we apply information to some decision or action, we have **knowledge**. **Wisdom** involves using knowledge for the greater good. Because of this, wisdom is deeper and more uniquely human. It requires a sense of good and bad, right and wrong, ethical and unethical.

One way to think about this is to consider how data, information, and knowledge form a hierarchy. Data are raw symbols (unconnected facts), information is data that have been processed so that they are useful, and knowledge is when information is applied to some decision or action. Russell Ackoff (1989) popularized this concept, although the root of the idea goes back to part of T. S. Eliot's poem "The Rock":

> Where is the Life we have lost in living?
> Where is the wisdom we have lost in knowledge?
> Where is the knowledge we have lost in information?

There are two reasons we bring up this hierarchy. First, it gives us a way to think about different kinds of systems and how they have evolved. Data processing is relatively straightforward. Handling information is messier. You need the capability to do ad hoc (e.g., specific, customized) queries to allow different views and combinations of data. Managing knowledge is messier still.

The hierarchy also parallels how businesses have applied **information technology**. In the early days, the focus was on processing data for relatively straightforward tasks, such as tracking payrolls or bookkeeping. This evolved into **information systems**, which enabled users to connect and process data to help with management decision-making. More recently, there has been movement toward being able to manage knowledge. Although it is hard to imagine, it is possible that we may see a day when technology helps us manage wisdom. This is debatable, since many consider wisdom to be uniquely human.

A second, and perhaps more important, reason to understand the hierarchy is that it helps frame two important concepts: connectedness and usefulness. The shift from data to information requires connecting data elements. Combining bits of data gives the data some context and meaning. This meaning may or may not be useful, though. To be useful, information is interpreted and applied (in other words, is used), leading to knowledge.

One last thought on the hierarchy is that when we reach the knowledge level, we often realize that we need more data and information. This creates a sort of cycle, as illustrated in Figure 1.2.

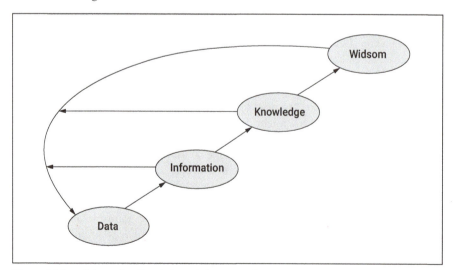

Figure 1.2 Data, Information, Knowledge, Wisdom Hierarchy

LEARNING ACTIVITY 1.4

How I Use Information

You might be surprised at the variety of different ways you use information every day. Take a few minutes to identify five ways you use information. For example, when deciding what to buy for lunch, you might use information about the restaurant's offerings, such as the descriptions and prices of the items. Try to come up with a range of decisions from simple to more complex. Then, for each one, rate the impact of having bad information as being a large, moderate, or small impact. After you have done this, compare your lists with another student. For each item, discuss why you rated the impact of bad information the way you did.

While it is useful to understand the relationships among data, information, and knowledge, it is more important to know how to deal with the tremendous amount of information available today. In other words, it is important to be information literate, which is the topic of the next section.

Information Literacy

What does it mean to be information literate? While there are a number of definitions of **information literacy**, it boils down to the ability to efficiently and effectively determine what information is needed and then access, evaluate, use, and manage that information in an ethical manner.

It may be hard to believe, but not that long ago, we had access to a much more limited amount of information. Then along came the Web, and today we are faced with a tremendous amount of information every day. While it is great to have access to almost unlimited information, this comes at a price; we must learn how to deal with this information. Think about it this way: if you do not have any money, understanding investments and money management is not very important. But once you start to make a good living, you need to pay a lot more attention to these matters. It is great to have money, but the price is that you have to pay more attention to using it effectively. We are faced with a similar situation with respect to information; we are information rich, which has many advantages but requires more skill and effort if we want to use that information effectively.

Let us look at an example. In the "old days" (before the Web), if you wanted to buy a car, you might look at a couple of magazines for reviews and then go to two or three dealers to check their prices. You would also need to get quotes on your trade-in. This was all very time consuming and inefficient, and the information you gathered was incomplete. Today, the situation is very different. You can access multiple automobile reviews, get a good estimate of the value of your current car, and see the pricing of dozens of dealers—all without leaving home. There is a catch, though—you have to know what you are doing. You have to know what information you need, where to get that information, and whether that information is trustworthy. Then, since you will have so much information, you need to be able to manage and use it effectively. In other words, you need to be information literate.

Understanding how to deal with information may well have serious benefits as you move through your business career. One way poor information literacy puts you at a disadvantage is through its effect on information asymmetries. Information asymmetries exist when one party in a transaction has more or better information than the other party. This additional information gives the better-informed party a serious advantage. So, you can see that being able to deal with information effectively is important. In fact, there are many who believe that information literacy is *the* critical skill for the future. Ilene Rockman (2004), a leading authority on information literacy, put it well:

> Individuals who are knowledgeable about finding, evaluating, analyzing, integrating, managing, and conveying information to others efficiently and effectively are held in high esteem. These are the students, workers, and citizens who are most successful at solving problems, providing solutions, and producing new ideas and directions for the future.

Uses of Information

All businesses, regardless of their size, use information for three main purposes: communication, process support, and decision-making. When properly applied, information technology enhances all of these.

The *Merriam-Webster's Dictionary* defines communication as "a process by which information is exchanged between individuals through a common system of symbols, signs, or behavior." This definition clearly shows that one use of information is to communicate. Organizations must exchange information for a variety of reasons, including sharing ideas, coordinating actions, and transmitting information to stakeholders.

Businesses also use information to support processes. A **business process** is a set of coordinated activities that lead to a specific goal or outcome. (Chapter 11 goes into much more depth regarding business processes.) Many business processes are quite complex and involve various parts of a business. For example, a customer

---

LEARNING ACTIVITY 1.5

How Businesses Use Information

As we have discussed, businesses use information in several ways. Team with another student and think about a business (one you have worked for, one you frequent, or one assigned by your instructor). Identify three decisions and three processes related to the business. Identify key pieces of information required to support each decision or process. Write them down and be prepared to share them with the class.

BUSINESS PROCESS

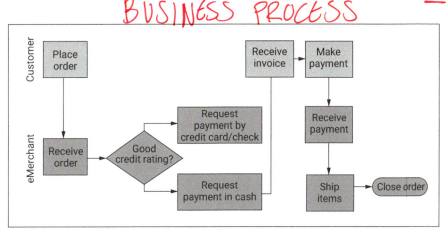

Figure 1.3 Order Process (adapted from http://www.readbag.com/media-visual-paradigm-media-docu ments-bpva20ug-pdf-bpva-user-guide-ch6)

order process might look something like Figure 1.3. (Note that this is a greatly simplified view.)

Information is used at each stage in the process. We need to know what products were ordered, who the customer is, where to ship the order, how much the products cost, and so on. This information must be communicated and properly processed throughout the process. Without information, the process simply cannot be completed. **Information systems** can help streamline and improve the quality of business processes.

Businesses (and individuals) also use information for **decision-making**. Decision-making involves choosing among alternatives. Making this choice requires information about the alternatives (as well as other information). Put simply, you cannot make decisions without information. For most of you, this is the main way you will use information. Because of this, quite a bit of this book is related to using information for decision-making.

Two other aspects of business information use are worth mentioning. First, some businesses actually use information as a product. For example, Google is a company that you are probably familiar with. Although they provide services such as a search engine and email, they are really in the business of information. They offer their products for free as a means to gather information. This helps them achieve their mission to "organize the world's information and make it universally accessible and useful."[1] While all businesses use information for communication, process support, and decision-making, only some businesses use information as a product. Second, information can be used to gain strategic value; this is the topic of Chapter 4.

Knowing how to successfully deal with the incredible and growing amount of information available to us is a key life and professional skill. Businesses use information in several important ways, including for communication, to support

---

1 https://about.google/intl/en/.

BUSINESS EXAMPLE BOX 1.1

Promising Information Careers

Careers related to information systems and information analysis are in high demand. They also tend to pay well. The table below shows a few examples.

	Median Salary ($)	Growth (%)
Information Systems Careers		
Computer systems analyst	90,920	9
Information systems manager	146,360	11
Computer support specialists	54,760	10
Software developer	107,590	21
Database administrator	97,750	9
Information Analysis Careers		
Financial analyst	85,660	6
Auditor	71,550	6
Information security analysis	99,730	32

Notes: Median salaries are from 2019; growth projections are from 2018 to 2028.

Source: Bureau of Labor Statistics, U.S. Department of Labor. 2020. *Occupational Outlook Handbook*. http://www.bls.gov/ooh/.

---

processes, and for decision-making. Some businesses also use information as a product. They gather and perhaps analyze information and then sell that information to other businesses that use it for a variety of purposes. Understanding information and how it is used and having good information literacy skills will help you be more successful professionally and personally.

Information and Your Career

Information systems and technologies are key elements of business. Because of this, knowledge of the systems and technologies you will learn about in this book can help you in your future career.

Individuals in information systems careers design, build, support, or manage information systems. These are the jobs that typically require a major in information systems or a related field. People in information analysis careers use these systems to retrieve, report on, and analyze the information contained in the systems. The results of the analyses are passed along to information consumers, who use the results to carry out transactions and make decisions. Information systems and information analysis offer a surprising variety of interesting, rewarding careers, as illustrated in Tables 1.1 and 1.2. Information-consuming careers include almost any business career, so we do not provide a table of examples for those.

Table 1.1 Information Analysis Careers	
Business intelligence professional	Uses various reporting and statistical tools to help organizations gain value from their informational assets
Data analyst	Collects, processes, and analyzes data using various data management and statistical analysis tools
Data scientist	Develops and implements systems and processes to gain knowledge and insights from data
Compensation analyst	Analyzes information related to the costs of paying employees
Economic analyst	Develops economic models that help organizations predict economic conditions and outcomes
Financial analyst	Provides guidance in how to best utilize financial assets
Forensic accountant	Analyzes accounting and financial information to determine whether an activity is illegal or fraudulent
IT auditor	Ensures that an organization's information systems are operating properly and producing accurate information
Marketing analyst	Analyzes information related to the marketing activities of an organization

Table 1.2 Information Systems Careers	
Business process analyst	Analyzes business processes to find ways to improve efficiency and/or effectiveness (which often involves applying information systems)
Chief information officer	Oversees all information and information systems–related activities of an organization
Database administrator	Manages systems that store and organize data used in information systems
Information security specialist	Ensures that the information contained in information systems is properly protected
Information system trainer	Designs or delivers educational programs that help users utilize information systems
IT project manager	Organizes and controls projects related to information systems
Software developer	Writes the computer programs that make up part of an information system
System analyst	Works with users to understand and document system requirements
System architect	Develops high-level designs of information systems, including the overall structure and essential design features

Chapter Summary

The overall goal of this chapter is to help you understand why information is valuable in business and your personal life. Understanding the value of information is important to understanding the importance of the various information systems presented in this book. Here are the main points discussed in the chapter:

- Emerging technologies such as 5G, the Internet of Things, and artificial intelligence are important because of their ability to gather, communicate, and process information.
- Data are raw symbols or unconnected facts.
- Information is processed data that is useful.
- Knowledge is information that is applied to a decision or action.
- Information literacy is an important professional and personal skill.
- Being information literate requires being able to efficiently and effectively determine information needs and then acquire, evaluate, use, and manage that information ethically.
- Businesses use information for communication, process support, and decision-making.
- Information-related knowledge is important to virtually all careers.
- There are many careers that involve designing, building, supporting, managing, and analyzing information.

Review Questions

1. Contrast data, information, and knowledge.
2. Briefly discuss why 5G, IoT, and artificial intelligence are important in enabling new information technology applications.
3. Apply the data, information, and knowledge hierarchy to the evolution of how information technology has been applied in businesses.
4. Describe the concepts of connectedness and usefulness as they relate to the data, information, and knowledge hierarchy.
5. Describe how the data, information, and knowledge hierarchy can be thought of as a cycle.
6. What are the key abilities related to information literacy?
7. Why is it important to be information literate?
8. Name three general ways businesses use information.
9. Name and briefly describe three information systems careers.
10. Name and briefly describe three information analysis careers.

Reflection Questions

1. What is the most important thing you learned in this chapter? Why do you think it is important?

2. Why is it important for business professionals to be information literate?

3. How does better information help businesses?

4. What impact do you think improving your information literacy skills will have on your career?

5. When has bad information or incomplete information lead to bad outcomes in your life? If you were facing that situation again, what would you do differently?

Additional Learning Activities

1.A1. Take a few minutes to browse the Smart City Press website (https://www.smartcity.press/). Pick a smart city application and identify the following: What information is being collected? How is the information being used? How might 5G and AI be used in this application?

1.A2. Do an Internet search for augmented reality smartphone apps. Pick three apps that you think would be useful to you now or in the future. Briefly describe the app and discuss why you think it would be useful.

1.A3. Think about a common situation that a business professional (such as one in marketing, accounting, or finance) might face. For example, a finance professional might be analyzing an investment opportunity. Identify five pieces of information that would be useful in that situation. For each piece, discuss how bad or incomplete information might lead to negative outcomes.

1.A4. Share your findings from 1.A3 with two other students who are in different majors. Compare your answers. How are the answers similar? How are they different?

1.A5. Consider one of the following major decisions: (a) where to go to grad school, (b) what car to buy, (c) which apartment to move to, (d) which job offer to accept. Identify five important pieces of information that you would need to make a good decision.

   a. For each piece, discuss why you think that information is important. Focus on consequences.

   b. Where would you find the information you need?

   c. How would you know whether or not to trust the information you gather?

1.A6. Consider the order process shown in Figure 1.3. For each stage in the process, what information is needed? What might be the consequences of bad information at that stage?

1.A7. Use a job-hunting website (e.g., Monster.com, Career.com, or your school's career website) to find job postings for two of the information systems and two of the information analysis careers discussed in the chapter. Be prepared to share your results.

References

Ackoff, R. L. 1989. "From Data to Wisdom." *Journal of Applied Systems Analysis* 16: 3–9.

Shuttleworth, J. 2019. "SAE Standards News: J3016 Automated-Driving Graphic Update." SAE International. https://www.sae.org/news/2019/01/sae-updates -j3016-automated-driving-graphic.

Rockman, I. F. 2004. *Integrating Information Literacy into the Higher Education Curriculum: Practical Models for Transformation.* San Francisco: Jossey-Bass.

Glossary

5G: Fifth-generation cellular network.

Artificial intelligence: Family of technologies that approximate human cognitive abilities such as reasoning, planning, learning, problem-solving, perception, and language processing.

Business process: Set of coordinated activities that lead to a specific goal or outcome.

Data: Raw symbols (unconnected facts).

Decision-making: Process of choosing among alternative courses of action.

Information: Data that has been processed so that it is useful.

Information literacy: The ability to know when information is needed, and to be able to locate, evaluate, and effectively use that information.

Information system (IS): A combination of technology, data, people, and processes that is directed toward the collection, manipulation, storage, organization, retrieval, and communication of information.

Information technology (IT): The hardware, software, and media used to store, organize, retrieve, and communicate information.

Internet of Things (IoT): Network of connected sensors embedded in physical objects that communicate over the Internet.

Knowledge: Information that is applied to a decision or action.

Wisdom: The use of knowledge for the greater good.

Chapter 2

Introduction to Information Systems

Learning Objectives

By reading and completing the activities in this chapter, you will be able to:

- Describe the major functions of an information system
- Explain why it is important for business professionals to understand information systems
- Explain key concepts related to systems
- Describe the information processing cycle
- Describe the critical elements of an information system
- Explain how information systems help managers deal with information
- Give examples of business rules
- Discuss how information systems facilitate organizational change
- Compare and contrast common information systems

Chapter Outline

Why All Business Professionals Need to Be Information Systems Managers
Focusing Story: Registering for Classes in 1981
    Learning Activity 2.1: Identifying My Information Systems
Overview of Systems
    Learning Activity 2.2: Components of a System
    Learning Activity 2.3: Components of a Business Information System
Foundations of Information Systems
How Information Systems Help Us Deal with Information
    Learning Activity 2.4: Information Systems and Change
How Information Systems Facilitate Organizational Change
Common Information Systems

Why All Business Professionals Need to Be Information Systems Managers

Information systems, such as the registration system in this chapter's focusing story, are an ingrained part of our daily lives. Each of us interacts with many information systems throughout the day. In fact, information systems are so prevalent that we often do not even notice them. Most of you probably use these systems as consumers. You check your course grades online, buy a song on the Web, and use your debit

Registering for Classes in 1981

When you registered for this class, you probably did so over the Web. Registration was very different in the 1980s. Let us describe how one of us registered for classes in 1981.

First, you get a copy of the printed schedule of classes. After spending quite a bit of time studying the course offerings, you finally come up with an ideal schedule. But wait . . . the chances of getting all your "first choice" classes are pretty slim. (Some things have not changed!) So you come up with some backup plans—alternative courses and sections that are less than ideal but will still work. Fortunately, you have a good registration appointment time this semester. A little before your appointment time, you head to the student union and join the very long line for registration. While you are waiting in line, you keep an eye on the bank of television screens that are scrolling through the classes and their statuses. You are constantly checking to see if any of your ideal classes have closed out. When this happens, you start turning to your backup plans. You finally make it into the registration room, which is just a big multipurpose room filled with computer terminals and people. As you stand in line, you keep a sharp eye on the screens displaying the most up-to-date class statuses. These screens are showing the transparencies of the entire class schedule. As you watch, a person with a transparency marker walks around, crossing off classes that have recently filled. Of course, if one of your classes closes, you turn to your backup, backup plan. You finally make your way to one of the computer terminals, which is operated by trained university staff. The staff person takes your desired schedule, enters the code numbers into the system, and in a minute or two you find out if you are successfully registered. If one of your classes happens to close between the time you checked out the transparencies and the time your data actually got entered, you are sent to the "problem table," where someone works with you to finally get a schedule. Once your schedule is successfully entered, you stand in another line, waiting for your printed schedule and tuition bill. Congratulations, after a couple of hours (if you are lucky), you are registered!

Focusing Questions

Compare the scenario above to your course registration process.

1. Name four ways the current process is better for students.
2. Name four ways the current process is better for universities.

---

or credit card to pay for a purchase. However, in the future, most of you will use information systems as part of your profession. Because of this, it is useful to start to think like an information systems manager.

Today, everyone in business needs to be an information systems manager to some extent. Let us be blunt about this; regardless of your major or your career path, you are going to be using information systems. Learning how to effectively use these systems can help you be more effective and successful in your career. Sales representatives who know how to use customer relationship management systems well, accountants who know the ins and outs of a company's enterprise system, and finance managers who know how to use a spreadsheet to analyze different investment scenarios will all be more successful than those who lack knowledge of these systems. To utilize these (and other) systems, it is helpful to know something about how they work. In this chapter, we lay the foundation by helping you understand concepts that are the basis of all information systems.

One reason it is useful to think like an IS manager is that information systems are increasingly ingrained in our business and professional lives. As you may have learned in the previous chapter, it is pretty hard to get through the day without interacting with an information system. How much of a pain would it be if you could not use these systems? To top it off, information systems are likely to become even more pervasive. So it is going to become even more important to understand and know how to use information systems. A big part of career success is knowing how to use important tools of the trade. Just as a carpenter needs to be skilled in using a hammer and saw, a businessperson needs to be skilled in using information systems. Our goal throughout this chapter and the entire book is to help you gain the knowledge you will need to make the best use of information systems in your career.

Overview of Systems

A **system** is a set of interacting components working together to form a complex, integrated whole to achieve some goal by taking inputs and processing them to produce outputs. Figure 2.1 shows a simple system and its key elements. Let us break down this definition to understand the key ideas related to systems:

- A system is made up of different pieces, called components. Components can take many different forms, ranging from human organs to computer software.
- These components work together; they are interrelated.
- A system has some purpose or goal.
- The goal is achieved by taking inputs and processing them to produce outputs.

There are a few systems-related concepts that are not apparent from the definition. First, a system is separated from its environment by the system's boundary.

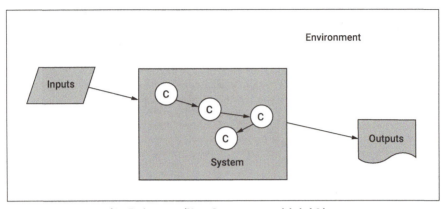

Figure 2.1 A System and Its Environment (*Note:* Components are labeled *C*.)

Most systems are **open systems**, which means that they interact with their environments. (Closed systems, in contrast, do not.) For example, your body reacts to hot weather (an element of the environment) by sweating. This is also an example of **feedback**. Your body senses its temperature rising, so the brain sends a signal to your sweat glands to activate. It is also important to note that systems are often made up of **subsystems**. A subsystem is a system that is part of a larger system. Your respiratory system is an example; it is a complete system in its own right but is also part of the larger system of your body. Finally, **equifinality** is the idea that in an open system there are many different potential paths to the final outcome. In other words, there are different ways to end up in the same place.

One way to think about systems is to consider the components as a set of transformation processes. In this view, systems follow an input-process-output sequence. Consider a manufacturing system. Raw materials are the inputs, which are processed by various machines to produce the final product (the output). Of course, this is a greatly simplified view of a manufacturing system. Information systems work simi-

---

LEARNING ACTIVITY 2.2

Components of a System

Part 1: Consider one of the following systems:

- Video game system
- Personal computer
- Home entertainment system

Sketch the different components (pieces) of the system. (For example, a television might be a component of a home entertainment system.) Draw lines to indicate which components interact.

Part 2: Repeat Part 1 using your school's course registration system (or another system assigned by your instructor).

LEARNING ACTIVITY 2.3

Components of a Business Information System

Pick one of the business information systems below (or another system assigned by your instructor). Identify at least five components of the system. Briefly describe the purpose of each component.

- Online ordering system (for example, ordering something from Amazon .com)
- Membership tracking system for a health club
- Hotel reservation system

larly. They take data as the raw material and transform it in various ways to produce usable information.

This brings us to our final general systems concept, feedback and **control**—a set of functions intended to ensure the proper operation of a system. Systems often have a feedback loop, where information from the environment is sensed by the system, which may change its behavior based on the feedback. Feedback is closely linked to the concept of system control. Many systems include a control element. Think about a thermostat, which is a subsystem of a heating and air conditioning system. The thermostat receives data (the current temperature) as feedback from its environment. When the temperature reaches a certain level, the thermostat sends a signal to the air conditioner or heater to begin operation (the control). Then, when the temperature reaches the desired level, the thermostat signals the air conditioner or heater to turn off.

Now that you have a basic understanding of systems in general, let us turn our attention to information systems. In the next section, we cover some important concepts regarding systems designed to deal with information.

Foundations of Information Systems

Earlier we defined an information system as a combination of technology (hardware, software, and communication media), data, people, and processes that is directed toward the collection, manipulation, storage, organization, retrieval, and communication of information. Later in the chapter, we discuss the six elements of an information system in more detail. The goal of this section is to help you understand some fundamental ideas related to information systems. Keeping these in mind will help you better understand the material in the rest of this book.

Before going further, it is important to realize that an information system does *not* require a computer. Information systems existed for hundreds of years before the invention of the electronic computer. In this book, we are primarily concerned with *computerized* information systems, which, of course, *do* require computers. Throughout this book, when we say "information systems," we are referring to computerized systems. Noncomputerized systems will be called "manual information systems."

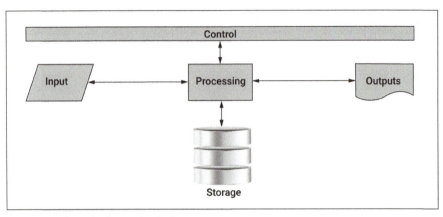

Figure 2.2 Information Processing Cycle

One final note before we get into the details: people are the most important component of any information system. We will discuss the role of people in an information system in more detail later. However, we want to make sure that you understand that people are critical to the success of any information system.

Information systems include the following operations, as also shown in Figure 2.2. This is sometimes called the information processing cycle. (*Note:* Some versions of this cycle do not include control as a separate operation.)

1. *Input:* Collection of data and their conversion into a form that allows processing.
2. *Processing:* Manipulation and transformation of data.
3. *Storage:* Holding place for data so that they can be retrieved at a later time.
4. *Output:* Transformation of processed data into a form that can be understood by its eventual user.
5. *Control:* Enforcement of correct processing procedures.

These operations are carried out by the various components of the information system. For example, the screen on your laptop computer may perform output operations. As is the case with any system, these components interact with one another in an organized manner. For instance, when you drag your mouse to select some text, the mouse provides input to the system, which processes the input and highlights the text on your screen (an output device). Your computer's central processing unit (CPU) carries out much of the processing and control activities, while your hard drive performs storage activities. These operations, which are things the information system does, can be distinguished from elements of an information system, which are the parts that make up the information system. The important elements of information systems are discussed next.

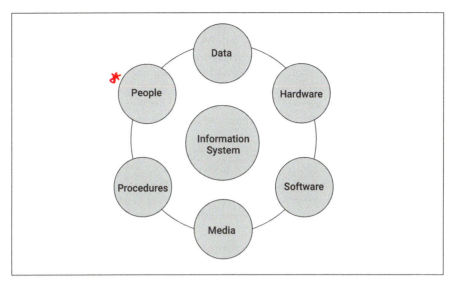

Figure 2.3 Elements of an Information System

Information systems have six critical elements, as shown in Figure 2.3:

- *Data:* Raw facts, text, numbers, images, and the like that serve as the inputs for producing information.
- *Hardware:* Physical devices, such as the processor, storage media, and peripheral devices (such as printers and displays).
- *Software:* Set of instructions that govern the operation of an information system.
- *Communication media:* Set of devices and protocols (rules) that enable computers to communicate with each other.
- *Procedures:* Instructions for the proper use of the information system.
- *People:* Individuals who use the information system.

You might be a little surprised to see people included in this list—even though we included them in our earlier definition—but they are critical to any information system. People are necessary to use and interpret the output of the information system. Even in information systems that seem like they are totally automated, people still need to monitor the system to make sure it is working correctly. Of course, people also build and maintain the system. Procedures for using the system are also necessary. Basically, the procedures tell the users how to properly utilize the system to support their tasks.

Before moving on, let us talk a little more about hardware and software. (There is more detail about this in Appendices A and B.) Software provides the instructions that govern the operation of a computerized system. Broadly speaking, there are two

systems software = operating systems

types of software: systems software and application software. Systems software includes operating systems, such as Windows, MacOS, Linux, Android, and iOS, and utilities such as antivirus and data-compression software. This is a little confusing, because when you install an operating system, numerous utility programs are often installed at the same time. For our purposes, it is enough to just think of system software as a "necessary evil." Most people would not buy a computer just to run the system software, but you have to have the system software for the computer to operate.

People buy computers to run application software, which lets you do what you really want to do with the computer. Examples include word processors, spreadsheets, accounting packages, games, and media players. One of the exciting aspects of today's information systems landscape is the dizzying array of applications that are available. As more and more applications are developed, our computers become more and more useful (and more fun!).

Some applications are installed directly on your computer, tablet, or other device. Others are "in the cloud," which means that they are accessed through a network, usually the Internet, rather than being stored on your computer. Cloud applications have many advantages, including being able to be accessed from virtually anywhere that has network access. One disadvantage to that, however, is that they typically require network access to operate, although some cloud applications allow limited offline operation. In Chapter 7 we discuss cloud computing in more depth.

There are also different types of hardware involved in an information system. In addition to the computers involved, there are storage devices (such as hard disks), printers, keyboards, and communication devices. You can learn more about hardware in Appendix A.

To understand how the elements of an information system fit together, let us consider a simplified view of an information system for hotel reservations. (To keep this discussion from being too long, we are omitting some details of the system.) The goal of this system is to allow reservation agents at a hotel to reserve rooms for guests. Let us look at the specific roles each element plays in this information system:

- *Data:* The data serve as the facts that are manipulated by the system to produce information that is used by the reservation agent and hotel management.
    - A list of rooms, their status (available or reserved), prices, payments, customer information, and name of reservation clerk
- *Hardware:* The hardware performs computations, stores the data and software used by the system, displays information, and provides the platform for users to interact with this system.
    - A reservation agent's personal computer, a computer that has the software for the system, hard drives that have the database that stores the data for the system, and printers for reports.

- *Software:* The software controls the operation of the computer, including how the data are retrieved, manipulated, and communicated.
  - •• The operating system of the various system computers and one or more applications specific to the task of making room reservations
- *Communication media:* This allows the various hardware components to communicate with each other.
  - •• The network cabling and other devices that facilitate communication, such as routers
- *Procedures:* The procedures govern how the reservation agent and hotel managers should interact with the system.
  - •• Instructions for how to get into the system, which data have to be entered, how to retrieve information, and how to generate reports
- *People:* In this system, people provide input for the system, control how the system is used, and interpret the information from the system.
  - •• Customer, reservation agent, and hotel managers

Now that you know the basics of information systems, let us turn our attention to how information systems help businesses deal with the tremendous volume of information they must cope with every day.

How Information Systems Help Us Deal with Information

Businesses face an ever-increasing amount of information. Given the increasing complexity of today's business environment, this is unlikely to change any time soon. We are fortunate to have information systems to help us cope with this tsunami of information. Let us look at how information systems accomplish this.

First, information systems let us gather large amounts of data quickly, easily, and reliably. Think about your last trip to the grocery store. Today, most stores have electronic scanners that read a bar code (called the universal product code, or UPC) that is printed on most products. The checkout clerk runs each product across this scanner, which reads the UPC and electronically enters the fact that the product is being purchased into the system that facilitates the checkout process. (Eventually, these data also go into other systems, such as the inventory control system.) The system retrieves the item's price from a database, and the amount is added to the order's total. (This is an example of a point-of-sale information system.) All this happens in a fraction of a second, and many different clerks can be performing the same basic operation at one time. In addition, the data are entered very accurately. A large grocery store has tens of thousands of transactions per day, so being able to gather these data quickly and accurately is important.

Second, information systems allow businesses to store and organize very large amounts of data. Databases with multiple terabytes (approximately one trillion

---

STATS BOX 2.1

Growth of Applications in Apple's App Store

By any measure, Apple's iPhone has been very successful. One reason for its success is the large and growing number of applications, or "apps," available in Apple's App Store. These applications make the phone more useful to the owner, whether the app is work related or just a fun game. Figure 2.4 shows the steady growth of the number of available apps. Google and Amazon have similar app stores for Android and Amazon devices. (The decline in the number of apps in 2020 is due to stricter standards that led to some apps being removed from the App Store.)

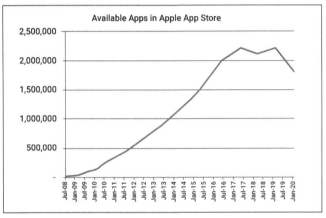

Figure 2.4 Available Apps in Apple's App Store Sources: http://en.wikipedia.org/wiki/App_Store_(iOS), https://www.lifewire.com/how-many-apps-in-app-store-2000252.

---

characters of data) are common. Information systems allow businesses to store volumes of data in an organized manner that allows for rapid retrieval. This simply would not be possible with manual systems.

Third, information systems perform their data manipulations quickly, accurately, and consistently. As long as the hardware is operating correctly and the software is designed and implemented correctly, an information system is very consistent in its manipulations. Using our grocery store example, the checkout system will retrieve the same price for a given item every time that item is scanned (unless the price changes). It will also correctly calculate any necessary sales tax and will correctly total the customer's order. Again, this degree of speed, accuracy, and consistency simply cannot be matched with manual systems.

Finally, information systems let us retrieve and output information in a variety of forms, depending on what is useful to the user. The same information can be displayed on a screen, printed, or graphed. Different purposes may require different forms of displaying data; charts and graphs are good for showing large amounts of data at a glance, while printed reports are often better for showing details.

BUSINESS EXAMPLE BOX 2.1

How Uber Provides On-Demand Service with Information Systems

Founded in 2009, Uber Technologies Inc. provides on-demand ride services in more than 600 cities worldwide. As the company name implies, technology is central to Uber's operations. In fact, technology is so important to Uber that the company describes itself as a "technology platform" (https://help.uber.com/riders/article/how-does-uber-work?nodeId=738d1ff7-5fe0-4383-b34c-4a2480efd71e). Uber uses a smartphone app to connect drivers with those wanting a ride. When you want a ride from Uber, you request the ride using the app. The request goes out to nearby drivers. Once a driver accepts the request, the app gives you information about the driver (such as the driver's name, a description of his or her car, and the car's license plate number) and provides an approximate wait time. The app also provides a quote for the cost of your ride. As your driver heads toward your pickup location, the app shows you a map with the driver's progress. When your ride is completed, Uber's information system automatically charges your preselected payment method for the ride.

Although Uber's app is simple to use, the supporting information systems are complex. For example, Uber uses geolocation technologies to locate riders and drivers. Google Maps and other mapping technologies provide routing information that shows progress during the ride. SMS (short message service) technologies allow Uber to inform you when a driver accepts your ride request and when the driver is approaching the pickup location. Uber uses dynamic, demand-based pricing. Other systems process your payment for the ride and pay the drivers. This is only a partial list of the various systems that must work together seamlessly for Uber to provide its services. For a more detailed view of Uber's technology, see https://eng.uber.com/tech-stack-part-one/.

Source: Abrosimova (2014).

One important function of many business information systems is to enforce **business rules.** A business rule is a statement that defines or constrains an aspect of a business with the intent of controlling behaviors within the business. All businesses have rules that govern their operations. The following are some examples:

- A hotel reservation system that will not allow assigning a room that is already occupied
- A course registration system that does not allow registering for a course that is full
- An order entry system that automatically calculates the proper sales tax and adds it to the customer's invoice

Information systems enforce business rules by not allowing violations to occur. Let us look at an example. Suppose your company had a business rule that said you could not accept an order from any customer who is more than 30 days past due on his or her bill. Only a supervisor can override this rule. The business's order entry

---

LEARNING ACTIVITY 2.4

Information Systems and Change

During the COVID-19 pandemic, many universities had to shift quickly to online classes. Various information systems helped make this shift possible. Identify two information systems that enabled the change to online classes. Briefly describe each system and discuss how it was used. Also, briefly comment on the effectiveness of each system and identify ways that the systems might be used to improve face-to-face classes.

(*Note:* If your school did not move classes online, research the technologies used by those schools that did.)

---

system enforces this rule by not allowing sales representatives to enter orders from past-due customers. When such customers want to place orders, the system refuses to accept the order until a supervisor gives approval. Later, we will tell you about how information systems are analyzed and designed. One critical aspect of systems analysis and design is to properly understand business rules, which are then carefully built into the design of the system.

How Information Systems Facilitate Organizational Change

There are many ways information systems help managers bring about organizational change. To help you understand how this occurs, we break down these ways into four categories: process improvements, automation, control, and information flow.

Process Improvements

The increasing use of information systems was partially responsible for the business process redesign movement that started in the 1990s and continues today. Information systems can help organizations improve both the efficiency and effectiveness of processes. For example, many companies use information to enable customer self-service, which means that individual customers interact with systems to perform their own customer-service activities. A classic example of this is the automated teller machine (ATM). In the past, when you needed cash, you would go into your bank and either cash a check or make a withdrawal. Now you simply find a convenient ATM, put in your card, push a few buttons, and get your cash. Some banks, such as Wells Fargo, even allow you to access the ATM using a smartphone app; you do not even need a card. The old process was less efficient from the bank's perspective because of the need for the teller's time. The ATM illustrates how information systems can bring about process improvements and control. Not only is the process more efficient, but it is also more effective because of the reduced number of errors and the increased convenience for the customer. Of course, there are many other examples of information systems improving efficiency and effectiveness, including the course registration example at the beginning of the chapter.

Automation

Some processes have been totally automated. A good example of this is online order-ing. Prior to the popularity of the Web, ordering something from a company typi-cally involved either calling the company and placing the order with a sales represen-tative or filling out and mailing in an order form, which was then entered into the company's order system by a clerk. Today, you simply go to the company's website, fill out an online form, and the order is directly entered into the order system. Even something as simple as buying gasoline is often fully automated. When you pull up to the pump, all you need to do is slide your credit or debit card through the slot, fill your tank, take your receipt, and go on your way. (Some gas stations even allow you to pay using a smartphone app.) In the past, an attendant would make an imprint of your card on a special form and would then fill out the relevant information. This form would go to the bank, which would do additional processing so that the gas station could be paid for your purchase.

Control

Information systems also enable organizational change by improving process con-trols. When properly designed and implemented, an information system can ensure that business rules are followed throughout a process. For example, an order entry system might refuse to accept an order for a customer whose account is past due. Returning to our gas station example, prior to automation, the station attendant was required to check a "hot list" of stolen or suspended credit cards before accepting a card as payment. As you might imagine, this business rule was often violated, espe-cially when the station was busy. Today, the information system checks the status of the account before allowing the customer to pump any gas. The course registration system from the focusing story is another example of enforcing rules. Under the less-automated system, it was relatively easy to register for a class even if you did not have the necessary prerequisites. (Do not ask us how we know this!) Today's systems typically check prerequisites before allowing a student to register for a course. (This assumes that the prerequisites have been properly entered into the system.)

Information Flow

Finally, information systems improve communication and information flow in orga-nizations. Workflow systems facilitate information flow throughout a work task. Consider a simple example of a grade change. The instructor enters the grade change into the information system. The system generates an email notifying the appropri-ate administrator that a grade change is pending her or his approval. The adminis-trator reviews and approves or denies the request. If the request is approved, the system notifies the instructor and student, and the grade change is recorded and made permanent. Throughout the process, the system manages the flow of informa-tion and controls the process. Information systems also facilitate information shar-ing. Examples include document repositories and communication systems, such as email.

BUSINESS EXAMPLE BOX 2.2

AgTech—Making Farming More Efficient and Sustainable

As the world's population continues to grow, providing adequate food supplies is an ongoing problem. At the same time, societies around the world are increasingly concerned about the environmental impact of food production. AgTech (the use of digital technology to improve agriculture) may be the key to increasing food production in an environmentally sustainable way. A wide array of information systems helps gather and process information to increase yields and improve efficiency, especially with respect to resource use. For example, geographic information systems combined with other technologies allow farmers to precisely apply fertilizer to specific areas based on soil composition, weather, and other factors. Sensors and drones combine to provide information that lets farmers better determine how specific areas of their fields should be irrigated. Both of these examples improve the sustainability of agriculture by reducing resource use.

Some dairy farms are making extensive use of robots to mix and distribute feed to cows with minimal human intervention. Automated milking machines not only milk the cows but also monitor cows' nutritional needs. The milking machine communicates with the feed robots to adjust the mixture of the feed to deal with various health issues. These systems help the dairy farm maximize milk production without harming the health of the cows. In fact, one of the goals of these systems is to keep cows stress-free, since stress reduces milk production. From more precise irrigation to stress-free cows, AgTech holds considerable promise for supporting the food needs of a growing population while at the same time reducing the environmental impact of agriculture.

Sources: CB Insights. 2017. "The Ag Tech Market Map: 100+ Startups Powering the Future of Farming and Agribusiness." https://www.cbinsights.com/research/agriculture-tech-market-map-company-list/; Kinnard Farms. 2016. "Kinnard Farms Dairy Tour." https://www.youtube.com/watch?v=v761DILatzg; Kite-Powell, J. 2016. "Take a Look at How Technology Makes Smart and Sustainable Farming." *Forbes*, December 31. https://www.forbes.com/sites/jenniferhicks/2016/12/31/take-a-look-at-how-technology-makes-smart-and-sustainble-farming/#6667d7bb3deb.

The ways information systems enable change combine to allow for changing organizational forms. To help you understand this, we discuss virtual teams and the role of middle managers. Members of virtual teams are not necessarily located in the same place; they may literally be spread around the world. Information systems allow these teams to work together by providing communication functions such as email and videoconferencing, which allow the members to interact even though they are not colocated (in the same location). Other tools, such as shared calendars and document repositories, also help virtual teams function.

On a larger scale, information systems help companies flatten their organizational structure. In the past, many midlevel managers existed to monitor the activities of lower-level workers and to facilitate communication between lower and upper levels of the organization. Today, information systems can serve some of these same functions by enforcing business rules and facilitating communication throughout various processes. In addition, communication technologies such as email and websites improve communication flow across organizational levels and provide employees with information that previously only existed at higher levels of the organization. As a result, many companies have flattened their organizational structures.

Our goal up to this point in the chapter has been to help you understand information systems and their role in organizations. In the next section, we provide brief overviews of some common information systems. Many of these are discussed in greater detail later in the book.

Common Information Systems

There are numerous different types of information systems in modern organizations; we will cover many of these in the rest of this book. For now, it is useful to get a glimpse of the range of information systems used by organizations. A useful way to do this is to classify information systems according to the impact or "reach" of the system, as shown in Figure 2.5. At the bottom of this triangle are personal applications, which are typically used by one person or a small group. At the top are global systems, the use of which spans multiple organizations around the world.

To further help you understand the range of information systems available today, we provide some brief examples of systems at each level of our triangle:

- *Personal applications:* These systems help make individuals' daily work more efficient and effective. Office automation systems (such as office suites like Microsoft Office) are included in this category. Other examples include note-taking systems such as Evernote or OneNote, to-do list systems, personal information managers, personal calendars, and drawing programs.

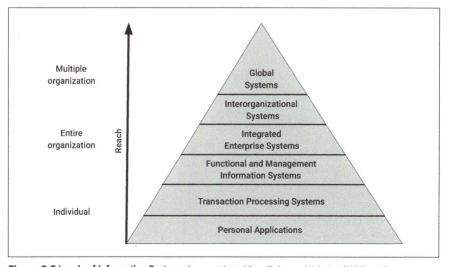

Figure 2.5 Levels of Information Systems Source: Adapted from Turban and Volonino (2009), p. 43.

- *Transaction processing systems (TPSs):* These systems collect, monitor, process, report, and store large volumes of data that are created by business processes. The grocery store point-of-sale system we mentioned earlier is an example. Many organizations have a large number of TPSs. Other examples include payroll processing systems, invoicing systems, and inventory control systems. (Note that many of these stand-alone TPSs have been replaced by integrated enterprise systems, which we discuss later.) In the earlier days of information systems, most TPSs operated in batch mode, which means that they gathered all the data to be processed and then processed the data in one batch. The modern trend is toward online transaction processing (OLTP) systems, which process the data as soon as they are gathered.
- *Functional and management information systems:* These systems monitor, control, and analyze the operation of functional areas. Examples include financial management systems, sales force automation systems, materials requirements planning (MRP) systems, and benefits administration systems. Collaboration systems such as Slack and Teams and videoconferencing systems such as Zoom are included in this category. As is the case with TPS, integrated enterprise systems have replaced some of these stand-alone systems. However, collaboration and videoconferencing systems are experiencing increased use.
- *Integrated enterprise systems:* Today, many organizations (particularly large organizations) have replaced many of their stand-alone TPS and functional systems with integrated, enterprise-wide systems that impact multiple functional areas. A stand-alone system is primarily used in isolation from other systems. An integrated system has multiple applications in a cohesive, interrelated system. For example, enterprise resource planning systems provide an integrated set of modules that carry out the information processing and reporting systems for the entire organization. Enterprise information systems are very important; we devote an entire chapter (Chapter 12) to them.
- *Interorganizational systems:* These span organizational boundaries to connect companies to suppliers and customers. **Electronic data interchange (EDI)** is at the heart of many of these systems. EDI allows the systems in one organization to directly interact with those in a partner organization. Many electronic business systems enable partner organizations to interact seamlessly and thus are considered interorganizational systems.
- *Global systems:* These are simply interorganizational systems that cross national boundaries. These systems are often more complex than other systems because of cross-national differences in language, currency, and culture.

Stand-alone systems are often integrated so that they can exchange data. For example, an accounting application may pull data from a transaction processing system. Such integration is very important to the smooth operation of an organization.

Chapter Summary

This chapter helped you learn how information systems help organizations operate effectively. This knowledge gives you a foundation that will be helpful in understanding the rest of the book. Here are the main points discussed in the chapter:

- Information systems combine technology, data, people, and processes to help collect, manipulate, store, organize, retrieve, and communicate information.
- In today's information-rich business environment, business professionals must be skilled at dealing with a variety of information systems.
- Systems are made up of components that work together to achieve a goal by taking inputs and processing them into outputs.
- The information processing cycle consists of input, processing, storage, output, and control.
- Information systems are made up of six critical elements: data, hardware, software, communication media, procedures, and people.
- Information systems help managers deal with information by improving how data are collected, organized, manipulated, and output.
- Information systems can help enforce business rules.
- Information systems facilitate organizational change through improving, automating, and controlling processes and improving information flow.
- Categories of common information systems include personal applications, transaction processing systems, functional and management information systems, integrated enterprise systems, and global systems.

Review Questions

1. Explain why business professionals need to understand information systems.
2. Define the term *system*.
3. Name the key system concepts discussed in this chapter.
4. Describe the information processing cycle.
5. Name and describe the six critical elements of an information system.
6. Why are people a critical element of an information system?
7. Contrast system and application software.
8. Name four ways information systems help managers deal with large volumes of information.

9. What is a business rule? How do information systems relate to business rules?
10. Name five ways information systems can facilitate organizational change.
11. Name and briefly describe six types of common information systems.

Reflection Questions

1. What is the most important thing you learned in this chapter? Why is it important?
2. What topics are unclear? What about them is unclear?
3. What relationships do you see between what you learned in this chapter and what you learned in Chapter 1?
4. How do information systems impact your life? Pick one information system and discuss how your life would change if it did not exist in a computerized form.
5. Think about your future career. How do you think you will use information systems? In what ways will better understanding information systems help advance your career?
6. In the text, we said that people are key components of information systems. How can people cause an information system to fail? Provide specific examples to back up your claims.

Additional Learning Activities

2.A1. Relate the information processing cycle to the class registration system described at the beginning of the chapter. Provide examples of the inputs required, the processing that occurs, outputs that are produced, data that are stored, and controls that should be in place.
2.A2. Use the information processing cycle to discuss a system that allows consumers to make purchases from an online store. (Your instructor may give you a different information system example to use.)
2.A3. Consider an information system that processes a company's payroll. Give specific examples of each of the critical elements involved in this information system. (The critical elements are data, hardware, software, communication media, procedures, and people.)
2.A4. Identify three business rules that you have encountered. For each rule, briefly discuss how information systems enforce the rule. (For example, a course registration system would block a student from registering for a course if he or she did not have the necessary prerequisites.)
2.A5. Find an example of how information systems facilitated organizational change. Using the categories presented in the chapter, discuss how the information system helped facilitate the change.

2.A6. Describe two examples of how an information system improved the efficiency or effectiveness of a business process. Be specific: What elements of the process were improved? How? Did the improvement benefit both the organization and its customers or suppliers?

2.A7. Find a specific example of an information system for each of the six levels of common information systems described in the chapter. For instance, Microsoft Excel would be an example of a personal application. (*Note:* Do not use the examples provided in the chapter!)

2.A8. Suppose a friend is starting a small coffee shop. Recommend five information systems that would help your friend run the shop. Briefly explain why each information system would be helpful.

References

Abrosimova, K. 2014. "Building an App Like Uber: What Is the Uber App Made From?" *Medium.* https://medium.com/yalantis-mobile/uber-underlying -technologies-and-how-it-actually-works-526f55b37c6f.

Pauli, D. 2010. "IT and Beer: A Love Story." *Computerworld,* June 4. https://www .computerworld.com/article/3470392/it-and-beer-a-love-story.html.

Turban, E., and L. Volonino. 2009. *IT for Management: Improving Performance in the Digital Economy.* New York: John Wiley & Sons.

Glossary

Business rule: A statement that defines or constrains an aspect of a business with the intent of controlling behaviors within the business.

Control: A set of functions intended to ensure the proper operation of a system.

Electronic data interchange (EDI): A B2B e-business model focusing on the electronic exchange of information among two or more organizations using a standard format.

Equifinality: The idea that in an open system, there are many different potential paths to the final outcome.

Feedback: A process by which a system regulates itself by monitoring its own output.

Information system (IS): A combination of technology, data, people, and processes that is directed toward the collection, manipulation, storage, organization, retrieval, and communication of information.

Open system: A system that interacts with its environment.

Subsystem: A system that is part of a larger system.

System: A set of interacting components that work together to form a complex whole by taking inputs and processing them to produce outputs.

Transaction processing system (TPS): A system that collects, monitors, processes, reports, and stores data generated by an organization's transactions.

Evaluating Information

Learning Objectives

By reading and completing the activities in this chapter, you will be able to:

- Discuss why it is important both personally and professionally to be an informed information consumer
- Describe information overload, its consequences, and approaches for dealing with information overload
- Discuss the relationship between information overload and information evaluation
- List and describe the dimensions of information quality
- List and describe the elements of an information evaluation framework
- Given an information-related task, evaluate information for its usefulness and believability

Chapter Outline

Being a Smart Information Consumer
Focusing Story: Biased Information in a Trusted Outlet
   Learning Activity 3.1: How Good Is This Information?
Information Overload and the Need to Evaluate Information
   Learning Activity 3.2: My Online Life and Information Overload
Dealing with Information Overload
Information Quality
   Learning Activity 3.3: Dimensions of Information Quality
Evaluating Information
   Learning Activity 3.4: Evaluating Information Sources

With the tremendous amount of information available today, information evaluation is an important skill. The goal of this chapter is to help you gain the knowledge and skills necessary for effective information evaluation.

Being a Smart Information Consumer

Being successful in today's knowledge society requires being a good information consumer. We are faced with an ever-increasing array of information. Being able to

The Personal Consequences of Misinformation

Although "fake news" has only recently become a widely used term, misinformation and disinformation spread through media have been around for a very long time. Publications such as *The National Enquirer* and *The Weekly World News* (often called "tabloids") have been a fixture of grocery store checkout lines for many years. The rise of online outlets coupled with the ease of sharing stories on social media has made fake news a matter of grave concern. Social media makes it simple to share a link to a news story, which can then be shared by others, and so on, as the story goes viral. Sometimes the stories cause little harm and fade away almost as quickly as they spread. However, false news stories can carry significant consequences. Consider the case of Kaci Hickox, a volunteer nurse for Doctors Without Borders, who served in West Africa during the Ebola hemorrhagic fever outbreak. (Ebola is a deadly disease that is largely untreatable. Ebola outbreaks have occurred several times, including the 2013–2016 West Africa outbreak.) When Ms. Hickox returned to the United States, she endured an 80-hour quarantine that was imposed by the New Jersey Department of Health. She was released after a negative Ebola test. According to news outlet NJ.com, New Jersey Governor (at the time) Chris Christie apologized for inconveniencing Ms. Hickox but went on to state that she was "obviously" still sick and might have to undergo another Ebola test. According to NJ.com, Governor Christie also stated, "There is no question the woman is ill, the question is what is her illness." In reality, a forehead thermometer at Newark airport showed a slightly elevated temperature, but an oral thermometer showed a normal temperature. She did not display any other symptoms. The governor's statements were picked up by other outlets as the story spread. When Ms. Hickox returned to her home in Maine, Governor Paul LePage tried to force Ms. Hickox into a three-week quarantine, although she refused to abide by the quarantine. Her landlord also asked her and her partner to move out of their home, purportedly over concerns about Ebola. Local police received calls from citizens concerned about the nurse being in their area. Ms. Hickox also received death threats, and her partner was forced to leave nursing school because of the school's concerns about him living with Ms. Hickox. Eventually, Ms. Hickox left Maine. She did sue and settled cases with the states of Maine and New Jersey over the incident. This nurse, who volunteered to help in the fight against Ebola, suffered significant personal consequences because of the widespread sharing of inaccurate information.

Governor Christie's statements might be classified as misinformation, since he likely did not intend any harm to Ms. Hickox and may not have even been aware that his statements were inaccurate. Although there is not universal agreement on these definitions, misinformation is generally considered to be information that is false but was not intended to cause harm. In contrast, disinformation is false information that is intended to cause harm (Spies 2020). Sometimes "fake news" stories are satire, such as articles in *The Onion*. These are intentionally false but are intended as amusement rather than being created to mislead. (Ethically, satirical news stories should be clearly identified as such—unfortunately, this is not always the case.)

Kaci Hickox's story is just one example of how misleading or inaccurate news stories can cause real harm. While most of us enjoy the rapid communication enabled by modern technologies, these same technologies can also speed the spread of harmful, false news.

Sources: Akpan, N. 2016. "The Very Real Consequences of Fake News Stories and Why Your Brain Can't Ignore Them." *PBS News Hour*, December 5. https://www.pbs.org/newshour/science/real-consequences-fake-news-stories-brain-cant-ignore; Arco, M. 2019. "Christie Defends Ebola Quarantine Announcement, Saying He's Trying to Protect N.J. Residents." *NJ.com*, March 29. https://www.nj.com/politics/2014/10/christie_defends_ebola_quarantine_announcement.html; Sullivan, G., and A. Ohlheiser. 2014. "Maine Gov. Paul Lepage Is Seeking Legal Authority to Enforce Ebola Quarantine on Nurse." *The Washington Post*, October 29. https://www.washingtonpost.com/news/morning-mix/wp/2014/10/29/after-fight-with-chris

-christie-nurse-kaci-hickox-defies-ebola-quarantine-in-maine/; Sherwood, D., and C. Jenkins, 2014. "Nurse Kaci Hickox and State of Maine Settle Quarantine Lawsuit." *Scientific American*, November 3. https://www.scientificamerican.com/article /nurse-kaci-hickox-and-state-of-maine-settle-quarantine-lawsuit/; Washburn, L. 2017. "Quarantined Ebola Nurse Settles Case against Gov. Christie." *NorthJersey.com*, July 28. https://www.northjersey.com/story/news/health/2017/07/28/ quarantine-ebola-patients-bill-of-rights/520480001/.

Focusing Questions

1. What are some other consequences of fake news?
2. How can you spot fake news? What aspects of a news story might cause you to question its truthfulness?
3. Think of a time you saw a fake news story online. Did you believe the story at first? What led you to realize the news was "fake"?

LEARNING ACTIVITY 3.1

How Good Is This Information?

Good decisions rely on good information, and bad information can lead to bad decisions. But what do we mean by "bad" information? In small groups, discuss the meaning of information quality. Develop a list of characteristics of "good" information.

After your discussion, your instructor will give you an information source, such as a blog or wiki. Evaluate the information contained in the source using the characteristics you developed as a guide.

---

deal with that information is a key life skill. In Chapter 1, we discussed the concept of information literacy. You may recall that being able to effectively evaluate information is an element of information literacy. The goal of this chapter is to help you improve your information evaluation skills.

As is often the case, the amount of readily accessible information available online has both advantages and disadvantages. The obvious advantage is that we now have easy access to information that would have been quite time-consuming and difficult to track down just a few years ago. Unfortunately, there are also a few downsides. One of the great things about the Internet is that there are almost no "gatekeepers" who determine what can be posted. However, this also means that there is no quality control. Almost anyone can post information about almost any topic. In more traditional media, evaluating the quality and correctness of information was the job of editors and publishers. Today, with respect to much of the information on the Internet, that responsibility shifts to the information consumer (in other words, you). So the ability to evaluate information is an important skill.

Often when we deal with something routinely, we tend to go into an autopilot mode; we act without really thinking. Being faced with so much information, we run the risk of using the information without thinking critically about it. When we are making small decisions, this is usually fine. However, when making higher impact decisions, it is worthwhile to be a more discerning information consumer, especially by critically evaluating the quality of the information.

STATS BOX 3.1

The Cost of Poor Information Quality

Poor-quality information is costly. One study estimated that 75% of organizations believe that customer service quality suffers because of inaccurate information. According to that same study, human error is the most common cause of inaccurate data (Experian Data Quality 2016). Another study, conducted by the Data Warehousing Institute, estimates that poor-quality data cost organizations in the United States more than $600 billion per year (http://www.crmbuyer.com/story/44711.html?wlc=1278446605).

Information evaluation is the systematic determination of the merit and worth of information. Information evaluation skills will be important to you both personally and in your business career. Personally, you need to be able to sift through and evaluate many kinds of information. As you go through life, you will have to make many decisions, some big and some small. When facing a big decision, such as choosing a job or making a major purchase, the consequences of relying on bad information can be quite severe. As you saw in the focusing story, acting on bad information can lead to bad outcomes.

The need to intelligently deal with information is also important to your career. As is the case with your personal life, you will face many decisions throughout your career. Knowing how to evaluate the information you must deal with is particularly important. Much of your reputation, and thus your career success, depends on the outcomes of the decisions you make. Making decisions based on better information usually leads to better outcomes.

In business, we often use information to reduce uncertainty. The more uncertainty there is surrounding a decision, the more we seek information to reduce that uncertainty. This makes intuitive sense, but there is an underlying assumption that the information is of good quality. Poor-quality information is ineffective at reducing decision uncertainty.

Consider the example of hiring a new employee. When a manager is evaluating potential new employees, he or she gathers information about the person to increase the odds of making a good hiring decision. Information on the applicant's experience, education, and character all come into play. If this information is accurate, then the manager is better able to hire the right person. But what if the information on the applicant's resume is false? In this case, the likelihood of hiring the right person decreases.

The bottom line here is that it is important to be able to intelligently evaluate all sorts of information. In this chapter, we help you understand information quality and provide some approaches to evaluating information.

BUSINESS EXAMPLE BOX 3.1

Business Impacts of Information Quality

As you will learn throughout this book, using information more effectively is the major reason businesses invest in information systems. So, you might wonder how good and bad information affects businesses. As noted earlier, high-quality information improves decision-making. For example, high-quality marketing and sales information allows businesses to price products correctly, target the right customers, and stock products that are in demand. Good-quality information also improves productivity, since workers spend less time checking on and fixing information errors. Quality information is also important to regulatory compliance. Many companies are required to make accurate reports on various aspects of their business. When these reports contain inaccurate information, significant legal troubles can result. Bad information can harm reputations, such as when Coca-Cola released "New Coke" based on poor-quality market research data. Poor-quality information can also lead to missed opportunities and lost revenue, among many other negative outcomes.

Source: Forbes Staff. 2017. "Poor-Quality Data Imposes Costs and Risks on Business, Says New Forbes Insights Report, May 31. https://www.forbes.com/sites/forbespr/2017/05/31/poor-quality-data-imposes-costs-and-risks-on-businesses-says-new-forbes-insights-report/#5479c128452b.

Information Overload and the Need to Evaluate Information

Herbert Simon (1971), the Nobel Prize–winning psychologist and economist, stated in 1971:

> In an information-rich world, the wealth of information means a dearth of something else: a scarcity of whatever it is that information consumes. What information consumes is rather obvious: it consumes the attention of its recipients. Hence a wealth of information creates a poverty of attention and a need to allocate that attention efficiently among the overabundance of information sources that might consume it.

Although he did not use the words "**information overload**," that is basically what he was talking about: being faced with more information than we can effectively process. The more information we have to sift through, the less attention we have to devote to other tasks. Information overload is a very real problem, in terms of both our business and personal lives. It reduces productivity, increases stress, and can actually lead to physical health problems.

Business managers bring some of this on themselves. Managers know the value of information and gather information for many different reasons (Butcher 1998):

• To improve decision-making
• To justify decisions
• To verify previously acquired information

LEARNING ACTIVITY 3.2

My Online Life and Information Overload

We all face a daily torrent of information. The more "connected" you are, the more information you face. As you go throughout tomorrow, pay attention to how you deal with the information you face. Prepare a one- or two-page report that addresses the following questions:

1. What strategies do you use to determine what information is important and to reduce the amount of information you deal with?
2. How successful are these strategies?
3. How could you improve your approach to information filtering?

- To "play it safe" by making sure they do not miss any relevant information
- To use the information later

Add to this the amount of unsolicited information that managers face, and you can see why information overload is a problem for many managers.

To better understand this, we will use the example of an information technology (IT) manager named Pat. Pat's company is getting ready to purchase a large number of laptop computers, and Pat has to decide which brand of computers to buy. (The company has already decided on the specifications needed.) Pat could easily become overloaded with information for this decision. She may begin reading reviews of the various brands, checking prices and specifications, and so on. She may reach out to her colleagues to see what their experiences have been with the brands being considered. While this is a big decision, Pat could easily become bogged down with more and more information. Being able to deal with these situations is important to being a successful manager.

Two major strategies for dealing with information overload are filtering and withdrawal (Savolainen 2007). *Withdrawal* essentially involves disconnecting from sources of information: not checking email, turning off the television, not surfing the Web, and so on. This may not always be a viable option for business professionals, so filtering may be a better choice for many of us.

Filtering information involves knowing what information we need and what information merits attention and use, which makes being able to evaluate information a critical skill in today's information-rich world. Most of us evaluate information almost constantly, but without really paying attention to how we do so. For example, we scan our email inboxes and choose which messages to open, which to save for later reading, and which to delete or ignore. We make these decisions, in part, based on an evaluation of the information we think is contained within the message. In most cases, we evaluate the information in email messages based on its relevance to whatever we are working on at the time. Relevance is one aspect of information quality, which we discuss later in the chapter.

If we dig a little deeper, we see that relevance is only one piece of the puzzle.

Dealing with Information Overload

How can you deal with the torrent of information you face every day? Fortunately, there are a number of techniques you can use to help avoid information overload. First, understand and identify your information priorities and needs. Once you have determined them, try to focus on the most relevant information. This can be difficult, especially when you see information that is not really relevant at the moment but may be later. To deal with these situations, it is a good idea to develop some organizational structure for this "may need later" information. There are some good technology solutions for this. If the information is mostly Web based, an application such as Pocket (https://getpocket .com) provides an easy means of saving items for later reading. You can also use a note-taking application such as Evernote (http://www.evernote.com) to save the Web page to read later and to make a quick note about the pages you have saved. Evernote lets you tag notes and organize them in notebooks, which can make recalling information easier. You can also use something like Google Docs to make a quick note or record a link for later use. Be sure to put something meaningful in the document so you can find it later. Finally, think about what information sources are the most useful and of the highest quality. Be sure to consider these sources first when faced with relevant information needs. This can save quite a bit of searching time.

Whether consciously or unconsciously, we evaluate other aspects of information when we decide what to do with it. If the information is too old, untrustworthy, or otherwise of questionable quality, we essentially ignore the information and do not use it. (At least, that is what we should do.) As you can see, improving your information evaluation skills will help you better deal with information overload.

Information Quality

"Garbage in, garbage out." This old saying succinctly sums up the importance of information quality. As we mentioned in previous chapters, businesses rely on information to carry out business processes and to make decisions, both of which require good information. If you use bad information as the basis for a decision, you are probably going to make a bad decision. But just what does "good" information mean?

The issue of information quality is surprisingly complex. As you may have noticed when doing your research for Learning Activity 3.3, there are many different opinions regarding the meaning of information quality. We like the definition of **information quality** as information that is fit for its intended use. In other words, the information is useful toward the achievement of whatever task is at hand. While we like this definition, it is not particularly useful in helping us understand information quality. A more useful approach is to think about the dimensions of information quality: the characteristics of information that make it useful or not useful.

Dozens of studies have investigated the issue of the dimensions of information quality. As a result, there are many different lists of information quality dimensions. (Some of these studies are listed in the References section at the end of the chapter.)

LEARNING ACTIVITY 3.3

Dimensions of Information Quality

Information quality is an important, complicated topic. Because of this, there are many views on what constitutes "high-quality" information. Search the Web to discover different views on the dimensions of information quality. What, in your opinion, are the three most important dimensions of information quality? Why do you think these are the most important?

Richard Wang and Diane Strong (1996) developed a framework that is useful for understanding and evaluating information quality. (They used the term *data quality*, but the concepts apply to information quality.)

Wang and Strong put quality dimensions into four categories:

- *Intrinsic quality* includes dimensions of quality that are important regardless of the context or how the information is represented.
- *Contextual quality* includes the dimensions that may be viewed differently depending on the task at hand.
- *Representational quality* concerns how the information is provided to the user.
- *Accessibility quality* has to do with whether authorized users can easily access the information.

In this chapter, we focus on the first two categories. Table 3.1 shows the dimensions in these categories and provides definitions for each dimension.

We need to make a few points regarding the information in Table 3.1. First, there is considerable disagreement regarding the dimensions of information quality. It is possible (even likely) that your instructor will have a different perspective on which dimensions are most important.

Second, we want to reemphasize the importance of considering context when thinking about information quality. Consider the example of stock price information. It is common for free information services (such as Yahoo! Finance) to delay stock price information by 15 minutes. If you are a casual investor, this usually is fine; the delayed information is acceptably current for keeping track of the value of your portfolio. However, if you are a professional stock day trader, a 15-minute delay is unacceptable (even a 15-second delay could be a problem) and would be considered poor-quality information.

Finally, information quality has a cost. On the surface, it may seem like we should want the highest quality information possible. However, this typically is not true. Few individuals or organizations are willing to invest the resources necessary to ensure the highest possible information quality. More commonly, we want information that is of sufficient quality to carry out tasks effectively. In other words, we want "good enough" information quality. This is not to say that information quality is not

TABLE 3.1 Information Quality Dimension Definitions	
Intrinsic Dimension	**Definition: Extent to which the information is . . .**
Accurate	Correct, free from error, and reliable
Believable	Regarded as true and credible
Objective	Free from bias
Consistent	Compatible with previous information
Understandable	Easily comprehended
Contextual Dimension	**Definition: Extent to which the information is . . .**
Relevant	Applicable and useful for the task at hand
Timely	Available in time to perform the task at hand
Comprehensive	Of sufficient depth and breadth for the task at hand
Current	Sufficiently up to date for the task at hand
Sources: Adapted from U.S. Department of Justice (2010) and Knight and Burn (2005).	

important; quite the opposite is true. Information quality is critical to business success. However, it is also important to consider the costs of information quality and what level of cost is justified.

At a more micro level, a good way to think about the costs of information quality is to consider the possible consequences of poor-quality information. For more important, higher impact decisions, it is worthwhile to pay much more attention to information quality than for lower consequence decisions. If you are buying a house, you want very high-quality information. If you are buying a pack of notebook paper, lower quality information is probably fine. We will talk more about this in Chapter 6, which covers decision-making.

It is easy to see how poor-quality information can negatively impact a business. Consider when you order a product from an online store. If the description of the product is inaccurate, you may end up with a product that does not meet your needs. If the inventory information is not current, you may not receive the product when expected. Biased reviews may lead you to order a poor-quality product. We could go on, but you probably have the idea. In almost any business situation, you can easily see the impact of poor-quality information.

Let us return to Pat, the IT manager who needs to pick a laptop brand. She needs to pay careful attention to the quality of the information she is using to make this decision. She clearly needs to have accurate information, but she also needs to get the information in a timely manner. Objective information is very important in this case. Pat probably would not want to rely on a vendor's sales representative for information on the durability of the laptops. One thorny issue in this situation is the completeness of the information. Pat needs to have enough information to make the decision while avoiding information overload. It is also important to have current

> ### BUSINESS EXAMPLE BOX 3.2
>
> ## Information Quality as an Ethical Issue
>
> In 1986, the journal *MIS Quarterly* published Richard Mason's important article "Four Ethical Issues of the Information Age." One of these four issues was information accuracy. In the article, Mason relates the story of the March family, who were the victims of information inaccuracy. Despite the fact that for years the Marches had faithfully made their mortgage payments on time, the computerized system used by the bank for some reason failed to record one of the payments. This ultimately led to a foreclosure proceeding. Unfortunately, Mrs. March learned of the foreclosure while recovering from a heart attack. The news was such a shock that she suffered a stroke that almost took her life. The good news is that after considerable time and effort, the Marches were able to keep their home and won a substantial settlement from the bank.
>
> Because of the degree to which our lives depend on information, Mason points out that the organizations responsible for the information have an ethical obligation to ensure and maintain the accuracy of the information.
>
> We have personal experience with this; one of us was once denied a loan because of incorrect information in a credit report. One of the negative entries in the credit report was a default on a credit card that occurred at the age of 12, which was clearly inaccurate. It took quite a bit of time and effort to eventually clean up the credit report. The loan was to make a substantial retail purchase, so in this case the merchant also lost a sale. There are many other examples of inaccurate information causing everything from minor inconveniences to death. We agree with Mason that information accuracy is an ethical issue and an obligation for organizations that store and provide information.

information. Specifications and pricing information change often in the computer industry, so Pat needs to ensure that she has current information.

Now that you have some understanding of information quality, it is useful to gain a better understanding of how to evaluate information, which is the topic of the next section.

Evaluating Information

Now that you understand a bit about information quality, the question of how to evaluate information comes into play. There are some elaborate methodologies for examining information that is internal to an organization's information systems. A full discussion of that topic is beyond our scope in this chapter but is discussed in more detail in Chapter 14. So, we focus on how to evaluate information obtained from third parties, such as websites and published articles. Although we frame the discussion around external sources, the basic principles can also be applied to internal information.

A quick Web search for information evaluation strategies reveals many different approaches to evaluation. There are common themes that run through most of these strategies. Basically, there are two questions you need to answer: (1) Is the information useful? and (2) Is the information believable? In the next few paragraphs, we discuss the various "questions within questions" that you need to answer to evaluate information. Figure 3.1 illustrates our information evaluation framework.

As we noted earlier, filtering is one strategy for dealing with information overload. The ability to effectively evaluate information is important to being able to effectively filter information. Being able to quickly disregard poor-quality information is a big help in dealing with information overload.

Before getting into the details, it is important to understand that the evaluation of information is highly context dependent. You can only evaluate information within the context of your information needs. Consider Wikipedia as an information source. Anyone can add information to Wikipedia, regardless of his or her credentials. If you are researching a topic for pleasure, Wikipedia is often a fine source of information. However, if you are researching a topic for a work task, we would caution against relying exclusively on Wikipedia. In the first situation, it really does not matter much if the information is slightly inaccurate, incomplete, or out of date. For a work-related task, however, inaccurate, incomplete, or old information may be completely unacceptable.

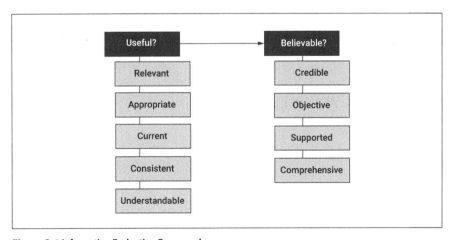

Figure 3.1 Information Evaluation Framework

Evaluating Usefulness

Before taking the time to assess the believability of information, you should determine whether the information is useful for the task at hand. If the information is not useful, then there is no need to assess its believability. To determine whether information is useful, evaluate whether the information is relevant, appropriate, and sufficiently current. Each of these is a "go, no-go" assessment. At any point, if you evaluate the information as not passing the test, there is no need to go further. For example, if the information is not relevant, you do not need to determine whether it is appropriate or sufficiently current.

Information relevance is the degree to which the information is pertinent to the task at hand. You can determine the relevance of information by asking yourself a simple question: Will this information help me accomplish my task? The answer to this question is often not a simple yes or no. There are varying degrees of relevance. Some information is highly relevant, while other information may only be tangentially relevant. With experience, you will be able to determine what degree of relevance merits further evaluation. To complicate matters further, relevance is context dependent. In some cases, information that does not seem relevant early in a task may become relevant as you move toward a solution. All this being said, relevance really does come down to figuring out whether the information helps you accomplish your task.

If the information seems relevant, then you should evaluate whether the information is appropriate: Is the information suitable for your purpose? For example, if you are researching a new technology when preparing a report for your manager, the information contained in a high school student's report may not be appropriate for your use. You will need to assess the level of detail and the depth of the information in light of your information needs. Again, this is a context-dependent question.

Currency is the final evaluation criteria related to usefulness. Assessing whether information is sufficiently current requires thinking about how current you need the information to be. If you are seeking information related to rapidly evolving topics, such as information technology, you may need information that is very up-to-date. For less volatile topics, older information may be fine. Unfortunately, determining the currency of Web-based information is often difficult. Many websites do not clearly indicate when the information was last updated. We caution you against relying too much on undated websites when you need up-to-date information.

Evaluating Believability

Once you establish that the information is sufficiently useful, you need to determine whether the information is believable. This requires assessing whether the information comes from a credible, objective source, is well supported, and is sufficiently comprehensive.

Evaluating the credibility of an information source can be tricky in many cases and relatively straightforward in others. For example, more formal publications, such as peer-reviewed journals, often include short author biographies, which can be helpful in assessing credibility. Often a quick search for information about the author can help you further evaluate credibility. If the author has written widely on the topic in reputable sources, he or she probably has sufficient expertise to merit using the information. In some cases, however, it is almost impossible to determine who wrote the information. We would be reluctant to trust information from such sources, unless it is provided by an organization with a solid reputation. If the information is provided by a respected organization, you can usually trust that the individual who created or compiled the information is knowledgeable on the topic.

The next step is to evaluate whether the information is objective. To determine whether information is objective, first consider the source of the information. An organization or individual who is credible may still have a bias or some agenda that makes the information less objective. For example, when an organization provides a review of its own product or service, that review may be biased. Using factual information from Dell's website is fine, but relying on Dell to provide an unbiased comparison of their computers to Hewlett-Packard's is a bad idea. You should also pay attention to the language used. If much of the information is presented using persuasive or emotional language, the information is probably not entirely objective. Language that is more fact based and neutral is more likely to be objective. Finally, if the topic is controversial, consider whether all sides of the issue are presented. If only one side is presented, the information is probably not entirely objective.

Evaluating the support for claims is also important. Begin by determining whether any support is offered. Claims without support should not be trusted. For example, we might say that this is the greatest information systems book ever written, but unless we present some evidence to back this claim, it is not believable. When support is offered, you should evaluate the quality of the support. Claims that are supported by credible sources, such as peer-reviewed journals, are usually believable. Also consider the reasonableness of the claim. There is a saying that extraordinary claims require extraordinary evidence. If a claim seems unreasonable or is widely different from other views of the topic, you should require more supporting evidence before believing the claim. Finally, think about whether the claim is testable. This does not mean that you have to actually test the claim, but if you can see no reasonable way to do so, be reluctant to rely on the information.

Our final evaluation criterion is whether the information is sufficiently comprehensive. Assessing comprehensiveness requires assessing the depth and breadth of the information. Breadth concerns whether all aspects of a topic are covered, while depth concerns the level of detail provided. When evaluating comprehensiveness, look for obvious gaps in the information. If it seems like something is missing, it probably is. Finally, consider whether there are unanswered questions. If there are, you will need to gather more information. Throughout your evaluation, keep in mind the context of your particular task; information that is sufficiently comprehensive for one task may be lacking for another.

All this probably seems like a lot of work. Keep in mind that the more information you evaluate, the more some of this will become second nature. Also, match the level of effort you put into evaluation against the importance of the task. The more important the task, the more worthwhile the effort put into evaluating the information.

Chapter Summary

In this chapter, you learned about information quality and information evaluation. The intent of the chapter was to give you an appreciation for the importance of information evaluation from both personal and professional perspectives and to help you gain the knowledge necessary to effectively evaluate information. Here are the main points discussed in the chapter:

- Being able to evaluate information is a key element of information literacy, which is an important skill for both our professional and personal lives.
- Career and personal success depend, in part, on the outcomes of the decisions we make. Our ability to evaluate the information we use to make these decisions affects the quality of our decisions.
- Information overload occurs when we are faced with more information than we can effectively process.
- Increasing our information evaluation skills helps us deal with information overload by reducing the amount of attention and time we devote to low-quality or nonuseful information.
- Intrinsic dimensions of information quality include accuracy, believability, objectivity, understandability, and consistency.
- Contextual dimensions of information quality include relevance, timeliness, completeness, and currency.
- Evaluating information concerns determining whether the information is useful and believable.
- Useful information is relevant, appropriate, and sufficiently current for the task at hand.
- Believable information is credible, objective, well supported, and comprehensive.

Review Questions

1. Briefly explain why information evaluation skills are important personally and professionally.
2. Explain the relationship between information and uncertainty in decision-making. Use a decision-making example to illustrate your answer.
3. What is information overload? How is information evaluation related to information overload?
4. Briefly describe five reasons managers gather information.

5. Name and briefly describe two major strategies for dealing with information overload.
6. Explain the phrase "garbage in, garbage out" as it relates to information quality and decision-making.
7. Name and briefly describe the four categories of information quality dimensions discussed in the chapter.
8. Contrast intrinsic and contextual information quality.
9. What are the two main questions that must be addressed when evaluating information?
10. Name and describe the elements of the information evaluation framework.

Reflection Questions

1. What is the most important thing you learned in this chapter? Why is it important?
2. What topics are unclear? What about them is unclear?
3. Compare and contrast misinformation and disinformation. How do these concepts relate to fake news?
4. What relationships do you see between what you learned in this chapter and what you have learned in earlier chapters?
5. How does information evaluation relate to information systems?
6. How do you think information evaluation skills will be helpful in your future career?
7. Describe a situation where, looking back on it, you spent more time gathering information than you should have. How do you know when you have gathered enough information?
8. Why is information quality an ethical issue?
9. How do you decide how much effort to put into information quality/evaluation?
10. Why is context important to information quality and evaluation?

Additional Learning Activities

3.A1. Think about a situation in which you relied on poor-quality information. Briefly describe the situation and what poor-quality information you relied on. Identify the dimension(s) of information quality where the information was not of adequate quality. (For example, you might have taken a course that you thought was required, but you were looking at out-of-date requirements, and the course was no longer required. This information was poor on the accuracy and timeliness dimensions of information quality.)

3.A2. A rubric is a set of rules for assessing something. For example, many instructors use rubrics for grading assignments. Develop a rubric for judging information quality. (For a rubric template, see http://template lab.com/rubric-templates/.)

3.A3. Many of you might have graduate school in your future. Choosing the right graduate school is a major decision-making task. Suppose you want to attend a full-time MBA program. Use the information evaluation framework to evaluate the information you would find in each of the following sources. For each information source, rate each dimension on a scale of 1 (lowest) to 5 (highest). Give a brief explanation of each of your ratings:

    a. Princeton Review's Graduate Business Schools website (https://www.princetonreview.com/business-school)

    b. A university's website about their MBA program

    c. A friend who recently got his or her MBA

    d. The admissions counselor for an MBA program

3.A4. For each of the following tasks, indicate on a scale of 1 (very little time) to 10 (a significant amount of time) how long you would spend evaluating the information you would use in completing the task. Briefly justify each of your ratings.

    a. Deciding where to go for a celebration dinner

    b. Deciding what pair of running shoes to buy

    c. Deciding what laptop to buy

    d. Deciding which of two postgraduation job offers to accept

3.A5. The COVID-19 pandemic was a time of great uncertainty. Many people sought information to help reduce some of that uncertainty. Think about the kinds of information that you or your friends may have sought to help you better deal with the pandemic. Briefly describe the information you sought and how you went about finding the information. Also, describe steps you took to evaluate the quality of the information. Finally, briefly discuss any poor-quality information you encountered.

3.A6. Think about a time when you had a bad experience with customer service. For example, you might have been misinformed about the length of time a repair would take. Briefly describe the situation. Also, write about any consequences for the company involved. What could the organization have done to avoid or improve your experience?

3.A7. Suppose you wanted to start a small business selling products on eBay or Etsy. List five pieces of information that would help in this effort. Briefly discuss how you would evaluate the information. For each piece of information, discuss the consequences to your business if that information is of poor quality.

3.A8. In this chapter, you learned about information quality and information evaluation. Create a diagram that links the information quality dimensions to the information evaluation criteria.

References

Butcher, H. 1998. *Meeting Managers' Information Needs.* London: Aslib.

Experian Data Quality. 2016. *The 2016 Global Data Management Benchmark Report.* https://www.edq.com/globalassets/white-papers/2016-global-data-management -benchmark-report.pdf.

Knight, S., and J. Burn. 2005. "Developing a Framework for Assessing Information Quality on the World Wide Web." *Informing Science* 8: 159–72. http://inform .nu/Articles/Vol8/v8p159-172Knig.pdf.

Lee, Y., D. Strong, B. Kahn, and R. Wang. 2002. "AIMQ: A Methodology for Information Quality Assessment." *Information & Management* 40: 133–46.

Mason, R. O. 1986. "Four Ethical Issues of the Information Age." *MIS Quarterly* 10(1): 5–12.

Savolainen, R. 2007. "Filtering and Withdrawing: Strategies for Coping with Information Overload in Everyday Contexts." *Journal of Information Science,* 33(5): 611–21.

Simon, H. 1971. "Designing Organizations for an Information-Rich World." In *Computers, Communications and the Public Interest,* edited by Martin Greenberger, 40–41. Baltimore: Johns Hopkins University Press.

Spies, S. 2020. "Defining 'Disinformation'." *MediaWell: Live Research Review.* https://mediawell.ssrc.org/literature-reviews/defining-disinformation/versions /1-1/.

Tuna, C. 2008. "How to Spot Résumé Fraud." *Wall Street Journal,* November 13. http://www.wsj.com/articles/SB122653695797922735.

United States Department of Justice. 2010. *Information Quality: The Foundation for Justice Decision Making.* http://www.it.ojp.gov/documents/IQ_Fact_Sheet_Final .pdf.

Wang, R., and D. Strong. 1996. "Beyond Accuracy: What Data Quality Means to Consumers." *Journal of Management Information Systems* 12(4): 5–34.

Glossary

Accurate (information): The degree to which information is correct and free from error.

Believable (information): The degree to which information is regarded as true and credible.

Comprehensive (information): The degree to which information is of sufficient depth and breadth for the task at hand.

Consistent (information): The degree to which information is compatible with previous information.

Current (information): The degree to which information is sufficiently up-to-date for the task at hand.

Disinformation: False information that is purposely created to cause harm to some person or group.

Information evaluation: The systematic determination of the merit and worth of information.

Information overload: Being faced with more information than one can effectively process.

Information quality: The degree to which information is suitable for a particular purpose.

Misinformation: False information that is not created with the intention of causing harm to some person or group.

Objective (information): The degree to which information is free from bias.

Relevant (information): The degree to which information is applicable and useful for the task at hand.

Timely (information): The degree to which information is available in time to perform the task at hand.

Understandable (information): The degree to which information is easily comprehended.

Chapter 4

Gaining Strategic Value from Information

Learning Objectives

By reading and completing the activities in this chapter, you will be able to:

- Present the main steps in the strategic planning process
- Identify competitive advantage frameworks and discuss their purposes
- Discuss methods for evaluating strategic initiatives
- Explain the concept of hypercompetition

Chapter Outline

Strategic Information Systems
Focusing Story: Health Care on My Smartphone
Strategic Planning Process
  Learning Activity 4.1: Trends in Top Management Concerns
Frameworks for Strategic Information Systems
Evaluating Strategic Initiatives
  Learning Activity 4.2: What Does This Company Need for Success?
Hypercompetition: Sustainability of Competitive Advantage
  Learning Activity 4.3: Competitive Advantage at TRIPBAM
  Learning Activity 4.4: Disrupting an Established Market!

Strategic Information Systems

Today, organizations need a wide variety of information systems for conducting their day-to-day business. For example, some information systems can be used for basic accounting services, such as payroll systems, or for communication support services, like electronic mail. In this chapter, we are interested in those information systems that are more specifically meant to provide organizations with competitive advantages. We call those business initiatives **strategic information systems**. Importantly, it is not necessary to have unique and proprietary information technology to make an initiative strategic; it is how the information systems are used that can provide the added value or strategic advantage organizations seek by implementing such initiatives.

Health Care on My Smartphone

In the business world, a key goal for many companies is to achieve some form of competitive advantage over rivals. Of course, creating a new product that no one has ever thought of before, or a product that is so much better than previous ones, definitely gives an organization a competitive advantage (think of Uber, for example). However, it is often possible to also develop competitive advantages from creative uses of existing technologies. One of these exciting newer trends in the world of medicine is mobile health, also called mHealth. mHealth offers health care services delivered via mobile technology, such as smartphones or tablets, and various fitness technologies. Mobile health care services can include compliance (ensuring that patients follow their medication plans), information dissemination (providing caloric information for obese patients or sugar content for diabetic patients), or monitoring (measuring hypertension or cholesterol or tracking activity).

The idea of mHealth is to provide health care services where they are needed, often to those who cannot access them. One application called text4baby from Voxiva helps prevent infant mortality by sending weekly text messages to young women about their babies' health. There are also mobile health screening programs at home for executives in Pakistan and at truck stops for truck drivers in Canada. mHealth smoking applications have been shown to help smokers quit smoking. Another company called Avacta developed a handheld device to detect dangerous flu diseases in the field without having to bring samples to laboratories for tests. Around the world, doctors' offices have started using less-expensive text messaging instead of phone calls to remind patients of appointments; results include improving attendance by 7% in China and reducing nonattendance rates by 40% in Malaysia. An even more exciting potential for mHealth is in providing health services to remote regions of developing countries, which are facing constant growth of life-threatening chronic diseases (e.g., hypertension, obesity, heart disease, and diabetes) and communicable diseases. In developing countries, there are limited health care infrastructures, limited hospital resources, and not enough health care workers, particularly in remote regions. This is where mHealth can make a difference, because even in remote regions of developing countries, mobile phones are often available. In 2016, for example, the International Telecommunications Union (ITU) and the World Health Organization (WHO) produced a report on how to use mobile phones to combat noncommunicable diseases like diabetes, titled "Be He@lthy, Be Mobile: A Handbook on How to Implement mDiabetes." By 2020, in light of the COVID-19 pandemic, the WHO had produced an app for global health, offering information on outbreaks and public health information. Even Apple and Google launched contact-tracing services on their mobile platforms to help quickly notify people who have been in contact with someone with a positive COVID-19 result.

mHealth also includes personal well-being, where individuals track their own health information with fitness trackers (like the Fitbit) or smart watches (like the Apple Watch). The devices are often worn 24 hours a day, seven days a week, collecting pulse rates, sleeping patterns, activities, and even food intake. Wrist wear is not the only type of personal health/fitness device, though it is the most common. Others include smart clothing and leg wear. Surveys show that as of early 2020, one in five Americans used an app to track their health data. Who would have thought people would start wearing watches again?

Sources: "Case Study: Organizations, Technology and PR Unite to Deliver Mobile Health Service Targeted at Moms and Moms-to-Be." 2010. *PR News* 66(45): 45; Contant, J. 2008. "Mobile Health Unit Rolls into Truck Stop." *OH & S Canada* 24(7): 21; "Dow University Launches Mobile Health Screening Program." 2010. *Financial*

Post, October 15; Kahn, J. G., J. S. Yang, and J. S. Kahn. 2010. "Mobile Health Needs and Opportunities in Developing Countries." *Health Affairs* 29(2): 252–58; World Health Organization. 2016. "A Handbook on How to Implement mDiabetes." *World Health Organization and International Telecommunications Union,* https://www.itu.int/dms_pub/itu-d/opb/str/D-STR-E_HEALTH.09-2016-PDF-E.pdf; 1 in 5 Americans Track Their Health Statistics Using an App. 2020. Marketing charts, January 7. https://www.marketingcharts.com/industries/pharma-and-healthcare-111492; World Health Organization. 2020. https://www.who.int/mediacentre/multimedia/app/en/.

Focusing Questions

1. What mHealth applications do you use?
2. What are the benefits of mHealth for existing health care organizations? For consumers?
3. How have fitness devices changed the health of individuals in recent years?
4. Would you participate in a contact-tracing program (or have you) for the benefit of community health? Why or why not?
5. What other day-to-day device could be transformed to change the competitive landscape?

Strategic Planning Process

The identification of strategic information systems should follow a structured set of steps, or a **strategic planning process**. Of course, there are situations when some tech-oriented individuals start using a new technology and believe everyone should have access to it to increase productivity of the workforce. When the BlackBerry and iPhone devices came out, many managers believed they needed them, even though they had no real idea why. Some companies gained real advantages from increased mobility, while others realized the devices did not really help their employees beyond being more accessible. Most large organizations, however, do have a strategic planning process in place. The goal of the strategic planning effort is to identify how the organization will use and manage its resources for strategic purposes (think competitive advantage). It provides a roadmap for decision-making related to information systems. Figure 4.1 shows an example of the steps involved in a typical information systems strategic planning process. Notice how the process is iterative, with later steps in the process requiring managers to revisit earlier steps. In addition, there may be several phases involved within each of the stages (rectangles).

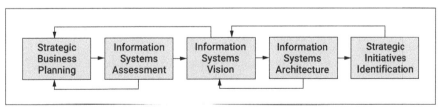

Figure 4.1 Sample Information Systems Strategic Planning Process

Step 1: Strategic Business Planning

Before an organization can decide what information technology (IT) initiatives can be used to gain a competitive advantage, it has to clearly identify what the mission of the organization is and how it is going to achieve this mission. The strategic business planning phase is often referred to as "Know Who You Are." Executives or leaders of the organization should perform strategic business planning on a regular basis. (How often this is varies substantially between companies.) The outcomes of strategic business planning sessions should include the mission and vision of the organization, its goals for the future, and the strategies that will be used to achieve those goals.

While strategic business planning is not specifically focused on information systems, it is essential to the IS strategic planning process because the organization needs to know what it is trying to achieve before it can identify how information systems are going to help achieve those strategic objectives. The goal is to achieve a high level of alignment between the business side of the organization and its information systems. With high alignment, there will be a high level of fit between the priorities and activities of the IS function and the strategic goals of the organization.

Step 2: Information Systems Assessment

Once managers know the main strategic goals of the organization, they need to identify the current state of information systems resources in the organization. This phase is sometimes referred to as "Know Where You Start." It is important to conduct a proper assessment of resources because these resources could enable information systems managers to meet or surpass some strategic objectives or, conversely, constrain the organization regarding what it can do. For example, if the organization has been collecting data about its customers for a very long time but never used those data, there might be some great potential for new systems that will make use of them. (We will discuss customer relationship management [CRM] systems in Chapter 12.) On the other hand, the lack of certain resources may also constrain what an organization can achieve. For example, if all individuals in the IT group are trained on a particular system, replacing the system with technology that is completely different may create a very disruptive situation for the organization, which may even impact profits for some time. The outcome of the **information systems assessment** is a picture of the current state of information systems resources in the organization. Depending on the assessment, it may be necessary for the organization to revise its strategic goals.

As you can see, information systems resources are not limited to just technology. There are three categories of resources: technical resources, which include hardware, software, and networks; data and information resources, such as databases; and human resources, which would include the skills and personal characteristics of the information systems employees, the user community, and the management, as well as the structure of the organization and its incentive systems.

LEARNING ACTIVITY 4.1

Trends in Top Management Concerns

The alignment of business and information systems objectives is a key concern for top managers. Every year, a survey is conducted among members of the Society for Information Management (SIM) about senior managers' top concerns about information technology. The most recent survey (Kappelman et al. 2020) highlights IT and business alignment as one of the top priorities, with all priorities being ranked in the following order:

1. Security/Cybersecurity/Privacy
2. Alignment of IT with the Business
3. Data Analytics/Data Management
4. Digital Transformation
5. Compliance and Regulations
6. Cloud/Cloud Computing
7. Agility/Flexibility (IT)
8. Cost Reduction/Controls (IT)
9. Innovation
10. Cost Reduction/Controls (Business)

For this learning activity, find previous IT Key Issues and Trends Studies by visiting MISQ Executive through your library's website, or use those provided to you by your instructor. Identify the trends in top management concerns, as these studies go back 10 or 15 years. Identify reasons you think might have led to which issues were of greater concern at different times. Be prepared to defend your reasons to the class.

Step 3: Information Systems Vision

In the first step, the executives of the organization developed mission and vision statements for the organization as a whole. Given those statements and the information systems resources identified in the second phase, information systems managers must now develop a vision specifically for information systems. For this and the following steps, the senior information systems person in the organization must gather not only employees of the information systems group, including individuals with technical skills, but also functional managers. Functional managers work in other areas of the organization, such as accounting or marketing, and are likely to be the end users of the information systems initiatives that will be identified later. They therefore need to be part of the information systems strategic planning process from the start.

The **information systems vision** should be a broad statement of how the organization should use and manage its information systems for strategic purposes. It basically suggests that organizations need to "Know Where They Want to Go." Clearly, the information systems vision has to be aligned with the organization's mission, vision, and strategies, which may require revisiting the organization's strategies. An example of an IS vision statement could be "We will strive to offer leading-edge but tested technologies to our functional areas, provide leadership in managing external and internal data, and promote technologies that will enhance the competitive advantage of our firm." Of course, each organization is unique, and the statement will be more specific to each company. Nevertheless, the statement should define what role the information systems should play in the organization.

Step 4: Information Systems Architecture

In most organizations, an information systems vision is likely to already exist. Therefore, the previous phase involves modifying it if needed. Similarly, most large organizations will already have selected an information systems architecture. **Information systems architecture** specifies how information systems resources should be used and how they should work together. It is the "Know How You Are Going to Get There" phase. We will discuss information systems architectures in more depth in Chapter 7. The development of an architecture may require managers to revisit the information systems vision if, for example, the architecture cannot support the vision that was established.

The selected information systems architecture will provide some guidelines for future uses, acquisitions, and the management of information resources. It is therefore general enough to provide guidelines for several years but specific enough to offer actionable guidelines. Consider the following architectural statement: "We will strive to provide a technology environment that will allow a variety of vendors to interact with our systems independently of their technology platforms." The statement clearly points to the need for interoperability of technology, which is the ability of heterogeneous systems to communicate with one another. Therefore, acquisition of software or hardware that limits interoperability is not consistent with the selected architecture. Architectural guidelines could include statements about all information resources, including hardware, software, and networks; the way data are stored, protected, and managed; and how information systems human resources are managed, such as whether information systems work can be outsourced.

Step 5: Strategic Initiatives Identification

The ultimate goal of the information systems strategic planning process is to identify strategic information systems initiatives that will provide competitive advantages to the organization. These initiatives tend to be longer term (two to five years), although some strategic initiatives can be launched rapidly. It should be noted that strategic initiatives can be about systems, but most of the time they involve much more than technology, which means they may result in a new strategic direction for the organization. For example, when an organization decides to implement customer relationship management systems, installing software and hardware is not enough; employees must be trained to be highly customer oriented if the implementation is to succeed.

There are many frameworks that can be used to identify **strategic information systems initiatives,** and we will discuss several of them in the next section of this chapter. As you learn how to use the frameworks to identify initiatives, however, keep in mind that the initiatives that are identified need to be aligned with the IS vision and the strategic goals of the organization.

Advantages of the IS Strategic Planning Process

The identification of strategic initiatives is in itself an important goal of the information systems strategic planning process. There are, however, several other benefits from using a structured approach to identify strategic initiatives:

- *Improved communication:* Having stakeholders involved in the planning process allows everyone to provide input into initiatives that will potentially affect individuals in different areas of the organization. By providing documented discussions of the advantages and disadvantages of each initiative, overall communication inside the organization is improved.
- *Improved coordination:* Since various stakeholders are involved in the planning process, they develop a shared mental image of the initiatives, their purposes, and their advantages and disadvantages. Furthermore, each member of the planning team gets a clear picture of everyone's responsibility. This improves overall coordination of the efforts related to the initiative, both before and during the implementation.
- *Improved decision-making:* When a structured approach is used for identifying strategic initiatives, a clear set of guidelines and criteria for selection of initiatives are established. For example, initiatives have to be aligned with the goals of the organization. The result is that decision-making is more consistent over time with respect to which strategic initiatives are supported and, more importantly, why.

Frameworks for Strategic Information Systems

The last phase of the strategic planning process is the identification of strategic initiatives. As previously discussed, each phase is actually composed of several steps. In this section, we look more specifically at a number of frameworks or tools that can be used to identify strategic information systems initiatives.

Information Systems SWOT Analysis

Most students will have discussed **SWOT (strengths, weaknesses, opportunities, and threats) analyses** in one of their management or marketing classes. SWOT analyses can be used for strategic information systems planning, just like for other areas of the organization. While the technique of a SWOT analysis has existed for some time now, the basic concepts remain the same. Table 4.1 summarizes the components of a SWOT analysis, with some example statements that could be applied to Amazon.com.

When considering the use of a SWOT analysis for the purpose of identifying potential strategic information systems initiatives, it is important to consider factors beyond technology in the strengths, weaknesses, opportunities, and threats. Say, for

TABLE 4.1 SWOT Analysis Components		
Component	**Description**	**Example for Amazon.com**
Strengths	What gives the organization advantages over others in its industry?	Leadership position in online retailing
Weaknesses	What creates disadvantages for the organization relative to others in its industry?	Increased size of the company requires more investments
Opportunities	What activities or factors could help the organization get new advantages over others in its industry?	Expanding to services instead of just retailing
Threats	What activities or factors could create disadvantages or troubles for the organization relative to others in its industry?	Increasing global competition from online retailers

SWOT Analyses in Practice

Does a SWOT analysis seem too basic to you? While it is true that SWOT analyses are more often discussed in academia, they do find their way into the real world. If you use a search engine and type "SWOT analysis examples," you will find many real-world companies that have been analyzed using this framework. For example, Sapphire Strategy explains how even tech companies like Amazon regularly use SWOT analyses. In 2017, when they acquired Whole Foods, they identified high-level strengths like their distribution network and that it would give them a brick and mortar presence, the small margins of the grocery business as a weakness, the opportunity to capture a share of the grocery business and use their Amazon Prime memberships to drive up market share, and that Walmart was a threat to this. SWOT analyses, however, are not solely for technology companies. For example, Scholefield Construction Law, a firm focusing on legal expertise related to construction and engineering, discussed in a *Business News Daily* article how they developed their SWOT to determine whether they should enter the mediation business, which they ended up doing.

Sources: Golding, J. L. 2018. "These Tech Companies Used SWOT Analysis to Come Out on Top of Competitors," Marketing Lifecycle, Sapphire Strategy, September 4. https://sapphirestrategy.com/these-tech-companies-used-swot-analysis-to-come-out-on-top-of-competitors/; Schooley. S. 2019. "SWOT Analysis: What It Is and When to Use It," Business News Daily, June 23. https://www.businessnewsdaily.com/4245-swot-analysis.html.

example, a firm has a great idea for implementing an information system to improve the sales cycle by providing mobile devices to sales representatives. If one of their weaknesses is that their customer base is limited to military agencies where mobile devices may not be allowed on the facilities for security reasons, then this may not be a good initiative for the organization. On the other hand, if one of their opportunities is to have a more physical presence with customers (which may seem contradictory to the use of electronic media), it may be that changes in technologies available at home can free up more time for sales representatives to visit their clients. The important point is that you should not limit yourself to thinking only about technology when you perform a SWOT analysis.

How do you identify information systems strategic initiatives from a SWOT analysis? Of course, you can look at the opportunities and see which ones can be achieved with the implementation of information systems. However, you can also look at weaknesses and see if you can remove or reduce those weaknesses via better use of information systems. Similarly, ask yourself, how can the organization reduce external threats by making better use of information systems? How can it make better use of its strengths through new information systems? All these ways to look at the SWOT elements can allow managers to identify a number of initiatives. Not all of them will be viable, but at this point, we are only focused on identifying them.

Porter's Five Competitive Forces Model

Porter's Five Competitive Forces Model is one of the most popular frameworks for analyzing a firm's competitive position by looking at the major forces that shape an organization's competitive environment. An adaptation of the framework is presented in Figure 4.2. The original purpose of the model is to analyze how competitive an industry is and therefore determine if a particular market (product, service, geography, etc.) could be attractive for an organization to consider. It could also be used to reevaluate a company's competitive position if one of the competitive forces changes—for example, if a major merger changes the level of rivalry in a particular industry.

The original model was developed back in 1979 but was later updated by Michael Porter in 2008. Some have argued that the model focuses too much on a perfect yet

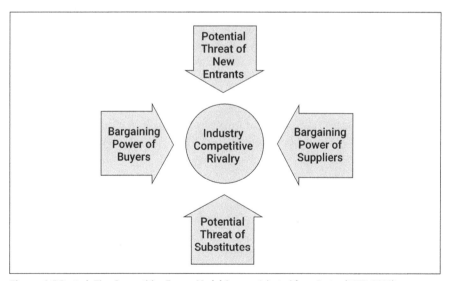

Figure 4.2 Porter's Five Competitive Forces Model Source: Adapted from Porter (1979, 2008)

TABLE 4.2 Porter's Five Competitive Forces Model and Strategic Information Systems			
Competitive Force	**Description**	**Desired Level**	
Potential threat of new entrants	How easy is it for new companies to enter the market in which the organization operates?	Low—no new companies can enter the market.	
Bargaining power of buyers	What is the ability (or market power) of the organization's buyers (customers) to reduce its competitive position (e.g., by bringing prices down)?	Low—customers cannot dictate lower prices or better conditions.	
Bargaining power of suppliers	What is the ability (or market power) of the organization's suppliers to reduce its competitive position (e.g., by bringing prices up)?	Low—suppliers cannot dictate higher prices or better conditions.	
Potential threat of substitutes	What is the likelihood that other products of equal or superior value will be available?	Low—no products can substitute for the organization's products (or services).	
Industry competitive rivalry	What is the current level of competition in the industry?	Low—there is limited competition.	

simple market situation where there is a relative static structure (where competition does not fluctuate very much; Recklies 2015). Yet while there have been some criticisms, the Five Forces model is still used as a way to analyze a firm's competitive position in today's business world. For example, Trefis, a stock analysis firm, has used the Five Forces model to analyze firms like Under Armour, Facebook, Nike, Coach, and Ralph Lauren (Martin 2019). Another example is the analysis of McDonald's by the Panmore Institute (Gregory 2018), which concluded that the company needed to focus its strategies on competition, the bargaining power of customers, and the threat of substitutes.

In the context of information systems, we can use Porter's Five Competitive Forces model to identify strategic information systems initiatives. Table 4.2 shows how each of the forces could be examined from a strategic information systems point of view.

Porter's Value Chain Analysis

Porter proposed another framework that could be useful in identifying strategic information systems initiatives. In a **value chain analysis**, managers identify all the activities that the organization must perform to conduct its business (Porter 1996). As each activity is performed, the organization adds value to the product or service it delivers. Clearly, all industries and organizations are likely to have their own value chain. Porter identified two broad categories of activities in the value chain: primary

Strategic Information Systems	Examples
How could you use information systems to increase barriers to entry in the organization's market?	Create barriers to entry—leverage data about customers that others cannot access, creating information asymmetry.
How could you use information systems to reduce the buyers' power of negotiation?	Create switching costs—give customers value-added services such as personalization or lower costs (through economies of scale) that will make them loyal to the organization.
How could you use information systems to reduce the suppliers' power of negotiation?	When there are few suppliers, this tends to be a high barrier. Use information systems to aggregate buying power with others (e.g., online exchanges) or reduce the cost of buying (electronic procurement).
How could you use information systems to make your products unique or make customers unwilling to use substitutes?	Create switching costs so customers will not be willing to use substitutes (personalization, lower costs, etc.) or use information systems to create unique or patented products such as proprietary systems (like Google applications or Apple products).
How could you use information systems to ensure that competition is limited in the industry?	Use information systems to differentiate the organization's products so that there are few or no competitors.

and support. *Primary* activities are directly related to the creation, processing, or delivery of the product or service. *Support* activities are those overall tasks that make it possible for the organization to function but that are not directly involved in the product or service. Figure 4.3 shows a typical value chain in a manufacturing firm. Note that many organizations today are in the service industry, and their value chain would be completely different. We will discuss service industry firms in the next section.

How could you use a value chain analysis to identify strategic information systems initiatives? Porter identified two major ways to create competitive advantages in the value chain: lower the costs of performing an activity or add more value to the final products and services in an activity. Examples of companies that have cost leadership are McDonald's and Walmart. Examples of companies with differentiation include Apple and Starbucks. An analysis of Starbucks' value chain can be found on the Investopedia website (https://www.investopedia.com/articles/investing/103114/starbucks-example-value-chain-model.asp).

To identify if a company can take advantage of either cost leadership or differentiation, for every activity in their value chain, managers can ask themselves: How can I use information systems to perform this activity at a lower cost? How can I use information systems to improve the value added from this activity to the final product or service? If we take the example of marketing a product, in what ways can information systems reduce marketing costs? The organization could use electronic

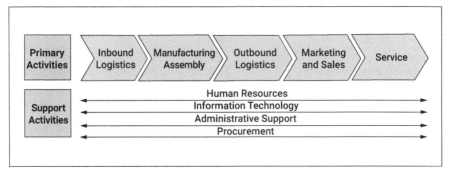

Figure 4.3 Sample Value Chain in Manufacturing Firm Source: Adapted from Porter (1996).

marketing through social networks or electronic mailings. What about adding value to the product? Well, the organization could make its products more personalized. For example, instead of selling cute teddy bears, the organization could create online profiles for the teddy bears that would make them have their own online "lives," making the product more attractive than the other teddy bears that do not have online lives.

There are a lot of opportunities for managers to examine the activities of their organization to identify potential strategic uses of information systems. But now think of extending this outside of the organization's value chain to its suppliers and customers. With more and more organizations linked electronically, are there possibilities for the activities of members of an organization's value system (upstream and downstream value chains) to be improved via technology? That is what the giant retailer Walmart is doing when it manages its inventory not only within its organization but also within its suppliers as well. (See Chapter 12, Enterprise Information Systems, for a story about this.)

Virtual Value Chain

As discussed in the previous section, many firms today are mainly information-based organizations. Think about what Google offers as a product. The company creates and offers software tools to individuals and companies. In addition, the company offers information to individuals through its search engine as well as information to organizations via its analytical software that provides an analysis of traffic on organizations' websites.

Instead of looking at activities that turn raw materials into a final product, as in manufacturing organizations, the **virtual value chain** looks at activities that turn raw data into useful information (Rayport and Sviokla 1995). An example of an information-based business is consulting. Consulting firms' main outputs are information and knowledge-based reports and documents they provide to their clients about specific questions of interest, specific analyses regarding their own business or certain markets they are considering, or even reports about general industry trends. In the business of consulting, professionals regularly gather substantial amounts of data and

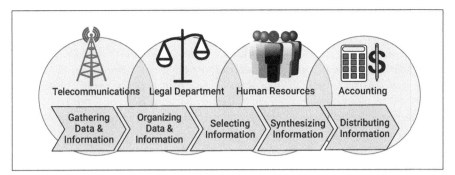

Figure 4.4 Virtual Value Chain: Handling Information Source: Adapted from Piccoli (2008).

information and then organize and store them in some useful way (retrievable, indexed, etc.). When the organization is asked to provide consulting on a specific topic, the professionals in the firm identify and retrieve the relevant information from a repository (where it is stored), synthesize the information (rendering it useful), and then distribute it to the client. Figure 4.4 shows an example of a virtual value chain.

Similar to the other frameworks discussed so far, an analysis of the virtual value chain can help identify strategic information systems initiatives. Managers can ask themselves in what ways information systems can make gathering, organizing, selecting, synthesizing, and distributing less expensive and add more value to the firm's information. For example, within the human resources (personnel) department, information could be distributed more rapidly to employees using an intranet (discussed in Chapter 7). Are there better ways to gather information and data for the human resources department? One could imagine, for example, collecting work hours (time sheets) for remote employees via website applications as opposed to paper forms. Similarly, are there better ways to organize and select information within the human resources department? It may be that a new system could be implemented for identifying information requirements for employees who travel internationally. As you can see, different questions can lead managers to the identification of potential strategic information systems. And we only discussed the human resources department! This can be extended to all departments that require gathering, organizing, selecting, synthesizing, and distributing information.

Today, as organizations have gone global, strategic information systems can also seek to leverage **global value chains**. Global value chains include production of goods and services on an international scale, with production broken down into activities and tasks carried out in different countries.

Evaluating Strategic Initiatives

In the previous section of this chapter, we used a variety of frameworks to identify potential strategic information system initiatives that the organization should consider. In theory, we should at this point have a significant number of initiatives identified. Not all of them will be implemented, as some simply cannot be done (for example, they might require technology that does not exist yet), and others are not worth doing (for example, they might be too costly, with only limited benefits). We therefore need to rank the initiatives, or at least eliminate those that cannot or should not be considered. There are several tools that can be used for doing this analysis, and we will discuss two: the critical success factors method and the priority matrix. Both methods can be used for evaluating the viability of strategic information systems initiatives, although the priority matrix provides a more in-depth analysis.

Critical Success Factors

Critical success factors (CSFs) are those few important considerations that must be achieved for the organization to survive and be successful (i.e., achieve its mission). These are not about technology (in general), but more about business objectives. In other words, CSFs must be in line with the organization's vision and mission. Organizations usually have just a few critical success factors, and top management typically identifies them.

Identification of CSFs is a process in itself that resembles brainstorming sessions where top managers sit together with a moderator who helps them develop a short list of the main factors necessary for the company's success. The key questions asked to start the discussion are similar to these: What needs to happen for our organization to increase its revenues? What are the most important actions we need to take to be more competitive? What is needed for us to be more successful? These questions just help start the discussion, which subsequently involves deliberations among the executives, identification of potential obstacles, and mostly the development of priorities among the most important factors for the organization's success.

Imagine that you are an executive at your school involved in developing strategies for the future of the school. Your group of school executives would have to ask themselves, what is needed for the school to be a leader in education? The group would probably identify factors such as the following:

1. Having the highest quality students possible (Having no students is not a good idea; having poor-quality students does not help the university be successful.)
2. Having the best possible faculty
3. Providing a close link to the community

Once the critical success factors are identified for the organization, managers can map their proposed strategic information systems initiatives to the CSFs. Initiatives that do not support the CSFs of the organization should not be considered as top priorities. This ensures that the organization achieves the business and IT alignment discussed earlier in the chapter and that the planning team is able to prioritize initiatives in a consistent way. In fact, consulting firms often suggest that businesses use CSF to ensure they meet their strategic objectives and highlight the importance of tracking the CSFs on a regular basis (Janse 2019; Lucco 2020).

Priority Matrix

The **priority matrix** allows managers to evaluate potential initiatives and prioritize them along two key dimensions: the effort to implement the system or project and the potential returns from this implementation. The effort dimension includes several factors, such as costs, time required, efforts required, and the actual complexity of the system. The potential returns can be in terms of revenues, market share, savings, reputation, and so on. Figure 4.5 shows a sample priority matrix.

Looking at Figure 4.5, you can easily identify which strategic information systems initiatives should be given higher priority by managers. **Imperatives** are initiatives that should be relatively easy to implement (i.e., low effort)—for example, because they cost little or the technology is already implemented—and have the potential to bring high returns to the organization. **Quick wins** represent initiatives that do not have much upside potential but are easy to implement with limited effort (low or no cost, low or no difficulties involved, not complex, etc.). They are called quick wins because the organization can get some visible impacts without much

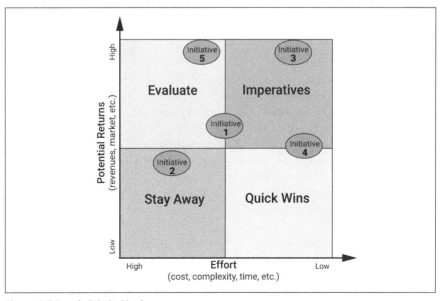

Figure 4.5 Sample Priority Matrix

S T A T S   B O X   4 . 1

CSFs for Improving Business in Australia

Critical success factors (CSFs) can be used not only for the organization as a whole, as we described in this chapter, but also sometimes for specific projects or industries. In this example, the Department of Agriculture and Food of the Government of Western Australia explains how businesses should ensure that they have clear CSFs and key performance indicators (KPIs) if they are to achieve their business goals. KPIs are the metrics used to ensure that the CSFs are working properly. They offer worksheets that businesses can use to establish their CSFs and offer examples for two industries: restaurants and manufacturing. The sample goal for a restaurant could be to increase its profits by 5% by a certain date via increasing throughput in lunches and dinners by 10% without reducing the gross margins. The table below lists the set of sample CSFs they propose.

CSF	KPI	Target
Market share	% of business within a 5 km radius	10%
Customer satisfaction	% of customers who are satisfied	95%
Meal quality	% of meals returned because of poor quality	2%

For the manufacturer, their example of a goal is to increase the operating profit that a washing machine manufacturer makes by 5% within a target number of years. They identify four CSFs related to product quality, process yields, production costs, and market share. Finally, they recommend three questions to help identify CSFs: "to achieve the goal it is critical to have . . . (these are the CSFs); the best indication of whether the CSF is working is . . . (this is the KPI); we will know we are on track to our goal when we have reached . . . (this is the target)" (https://www.agric.wa.gov.au/improvement-tools -critical-success-factors-and-key-performance-indicators).

Source: Department of Agriculture and Food. 2017. "Improvement Tools: Critical Success Factors and Key Performance Indicators." Government of Western Australia. https://www.agric.wa.gov.au/improve ment-tools-critical-success-factors-and-key-performance-indicators.

---

effort. The initiatives may not be the top priorities, but since they can be implemented rapidly and easily, it is worth considering them. The initiatives in the **evaluate** quadrant need to be discussed and evaluated thoroughly by managers. They have the potential to provide high returns, but they require substantial effort. Some of them will be implemented and others will not. In the end, it may depend on how many initiatives are on the table and what resources are available to the organization to implement these initiatives. Finally, the **"stay away"** quadrant includes what many say are lost causes—initiatives that would require substantial effort to implement and yet would have limited potential returns.

Hypercompetition: Sustainability of Competitive Advantage

Most of the frameworks discussed previously to help identify potential strategic information systems initiatives were developed with the idea that firms can create

LEARNING ACTIVITY 4.3

Competitive Advantage at TRIPBAM

This activity requires you to complete a competitive analysis of TRIPBAM (Piccoli and Pigni 2016). The case may be provided to you by your instructor. Use one of the frameworks discussed to (1) identify the competitive position of TRIPBAM, and (2) identify three strategic information systems initiatives you believe the company should consider.

Source: Piccoli, G., and F. Pigni 2016. "TRIPBAM: Leveraging Digital Data Streams to Unleash Savings." *Communications of the Association for Information Systems* 39(25). https://aisel.aisnet.org/cais/vol39/iss1/25.

competitive advantages that differentiate them from other organizations over time. Today, however, the pace of change in information technologies and other innovations is faster than ever. This has led some individuals to question the sustainability of any competitive advantage a firm obtains through its initiatives. This is the concept of **hypercompetition**.

In his 1994 book *Hypercompetition*, Richard D'Aveni suggests that any competitive advantage an organization has will be rapidly eroded by competitors and that focusing on trying to sustain competitive advantages could be a deadly distraction for organizations. Instead, he suggests that organizations should seek to create disruptions in the market through a series of initiatives. He identifies seven strategic moves that organizations should consider to compete in hypercompetitive markets; these are called D'Aveni's 7 *S*s, which are summarized in Table 4.3.

While many companies have not adopted the 7 *S* framework of hypercompetition, it can still be used to identify strategic information systems initiatives. Think

TABLE 4.3 D'Aveni's 7 *Ss*	
Strategic Move	**Description**
Superior stakeholder satisfaction	Maximizing customer satisfaction by adding value strategically
Strategic soothsaying	Using new knowledge to predict or create new windows of opportunity
Positioning for speed	Preparing the organization to react as fast as possible
Positioning for surprise	Preparing the organization to respond to the marketplace in a manner that will surprise competitors
Shifting the rules of competition	Finding new ways to serve customers, thereby transforming the industry
Signaling strategic intent	Communicating intentions to stall responses by competitors
Simultaneous and sequential strategic thrusts	Taking steps to stun and confuse competitors to disrupt or block their efforts
Source: D'Aveni (1994).	

> ## LEARNING ACTIVITY 4.4
> ### Disrupting an Established Market!
>
> The purpose of this activity is to practice some of the concepts discussed in this chapter. Your instructor will assign you to groups, ideally the same groups used for Learning Activity 4.2. Each group represents a senior management team at Lyft (http://www.lyft .com) or Spotify (http://www.spotify.com). Your team is tasked with selecting one of the 7 *S*s from D'Aveni's framework and specifically targeting it to find a way for Lyft to better compete with Uber (http://www.uber.com) or Spotify to better compete with Pandora (http://www.pandora.com). The team must then identify a specific strategic initiative that will be used to implement the selected strategy.
>
> Once this activity is completed, all initiatives proposed by the groups in the class should be prioritized using the critical success factors identified in Learning Activity 4.2.

about new ways that an organization can try to maximum customer satisfaction (**superior stakeholder satisfaction**) using information systems. For example, a new customer relationship management system (Chapter 12) can be used to determine what customers prefer in their interactions with the organization. Similarly, new Web 2.0 technologies (Chapter 7) can be used to seek new knowledge directly from potential customers (**strategic soothsaying**). As you learn about the various technologies and information systems available today and in the near future in the next chapters, remember how they can be used for strategic purposes.

Chapter Summary

In this chapter, we discussed how organizations can obtain competitive advantages by implementing strategic information systems. We started by discussing the strategic planning process, which involves more than just planning about information technology, and then explored various frameworks that can help managers identify strategic information systems initiatives. We then discussed how managers can evaluate which strategic initiatives to implement and concluded with a discussion of hypercompetition, or the idea that competitive advantage cannot be sustained but that organizations should try to achieve market disruptions instead.

Here are the main points discussed in the chapter:

- There are five main steps in the strategic planning process:
  1. Strategic business planning
  2. Information systems assessment
  3. Information systems vision
  4. Information systems architecture
  5. Strategic initiatives identification
- We discussed four frameworks that can be used to identify strategic information systems initiatives. The SWOT analysis is used for managers to identify strengths, weaknesses, opportunities, and threats for the orga-

nization. Porter's Five Competitive Forces Model helps managers analyze the organization's competitive position by looking at five major forces in the firm's competitive environment: threats of new entrants, threats of substitutes, bargaining power of buyers, bargaining power of suppliers, and industry rivalry. Porter's value chain helps managers identify all the activities that the organization must perform to conduct its business. The virtual value chain looks at activities that turn raw data into useful information instead of looking at activities that turn raw materials into a final product like in manufacturing organizations. Global value chains involve production of products on an international basis.

- We discussed two methods for evaluating strategic initiatives: the critical success factors method and the priority matrix. Critical success factors (CSFs) are those few important considerations that must be achieved for the organization to survive and be successful (i.e., achieve its mission). The priority matrix allows managers to evaluate potential initiatives and prioritize them along two key dimensions: the ease of implementation and the potential returns.

- Hypercompetition is when competitors rapidly erode competitive advantages; in this case, organizations should focus on market disruptions instead of trying to sustain competitive advantages.

Review Questions

1. What is a strategic information system? How is it different from a strategic information system initiative?
2. What is the purpose of the strategic planning process, and what are its main steps?
3. What are the main advantages of using an information systems strategic planning process?
4. What do the SWOT analysis, Porter's Five Competitive Forces Model, and Porter's value chain model have in common?
5. What are the key differences between the traditional value chain model and the virtual value chain?
6. How can a SWOT analysis be used for identifying strategic information systems initiatives?
7. What is the purpose of Porter's Five Competitive Forces Model, and how can it be used for information systems strategic planning?
8. What are critical success factors (CSFs), and how should they be used to evaluate strategic initiatives?
9. What is the priority matrix, and how should it be used to evaluate strategic initiatives?
10. What is hypercompetition, and how can D'Aveni's 7 Ss framework be used in that context?

Reflection Questions

1. What is the most important thing you learned in this chapter? Why is it important?
2. What topics are unclear? What about them is unclear?
3. What relationships do you see between what you learned in this chapter and what you have learned in earlier chapters?
4. Why do you think the information systems strategic planning process is an iterative process?
5. If the strategic planning process has the benefits of improved communication, coordination, and decision-making, could it be used or applied by students for group projects? How?
6. In what situations in your personal life could you apply a SWOT analysis (not for a specific course)?
7. How are the virtual value chain and the traditional value chain similar and different?
8. How does the priority matrix relate to critical success factors?
9. Do you believe that hypercompetition exists today? If so, do you think it will continue to happen in the future?
10. Looking back over the content of the chapter, why do you think it is important to discuss strategic initiatives in the context of introducing information systems?

Additional Learning Activities

4.A1. Use the priority matrix to rank initiatives identified in Learning Activity 4.4. Compare the results of your analysis with the results of the analyses conducted in Learning Activity 4.4. Be prepared to discuss the differences.

4.A2. This activity requires you to complete a competitive analysis of Uninor. The case may be provided to you by your instructor or found online. Bose, I. 2017. "Persevere or Exit: What Is the Right Strategy?" *Communications of the Association for Information Systems* 41(12). https://aisel.aisnet.org/cais/vol41/iss1/12.

4.A3. Find the mission statements for Amazon.com and your school. Compare the two statements and identify similarities and differences. Why are there differences?

4.A4. This activity requires you to complete a competitive analysis of Lenovo. The case may be provided to you by your instructor or found online. Zwanenburg, S. P., and A. Farhoomand 2018. "Lenovo: Being on Top in a Declining Industry." *Communications of the Association for Information Systems* 42(17). http://aisel.aisnet.org/cais/vol42/iss1/17.

4.A5. Conduct a virtual value chain analysis to identify potential strategic information systems in the hotel industry.

4.A6. This activity requires you to complete a competitive analysis of an Italian government initiative. The case may be provided to you by your instructor or found online. Datta, P. 2020. "Digital Transformation of the Italian Public Administration: A Case Study." *Communications of the Association for Information Systems* 46(11). https://aisel.aisnet.org/cais/vol46/iss1/11/.

References

D'Aveni, R. 1994. *Hypercompetition.* New York: Free Press.

Gregory, L. 2018. "McDonald's Five Forces Analysis (Porter's Model) & Recommendations." *Panmore Institute.* http://panmore.com/mcdonalds-five-forces-analysis-porters-model.

Janse, B. 2019. "Critical Success Factors." *toolshero.com.* https://www.toolshero.com/strategy/critical-success-factors/.

Kappelman, L., V. L. Johnson, C. Maurer, K. Guerra, E. McLean, R. Torres, M. Snyder, and K. Kim, 2020. "The 2019 SIM IT Issues and Trends Study." *MIS Quarterly Executive* 19(1): 69–104. https://aisel.aisnet.org/misqe/vol19/iss1/7/.

Lucco, J. 2020. "How to Determine Critical Success Factors for Your Business." *ClearPoint Strategy.* https://www.clearpointstrategy.com/how-to-determine-critical-success-factors-for-your-business/.

Martin, M. 2019. "How Porter's Five Forces Can Help Small Businesses Analyze the Competition." *Business News Daily,* December 3. https://www.businessnewsdaily.com/5446-porters-five-forces.html.

Porter, M. E. 1979. "How Competitive Forces Shape Strategy." *Harvard Business Review* 57(2): 137–45.

———. 1996. "What Is Strategy?" *Harvard Business Review* 74(6): 61–78.

———. 2008. "The Five Forces That Shape Strategy." *Harvard Business Review* 86(1): 78–93.

Rayport, J. F., and J. Sviokla. 1995. "Exploiting the Virtual Value Chain." *Harvard Business Review,* November–December: 75-85. https://hbr.org/1995/11/exploiting-the-virtual-value-chain.

Recklies, D. 2015. "Porters Five Forces—Content, Application, and Critique." *TheManager.org,* November 20. http://www.themanager.org/2015/11/porters-five-forces/.

Glossary

Bargaining power of buyers: In Porter's Five Competitive Forces Model, determines the ability of the organization's buyers (customers) to reduce the organization's competitive position.

Bargaining power of suppliers: In Porter's Five Competitive Forces Model, determines what is the ability of the organization's suppliers to reduce its competitive position.

Critical success factors (CSFs): The few important considerations that must be achieved for the organization to survive and be successful (i.e., achieve its mission).

Evaluate: In a priority matrix, the initiatives that need to be discussed and evaluated thoroughly by managers.

Global value chains: Value chains where goods are produced on an international scale, with production broken down into activities and tasks carried out in different countries.

Hypercompetition: When competitive advantages are rapidly eroded by competitors and organizations should focus on market disruptions instead of trying to sustain competitive advantages.

Imperatives: In a priority matrix, the initiatives that should be relatively easy to implement and have the potential to bring high returns to the organization.

Industry competitive rivalry: In Porter's Five Competitive Forces Model, determines the current level of competition in the industry.

Information systems architecture: Specifies how information systems resources should be used and how they should work together.

Information systems assessment: The step of the planning process where the organization identifies the current state of information systems resources in the organization.

Information systems vision: A broad statement of how the organization should use and manage its information systems for strategic purposes.

Opportunities: In a SWOT analysis, determines the activities or factors that could help the organization get new advantages over others in its industry.

Porter's Five Competitive Forces Model: Framework to help managers analyze the organization's competitive position by looking at five major forces in the firms' competitive environment: threats of new entrants, threats of substitutes, bargaining power of buyers, bargaining power of suppliers, and industry rivalry.

Positioning for speed: In D'Aveni's 7 Ss framework, a strategy about preparing the organization to react as fast as possible.

Positioning for surprise: In D'Aveni's 7 Ss framework, a strategy about preparing the organization to respond to the marketplace in a manner that will surprise competitors.

Potential threat of new entrants: In Porter's Five Competitive Forces Model, determines how easy it is for new companies to enter the market in which the organization operates.

Potential threat of substitutes: In Porter's Five Competitive Forces Model, determines the likelihood that other products of equal or superior value are available.

Priority matrix: Framework that allows managers to evaluate potential initiatives and prioritize them along two key dimensions: the ease of implementation and the potential returns.

Quick wins: In a priority matrix, the initiatives that do not have much upside potential but are easy to implement.

Shifting the rules of competition: In D'Aveni's 7 *S*s framework, a strategy about finding new ways to serve customers, thereby transforming the industry.

Signaling strategic intent: In D'Aveni's 7 *S*s framework, a strategy about communicating intentions in order to stall responses by competitors.

Simultaneous and sequential strategic thrusts: In D'Aveni's 7 *S*s framework, a strategy about taking steps to stun and confuse competitors to disrupt or block their efforts.

Strategic information systems: Information systems specifically meant to provide organizations with competitive advantages.

Strategic information systems initiatives: Detailed (usually two- to five-year) plans for implementation of systems that may result in a new strategic direction for the organization.

Strategic planning process: Structured set of steps to identify strategic information systems.

Strategic soothsaying: In D'Aveni's 7 *S*s framework, a strategy for using new knowledge to predict or create new windows of opportunity.

Stay away: In a priority matrix, the initiatives that would be difficult to implement and would have limited potential returns.

Strengths: In SWOT analysis, asks what gives the organization advantages over others in its industry.

Superior stakeholder satisfaction: In D'Aveni's 7 *S*s framework, a strategy about maximizing customer satisfaction by adding value strategically.

SWOT analysis: Framework used by managers to identify strengths, weaknesses, opportunities, and threats for the organization.

Threats: In SWOT analysis, determines what activities or factors could create disadvantages or troubles for the organization relative to others in its industry.

Value chain analysis: Framework to help managers identify all the activities that the organization must perform to conduct its business.

Virtual value chain: Framework that looks at activities that turn raw data into useful information instead of looking at activities that turn raw materials into a final product like in manufacturing organizations.

Weaknesses: In SWOT analysis, determines what activities create disadvantages for the organization relative to others in its industry.

Storing and Organizing Information

Learning Objectives

By reading and completing the activities in this chapter, you will be able to:

- Discuss the purpose of a database management system
- Decide whether it is better to store data using a database management system or another alternative, such as a spreadsheet
- Explain the basic structure and components of relational databases
- Describe the purpose of foreign keys in a relational database
- Discuss the purpose of a relational database schema and explain its notation
- List and describe a number of online databases
- Understand what Big Data is and how businesses can use it to make more informed decisions

Chapter Outline

Focusing Story: The Database behind Facebook
  Learning Activity 5.1: Data for an Amazon Order
Overview of Relational Databases
Databases versus Spreadsheets: When to Use a DBMS
  Learning Activity 5.2: Connecting Data Elements
Database Diagrams
  Learning Activity 5.3: Finding Business Databases Online
Online Databases
Big Data

Throughout this book, the focus has been on information. Much of the information used by businesses is stored in databases. While there are different schemes for storing data, much of the data you will deal with is stored in relational databases. Databases may seem pretty mysterious, but they are actually not hard to understand at a basic level. In this chapter, we will help you learn about how relational databases store data. We will help you learn how to construct, populate, and retrieve data from databases using Microsoft Access in Appendices C and D.

The Database behind Facebook

Odds are you are one of the more than 3 billion active users of Facebook. If so, you probably know how to "friend" people and post photos, videos, and status updates. What you may not know is that there is a giant database that keeps track of all this information. Here is a partial list of what Facebook's database must track:

- Almost 1 billion objects, such as pages, groups, events, and communities
- More than 30 billion pieces of content, including links, posts, photos, notes, videos, and new stories
- Friend connections among the more than 3 billion active users (The average user has well over 100 friends.)

Keeping track of all this information requires a very complex database design, in addition to a robust infrastructure. Facebook uses a variety of tools to create and manage its data, including Apache Cassandra, which manages data across hundreds of servers; Apache Hive, which facilitates summarizing and retrieving data in very large databases; and Scribe, which reliably delivers billions of Facebook messages each day.

While you may never have to deal with databases this large, you will probably have to use databases throughout your work life. Much of the information you will need to access to do your job will be stored in relational databases. Relational databases underlie many of the applications discussed throughout this book, including enterprise resource planning, customer relationship management, and supply chain management systems, which you will learn about in Chapter 12. (Note that Facebook uses other types of database technologies in addition to relational databases.)

Source: https://about.fb.com/company-info/.

Focusing Questions

1. Identify the information elements that are part of a Facebook profile.
2. What information do you think Facebook uses to determine what friend suggestions to make?

Digital Data

How much digital data exists? This is a big, complicated question, given that Google, Facebook, Microsoft, and Amazon alone store 1,200 petabytes of data. A petabyte is a 1 with 15 zeros after it. In fact, 90% of the data stored in the world was created in the past two years. (This includes business and personal data, such as digital photos, music, and video.) According to World Economic Forum, there are currently 44 zettabytes of data stored worldwide. A zettabyte is a 1 followed by 21 zeros. That number is expected to grow even more, as it is expected that 0.5 zettabyte of data will be created per day by 2025. That would mean over 180 zettabytes of data will be created in the year 2025. The amount of data collected is expected to grow as more sensors, websites, and other means of collecting data increase by the day. With the ever-expanding glut of digital data, you can understand why efficient ways to store, organize, and retrieve the data are increasingly important.

Source: https://seedscientific.com/how-much-data-is-created-every-day/.

LEARNING ACTIVITY 5.1

Data for an Amazon Order

Suppose you are ordering something from Amazon.com. What pieces of data (for example, your name) does Amazon need to fulfill your order?

Overview of Relational Databases

Databases and Database Management Systems

A **database** is an organized collection of data. In this chapter, we focus on a particular type of database called a relational database, because this is the dominant type of database for business applications. These databases can store different types of "information," including text, numbers, documents, images, and videos. Given the amount and variety of information that businesses deal with, well-structured databases are a must for any business.

DBMS

Typically, a database is managed by a **database management system (DBMS)**. A database management system provides the means for creating, maintaining, and using databases. For most of you, the "using" part is the most important aspect of a DBMS, although you may create small databases. Professional database designers and administrators handle the creation and maintenance tasks in most organizations.

There are many database management systems used by businesses. Smaller database-oriented tasks can be handled by a personal DBMS, such as Microsoft Access or OpenOffice Base. Larger, more complex databases require an enterprise-level DBMS, such as Oracle, MySQL, Microsoft SQL Server, or IBM's DB2. Each of these databases has its advantages and disadvantages, but all are quite capable. Oracle has been the market leader for many years, but the others are widely used in business.

Databases are integral elements of information systems. Most information systems today use multitiered architectures that divide processing into different elements. **Multitiered architecture** refers to the database server, the Web server, and applications, all existing on different computers. You will learn more about this in Chapter 7. Without going into too much detail, applications handle the processing of data, while the DBMS is responsible for managing the data. As shown in Figure 5.1, the application requests data from the DBMS, which provides the requested data. The application can then manipulate the data (such as reducing inventory when a product is ordered) and send updated data to the DBMS. The DBMS updates the actual databases according to the update sent by the application. It is possible to do all this within the application, but most large systems use the multitier approach.

Applications = processing data
DBMS = managing data

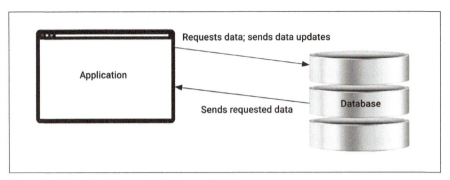

Figure 5.1 Interaction between Applications and Databases

Databases versus Spreadsheets: When to Use a DBMS

Many people get a little confused about whether to use a spreadsheet (such as Microsoft Excel) or a DBMS (such as Microsoft Access) for storing and organizing information. One reason for this confusion is that both store information in tables of rows and columns. This similarity is a bit misleading, however. The two applications have very different purposes, strengths, and weaknesses.

Spreadsheets are fine for very simple data storage tasks, such as keeping simple lists, like a personal contact list or a home inventory. Unfortunately, spreadsheets have a number of limitations that make them unsuitable for more complex data storage. Some of the problems you might run into if you use a spreadsheet for data storage include the following:

- Unnecessary data duplication
- Inconsistent data
- Difficulty in data retrieval and search
- Poor data integrity
- Difficulty in relating different data elements (such as customers and orders)

The problems of unnecessary data duplication and poor data integrity are particularly troubling. (Data integrity problems include inaccurate, inconsistent, and out-of-date data.) Some studies indicate that most complex spreadsheets have errors. (See http://www.panko.com/ssr/index.html for a good overview of these studies.) While not all of these errors are due to data issues (for example, formulas could be incorrect), many are. Most database management systems have methods for keeping data up to date and consistent. While data stored in databases are not perfect, they typically are of higher quality than data stored in spreadsheets (or in other formats, such as documents, pictures, or videos).

Using an out-of-date version can also be a problem with spreadsheets. Businesses sometimes make bad decisions because someone has relied on an old version of a

LEARNING ACTIVITY 5.2

Connecting Data Elements

In Learning Activity 5.1, you came up with a list of data that Amazon would need to process an order. Take those elements, group similar items together, and name each group of elements. For example, the customer's name and address might be in a group called Customer.

spreadsheet. Often this is due to data not being updated, but there can be other changes that make a spreadsheet out of date. Reports from database management systems draw data directly from databases, which typically contain current information. It is worth noting that spreadsheets can be built to connect to a database to avoid the problem of outdated data.

Because spreadsheets are relatively easy to use and require less planning than database management systems, it is tempting to use a spreadsheet for managing information. For all but the simplest tasks, this is a mistake. In the long run, you are usually much better off taking the time to plan and design a proper database, especially as the amount of information being stored increases.

Databases are good at storing and organizing information. Spreadsheets are good for analyzing and displaying information visually. A complete information solution requires both. Fortunately, it is relatively easy to extract data from a database and import them into a spreadsheet for analysis, so you do not have to make an "either/or" choice. Use a database to store the information; use a spreadsheet to make sense of the information.

Now that you have a general idea of the purposes of databases and their relationship to information systems, we can get into more detail about how relational databases function.

Relational Databases

A **relational database** stores data in the form of connected tables. Tables are made up of records (rows) and fields (columns). A **record** is a set of fields that all pertain to the same thing, while the **fields** represent some characteristic of the thing. An example may make this easier to understand. Consider Table 5.1.

Each row represents a single instructor. Each column represents a characteristic. The first row of data (not counting the headings) contains data about the

TABLE 5.1 Instructors Table			
InstructorID	**Last Name**	**First Name**	**Email**
1	Smith	Sadie	ssmith@school.edu
2	Jones	Maggie	mjones@school.edu
3	Thurman	Richard	rthurman@school.edu
4	Wilson	Fred	fwilson@school.edu

instructor named Sadie Smith. The table stores four characteristics (fields) about the instructors:

- Identification Number (InstructorID)
- Last Name
- First Name
- Email Address

Fortunately, this structure makes intuitive sense for most of us. Things get a little more complicated when we need to store data about multiple things. Take a look at Figure 5.2, which shows a small database about course offerings. This database stores data about three "things" that are related to each other: sections, courses, and instructors. Since we want to store data about three things, we have three tables. This is a key feature of relational databases. Each table stores data about a separate thing. This structure allows for flexibility when retrieving data while minimizing redundancy. When storing data, unnecessary redundancy can be a bad thing because redundancy can lead to inconsistencies. By storing each piece of information only once, we ensure that the data are consistent. We explain more about redundancy and consistency later in this chapter.

For a database to work, we need some way to uniquely identify each record. You might think that we could use the instructor's name for this purpose, but this will not work because two instructors might have the same name. Sometimes there is a natural choice for this unique identifier called the **primary key**. For example, student ID numbers might be a good choice for a database that tracks members of a student organization. Sometimes, however, you might simply make up an identifier. Since the identifier typically does not store any meaningful data (other than to identify a row), making up a number is not a problem. Each table in a database has a primary key. Sometimes a primary key is made up of more than one field. We call this a *composite primary key*.

You might have noticed that earlier we said that the tables in our database are related to one another. For example, the Sections table is related to the Instructors table and the Courses table. In a relational database, these connections are implemented by **foreign keys**, which are fields that reference a primary key in a related table. For example, the CourseID field in the Sections table references the CourseID field in the Courses table. This cross-referencing is called a *relationship*. The arrows in Figure 5.2 show how the foreign keys reference the primary keys.

These cross-referencing foreign keys make it easy to combine data contained in multiple tables. See if you can answer this question: What is the title of the course being taught in Section 1001? This is a pretty easy question to answer. Section 1001 is Introduction to IT. All you have to do is look at the CourseID in the Sections table and then find the matching CourseID in the Courses table. There is another foreign key in the Sections table. InstructorID references the primary key of the Instructors table. While there are a number of very formal, precise rules that govern

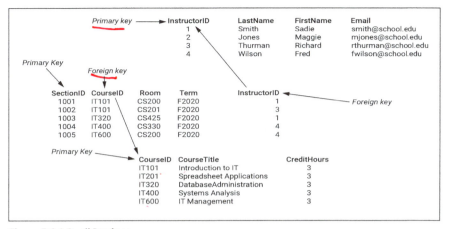

Figure 5.2 A Small Database

relational database design, if you understand this cross-referencing scheme, you understand the basics of relational databases. Even very large, complex databases follow this same basic structure. They just have more tables and fields.

There is an additional characteristic of relational databases that we need to discuss. For reasons that are beyond our scope, in a relational database, each row/column intersection can store at most one item of data. For example, if there is more than one section of a course, you must have a separate record for each section.

There are some other rules that govern relational database design. These rules, which are called *normal forms*, are in place to ensure data consistency by eliminating unnecessary redundancy. While the particulars of these rules are beyond our scope, we want to use a simple example to show the relationship between redundancy and data consistency. Consider the data shown in Table 5.2.

Answer the following question about Table 5.2: What is Instructor #1's last name? You really cannot be sure of the answer to this question. This instructor's last name is Smith in the first row and Jackson in the third row. We do not really know which one is correct. This inconsistency, which we call an *anomaly*, results from storing the instructor's name in multiple rows. Compare this structure to the one in Figure 5.2. The earlier structure prevents these inconsistencies by storing each item of data only once. (The exception is the foreign key.) This is why each table stores data about one "thing." To summarize simply, if you only store a data element in one place, you cannot have inconsistent values. One of the values could be incorrect (for example, an instructor's name could be misspelled), but that is a different problem from inconsistency.

Sometimes a particular row in a table can be related to at most one row in a related table. For example, in most businesses, a specific order can only be related to one customer. If we look at this from the other direction, a specific customer can be related to more than one order. This is an example of a *one-to-many* relationship.

SectionID	CourseID	CourseTitle	InstructorID	LastName	FirstName
1001	IT101	Introduction to IT	1	Smith	Sadie
1002	IT101	Introduction to Computers	3	Thurman	Maggie
1003	IT320	Database Administration	1	Jackson	Sadie
1004	IT400	Systems Analysis	4	Wilson	Fred
1005	IT600	IT Management	4	Wilson	Fred

TABLE 5.2 Redundancy Example

Many-to-many relationships also exist. In these relationships, a specific row can be related to multiple rows in a related table. But in contrast to the one-to-many relationship, this is true in both directions. Consider the example of students registering for course sections. A specific student can be registered for more than one section (each for a different course), and a specific section can have more than one student registered for it. This is known as a many-to-many relationship.

Because of the structural rules of relational databases, many-to-many relationships require creating a new table that links the two related tables. Not surprisingly, these are called *linking* or *intersection* tables. Figure 5.3 shows a linking table for student registrations. The purpose of the Registrations table is to link a student to a section. You can see that Student #1 (Pat Johnson) is registered for four sections (1001, 1002, 1004, and 1005). Section 1001 has three students registered (1, 2, and 4).

Although they are more unusual, one-to-one relationships also exist. In these relationships, a specific row in a table can be related to at most one row in a related table. This is true in both directions of the relationship. For some businesses, the relationship between customers and accounts could be a one-to-one relationship. (It could be a one-to-many relationship for other businesses.)

As you might imagine, using the actual data to show the structure of a database only works for very small databases. For larger databases, we illustrate structure using database schema diagrams, which we discuss next.

Database Diagrams

Database designers use several different diagrams when coming up with the design of a database. Here we are interested in a view that shows tables, fields, and relationships. (Note that we often call fields *attributes* when discussing database models. We are going to stick to *fields* to make things less confusing.) Some people call these diagrams *entity-relationship diagrams (ERDs)*, while others call them *database schemas*. There are technical differences between the two, but both essentially show the data elements and the relationships among them, which is what we care about here.

Figure 5.4 shows a diagram for a database that stores wish lists for customers. Each rectangle shows the name of the table at the top. Each field is included inside the rectangle. The primary key is indicated by the PK in the left-hand column. For-

Students

StudentID	LastName	FirstName	Email
1	Johnson	Pat	pjohnson@school.edu
2	Gonzales	Jorge	jgonzales@school.edu
2	Andrews	Annie	aandrews@school.edu
4	Goldman	Beau	bgoldman2school.edu
5	Simpson	Lois	lsimpson@school.edu

Sections

SectionID	CourseID	Room	Term
1001	IT101	CS200	F2020
1002	IT101	CS201	F2020
1003	IT320	CS425	F2020
1004	IT400	CS330	F2020
1005	IT600	CS200	F2020

Registrations

Registrations

SectionID	StudentID
1001	1
1001	2
1001	4
1002	5
1002	1
1003	3
1003	2
1004	1
1004	3
1004	5
1004	2
1005	1
1005	5

link

Figure 5.3 Many-to-Many Linking Table Example

eign keys have FK beside them to the left. Since tables can have more than one foreign key, the FK notations are numbered. Recall that the foreign keys relate a row in one table to a row in a related table. For example, the field "CategoryID" in the Products table refers to a CategoryID value in the Categories table.

You might find the notation for the WishList table a little confusing. This is an example of a linking table. It links together two other tables. These types of tables are needed to control unnecessary redundancy. Many linking tables have composite primary keys (primary keys composed of more than one field). In our notation, each field in the primary key is boldfaced, underlined, and has PK next to it, as is the case with CustomerID and ProductID in the WishList table. Together, CustomerID and ProductID serve as the primary key of the WishList table. Separately, however, each of these fields acts as a foreign key. You can tell this because of the FK1 and FK2 next to CustomerID and ProductID, respectively. The CustomerID value in the WishList table refers to the CustomerID of the specific customer to whom the WishList item is assigned. ProductID in the WishList table serves a similar purpose. An arrow between the tables shows relationships between tables graphically. The direction of the arrow indicates the direction of referencing (from the foreign key to the corresponding primary key).

Note that this is only one way to show a database diagram. Your instructor may prefer another, such as crow's foot notation, IDEF1X, or Chen's notation. We use this particular notation simply because it is well implemented in Microsoft Visio and other programs. Once you have the logical design, you have most of the information necessary to create a database, especially for simple applications. More complex applications require additional information, but that is beyond our scope here.

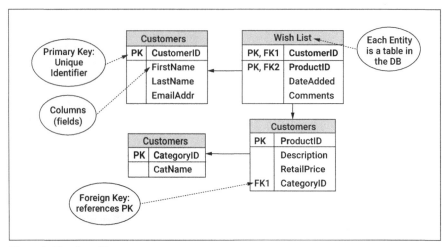

Figure 5.4 Database Diagram

Being able to read a database diagram is important, even if you never design or build a database. The diagram is a key to understanding the structure of the database, which is useful when retrieving data from a database or if you have the job of a programmer or systems analyst and you need to work with a database administrator to create or modify the database as part of a project. Learning how to create a database design is beyond the scope of this chapter but is covered in Appendix E.

iTunes Match

As people purchase more and more electronic devices to play their digital music, it has become necessary to have a central repository in which to store these songs. One such repository, provided by Apple, is iTunes Match. This service catalogues all the digital songs that you have purchased and stores them in the cloud. "Stored in the cloud" means that rather than being stored on one of your devices, such as a laptop, the songs are stored online, which makes them available to all of your devices. When logged onto a device that utilizes iTunes Match, it then allows for the music to be streamed directly from the cloud or downloaded onto the device.

With a number of people utilizing iTunes Match, how does Apple go about storing all this information? Apple's iTunes database has 4.3 million songs stored. If the song that a person owns is part of this database, an indicator that the person owns that song is stored rather than the song itself. The only copies of songs that have to be stored separately are those that Apple does not already have in its database. This centralized database approach allows Apple to keep its database storage needs low while providing a centralized service for the many people who want to use it.

It should be noted that Amazon, Google, Spotify, and others utilize a similar process to catalogue their music libraries as well.

Sources: https://support.apple.com/en-us/HT204146; https://www.lifewire.com/itunes-match-faq-199 9186.

Online Databases

So far in this chapter, we have primarily discussed relational databases that store text and numeric data related to business operations. There are other types of databases that are quite useful for performing research; many of them are accessible through the Internet. You may have used some of them when working on a research paper for class. These databases can also be quite useful in your professional life. For example, if you are a marketing professional, you can use these databases to gather information about prospective clients or competitors. Demographic databases are important for understanding markets. For example, retail organizations use demographic databases to gather data to help decide where to locate new stores. These databases serve a different purpose than the databases we discussed earlier. Rather than being containers for raw data, these online databases contain pointers to sources of information.

Knowing what databases exist and how to use them can be very beneficial to you, both personally and professionally. Here are just a few examples of situations you might face and databases that might be helpful. Most of these resources are available through libraries.

- Researching a career
  - The United States Bureau of Labor Statistics (http://www.bls .gov/data/) has a searchable database of employment, pay, and benefit statistics and occupation growth projections.
  - Job search sites such as AllianceQ, Monster.com, CareerLink, and CareerBuilder.com maintain databases of job openings. You can usually search these databases by a variety of criteria, including occupation and location.
- Preparing for a job interview
  - Article databases such as ABI/Inform and LexisNexis Academic (often available through your university library's website) are useful for finding news articles about a company and its executives.

LEARNING ACTIVITY 5.3

Finding Business Databases Online

Your campus library likely has a variety of business-related online databases. (See http://guides.lib.vt.edu/subject-guides/bus for an example.) Visit your library's website and find its list of business databases. Identify and briefly describe one database for each of the related purposes:

- Economic statistics
- Industry information
- Demographic statistics
- Business-related news articles

- •• The Business and Company Resource Center (often available through your university library's website) maintains databases of company- and industry-related information.
- • Gathering information on customers or competitors
    - •• The Business and Company Resource Center is also useful for learning more about customers and competitors.
    - •• Hoovers.com provides extensive information about companies and industries.
- • Conducting market research
    - •• In addition to many of the above, the United States Census Bureau (http://www.census.gov) provides access to a variety of demographic and economic data.
- • Economic forecasting, investment planning, and portfolio management
    - •• Wharton Research Data Services (WRDS) is essentially a portal that provides a Web-based interface to databases of financial information, including the following:
        - ••• *Compustat:* Company financial items such as income statements and balance sheets
        - ••• *CRSP:* Historical stock and mutual fund prices and U.S. Treasury interest rates
        - ••• *Global Insight:* United States and international macro-economic data
    - •• Federal Reserve Economic Data (FRED) (https://fred.stlouis fed.org/) provides more than 21,000 time-series data sets, including interest rates, inflation, and economic growth.

Big Data

A chapter covering databases would not be complete without a discussion of Big Data. **Big Data** is something of a buzzword that refers to the vast amount of data (often measured in terabytes or petabytes) that are created and stored that have grown beyond the capabilities of traditional data processing tools and applications. The meaning of Big Data is not rigorously set and will be subject to change over the years as technology evolves. However, it is commonly believed that there are at least 3 V's that define Big Data—Volume, Velocity, and Variety. Volume refers to how much data is being stored by the organization. Velocity refers to how quickly the data is moving through the system. Variety refers to the number of different sources that provide data that needs to be analyzed and stored. For example, an organization may utilize data from Facebook, customer service message boards, and internally created transactions, and the data from Facebook may be coming in at a high velocity. In addition to the data coming from the three different sources, the transaction logs also create a high volume of information that needs to be stored. For the organization to utilize all of this information, it must create new methods and processes

LEARNING ACTIVITY 5.4

Fitness Trackers

You likely have used a fitness tracker or know somebody who has. Most people who have a fitness tracker also use an app on their smartphone that collects and stores the fitness information. This information is utilized to identify trends over time, make recommendations, and encourage users to continue their exercise habits. For this activity, do the following:

- Identify and list the information captured by fitness trackers.
- Identity and list the features that are provided because a smartphone app has access to this fitness information.
- List potential issues that could arise because fitness trackers are collecting this information.

to analyze the data to identify trends that will help it make better business decisions. However, there are some challenges that must be overcome to capitalize on the increased availability of data. Such challenges include the following:

- How should the data be stored?
- How do data from various sources integrate with one another?
- How are data retrieved and disseminated?

The answers to all these questions were straightforward with a relational database. But in the case of Big Data, new solutions may be required. It should be noted that storing the data necessary to conduct Big Data processing can be done using traditional relational databases. However, in this section, we discuss an alternative method to relational databases along with some of their strengths.

Storing Data

In a relational database, data are stored in a way that ensures data integrity as records are inserted, updated, and deleted. However, one of the trade-offs with storing data in this fashion is performance. As the amount of data being stored, and then ultimately retrieved, grows as large as it has in this era of Big Data, it may be necessary to sacrifice some levels of data integrity to keep performance at an acceptable level. While a normalized database is stored in a highly structured way, in a Big Data implementation, data can be stored unstructured. This is done using either a network-attached storage (NAS) model or a direct-attached storage (DAS) model. The NAS model utilizes a series of file servers that can easily expand to grow capacity and high-speed connections between them. The DAS model keeps the data more centralized. This approach allows for faster access time when it comes to processing but limits the scalability of data size. Hybrid approaches can also be used. Data stored permanently could utilize the NAS model, while temporary data that needs to be accessed more quickly could be stored using the DAS model.

Big Data at Facebook

Many of you probably enjoy using Facebook (or at least know people who do). A single user alone has hundreds of friends, and when they log on to Facebook, they want to see an update specific to them. Imagine now that there are more than 3 billion Facebook users, and each of these users has a unique set of friends. Facebook is successful at selecting the correct posts and making recommendations based on each individual's interactions with the website. This success is due to utilizing Big Data in such a way that it can very quickly select the updates that a given user will want to see as well other recommendations customized for him or her based on previous visits to Facebook. Thanks to advances in Big Data, you are able to enjoy this and many other services that bring together the interaction of millions, if not billions, of users in one service.

Data lakes provide one approach to storing the massive amounts of data generated and collected by organizations. Data lakes store huge volumes of data in their original formats by giving each data element a unique identifier and tagging it with descriptors that tell users what it is about. In most data repositories, data structure and requirements are defined up front. In a data lake, data structures and requirements are not defined until the data are needed. The bottom line is that the best ways to store, organize, and use large volumes of data are still emerging. Time will tell if data lakes are important long term or just an interesting method that will lead to further advancements.

Integrating Data

With data coming from a variety of different sources, it is necessary to ensure that records from various data sources can be combined to make sense of the data. This process is referred to as the E/T/L process, which stands for extract, transform, and load. We will go into more detail about the E/T/L process in Chapter 14, but we will briefly discuss it here for the purpose of Big Data. The first step in this process is to extract, or retrieve, the data from the various locations in which it was stored or gathered. This involves identifying the sources of the various data that are needed to make the decisions required. Because of the vast amount of data that could be stored, it is important to strategically identify what data are needed for making business decisions. The recommendations discussed in Chapter 4 would be followed.

Once the necessary data have been identified and extracted, it is then necessary to transform the data. This is the process where the data are modified and combined in such a way that records from various sources can be combined. For example, one data source may provide data that are stored at the second-by-second transaction level. However, this may need to be combined with other data that are stored in a daily summarized manner. To combine them, the data will need to be transformed in such a way that the two different levels will make sense when combined. It may make sense to summarize the transaction-level data to daily summarized data and analyze at that level. Or it may make sense to take the summarized data and create a way for it to be com-

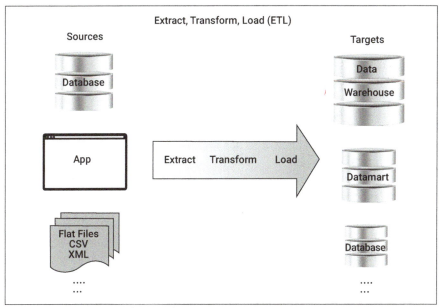

Figure 5.5 The Extract, Transform, Load Process Source: https://commons.wikimedia.org/wiki/File:Extract,_Transform,_Load_Data_Flow_Diagram.svg.

bined at the transaction level. Whatever decisions are made should be done in such a way as to be in alignment with a business's strategy. However, once data are transformed to be summarized at a level higher than the transaction level, it is impossible to desummarize them. Much caution at this decision-making stage is recommended. Once the transformation of the data has occurred, the data must then be stored. This is the load process. Loading is a matter of writing the data to a disk in a fashion that can be readily retrieved as necessary. An illustration of this process is provided in Figure 5.5.

Retrieving and Disseminating Data

One of the advantages of Big Data is the ability to analyze the data that are being stored to make informed business decisions. If the data are stored in a relational database, then SQL will be utilized to query the data. However, if it is stored in an unstructured format, it is necessary to retrieve the data in a manner other than SQL. One way this is done is by utilizing what is called NoSQL, which some interpret as "Not only SQL." NoSQL allows for greater scalability and efficiency in storing and retrieving data. With NoSQL, one approach to querying data is to assign all attributes keys. This allows for the dynamic creation of a table for a given query. Then, when that query is over, a new dynamic creation can be utilized depending on the question being answered. This method of retrieving data then results in the ability to analyze data in an ad hoc fashion without necessarily needing to know what questions will be asked prior to the creation of the database structure.

Regardless of how the data are stored or queried, as discussed above, it is still necessary to interpret and share these results. One manner for doing so in an easy-to-utilize fashion is through data analytics. Data analytics utilizes the data that the company has available and uses statistical analysis through various tools to make business-related decisions. This could include forecasting future sales, graphically displaying information about the organization through charts and graphs, and tracking metrics that the organization has deemed important for its success. The important aspects necessary for a company to use data analytics are to possess the necessary data, understand how it relates to the business problems, and then statistically interpret the data.

Chapter Summary

In this chapter, you learned about databases, which are critical to the operation of information systems. The intent of the chapter was to help you gain an appreciation of the importance of databases, with a special emphasis on relational databases. Here are the main points discussed in the chapter:

- Database management systems provide the means for creating, maintaining, and using databases.
- Spreadsheets are good for storing simple lists of information. However, they have a number of limitations that render them less effective for more complex data management.
- A relational database stores data in the form of connected tables. Tables are made up of records and fields.
- In a relational database, a record is a set of fields that all pertain to the same thing, while the fields represent some characteristic of the thing.
- In a relational database, foreign keys are fields that reference the primary keys in related tables.
- A database diagram shows the logical structure of a relational database, including its tables, fields, and relationships among tables. Primary and foreign keys are also indicated.
- There are many online databases that store a vast array of information. These databases include article databases, market and economic databases, and databases of demographic and governmental information, among other topics. Examples include LexisNexis Academic (article database), Federal Reserve Economic Data service (economic data), and Monster.com (employment database).
- Big Data provides a new opportunity to analyze information and identify trends that can help businesses make better-informed decisions.

Review Questions

1. What are the main functions of a database management system?
2. Name five potential problems that might arise from using a spreadsheet for data storage.
3. Briefly describe how a software application interacts with a database management system in a multitiered architecture.
4. Briefly describe how relational databases store information.
5. What are the main elements of a relational database?
6. In a relational database, what is the function of a primary key?
7. In a relational database, what is the function of a foreign key?
8. What is the purpose of a database diagram?
9. What are the main elements of a database diagram?
10. Name and briefly describe four online databases.

Reflection Questions

1. What is the most important thing you learned in this chapter? Why is it important?
2. What topics are unclear? What about them is unclear?
3. What relationships do you see between what you learned in this chapter and what you learned in previous chapters?
4. Compare and contrast data storage using electronic spreadsheet software and database management systems.
5. Suppose you need to keep track of some data. How would you decide whether to use a spreadsheet or database management system to store the data?
6. As a user of databases, why is it useful to understand how relational databases are structured?
7. What challenges do you think arise as organizations collect the amount of data necessary to conduct the analyses provided by Big Data?

Additional Learning Activities

5.A1. You have just been elected chair of one of your clubs. Your first task is to design a form to keep track of information about your members. For now, you just need to track basic information: member's name, email address, mobile phone number, and join date. You want to be able to see the information about multiple members at one time. To keep things simple, you are going to keep the information on paper (at least for the moment). Sketch out a format for how you would store this information.

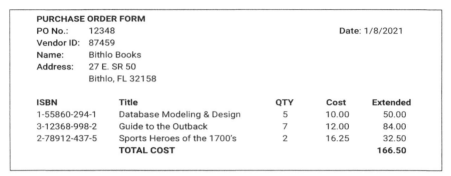

Figure 5.6 Purchase Order Form

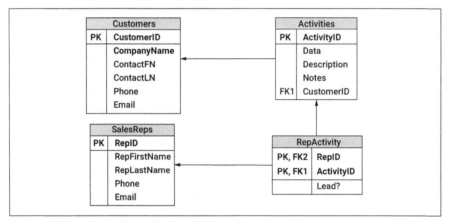

Figure 5.7 Customer Activity Database Design

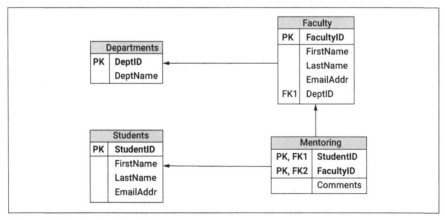

Figure 5.8 Mentoring Database Design

5.A2. Suppose that you want to store the club member information from the previous question on a computer. Would you use a database (such as Microsoft Access) or a spreadsheet? Explain your choice.

5.A3. Identify the data elements necessary for you to register for a course. Group these elements into tables. Draw a sketch showing how these tables might relate to one another.

5.A4. Identify the tables and fields that would be a database that stores information for the purchase order form shown in Figure 5.6. (*Note:* "Extended" is the extended cost, or quantity multiplied by cost.) Draw a sketch showing how these tables might relate to one another.

5.A5. Use the database shown in Figure 5.7 to perform the following tasks:
   a. Name the tables in this database.
   b. For each table, list its fields and primary key.
   c. List the foreign keys.
   d. For each foreign key, list the table to which it refers.

5.A6. Use the database shown in Figure 5.8 to perform the following tasks:
   a. Name the tables in this database.
   b. For each table, list its fields and primary key.
   c. List the foreign keys.
   d. For each foreign key, list the table to which it refers.

5.A7. You are probably thinking about your future career. One element of your career choice is finding out what industry (or industries) interests you. Visit your library's website and identify three online databases that provide information that can help you evaluate an industry. Briefly describe each database and explain why you think it would be useful in this task.

5.A8. Big Data has become a part of almost every business career. Think about a company that you would want to work for upon graduation. Do an Internet search with the name of that company and Big Data and see what you can find out about how it is using Big Data. Identify the company for which you searched, briefly describe how it uses Big Data, and speculate on how you could see the use of Big Data impacting the work you would do for this company.

Glossary

Big Data: The vast amount of data that are created and stored that have grown beyond the capabilities of traditional data processing tools and applications.

Database: An organized collection of data.

Database management system (DBMS): A set of programs that control the creation, maintenance, and use of databases.

Field: In a relational database, stores data about a single characteristic of a record.

Foreign key: In a relational database, a field (or fields) that references the primary key of a related table.

Multitiered architecture: Refers to the database server, the Web server, and applications, all existing on different computers.

Primary key: In a relational database, the unique identifier of a record.

Record: In a relational database, a set of related fields.

Relational database: A data store that organizes data in connected two-dimensional tables.

Chapter 6

Analyzing Information for Business Decision-Making

Learning Objectives

By reading and completing the activities in this chapter, you will be able to:

- Discuss the importance of having good decision-making skills
- Explain how decision-making changes depending on organizational level
- Discuss the role of information in decision-making
- Contrast structured, semistructured, and unstructured decisions
- Apply a decision-making methodology
- Choose the appropriate technology tool for a given decision-making task

Chapter Outline

Focusing Story: How a Spreadsheet Helped Me Make a Hard Decision
The Importance of Good Decision-Making Skills
    Learning Activity 6.1: Information and Decision-Making
    Learning Activity 6.2: What Kind of Information Do I Need?
Using Information for Decision-Making
Types of Decisions
A Decision-Making Process
    Learning Activity 6.3: What Kind of Information Do I Need?
Information Retrieval and Analysis Tools
    Learning Activity 6.4: "What If" I Get a 75% on the Final?

Making good decisions requires information, but that is not enough. You must also be able to analyze information. In this chapter, we want to help you understand the basics of business-oriented decisions and the process by which they are made. In Chapter 14, you will learn about systems specifically designed to support decision-making. Here the focus is on understanding more about the decision-making process and the tools that can help you retrieve and analyze information as part of that process.

How a Spreadsheet Helped Me Make a Hard Decision

Several years ago, I decided to apply to graduate school so I could earn a PhD and become a professor. As I began the research process of where to apply, I discovered that there were hundreds of schools that offered doctoral degrees in Information Systems. Complicating matters further, I had a wife and two young children whose needs and desires needed to be taken into consideration. I quickly found that there were too many factors to consider for me to feel comfortable making a fully informed decision. My wife and I sat down and created six factors that were important in considering where to relocate to for the next four years of our lives. Included in this list was evaluating the quality of education I would be pursuing in exchange for quitting the well-paying job I already had. After making this list, we then assigned percentages to each of these categories. The resulting list and percentages assigned were quality of school (20%), amount of financial support provided by the school (20%), impression of faculty (20%), cost of living (20%), crime rate (10%), and could we see ourselves enjoying this school and city (10%). Notice that these percentages all add up to 100. I created a Microsoft Excel spreadsheet with each of these categories and rankings listed in separate columns. We then did research to rank each possible school on these various categories using a ranking of 1 to 10. These rankings for each school were inserted in separate rows within the spreadsheet. After completing this task, it was a matter of using an Excel formula to multiply the rank with the percentage and then adding these scores across categories for each school. At the end of this process, the schools at the top of the list were best aligned with what I was looking for in a graduate school. I then applied to the top schools. As result of using a Microsoft Excel spreadsheet, I was able to make a difficult decision much easier to resolve.

Focusing Questions

1. What role did information play in this decision?
2. An electronic spreadsheet was helpful in making this decision. What would have been different if spreadsheet software was not available?

A **decision** is simply a choice among alternatives. We make many decisions each day. Sometimes these decisions are trivial, such as deciding what to eat for lunch, while others have enormous consequences, such as deciding whether to accept a job offer. Regardless of the importance of the decision, making the decision requires retrieving and analyzing information.

The Importance of Good Decision-Making Skills

Good decision-making requires good information along with good decision-making skills, processes, and tools and is the primary focus of this chapter. The ability to make good decisions is one of the skills that sets apart successful business professionals. There is no escaping decision-making. Being able to make sound decisions that result in positive outcomes is a critical skill for a successful business career. Analytical reasoning skills and the ability to solve complex problems are among the most highly sought-after qualities for potential employees. These skills are directly related to being able to make good decisions.

LEARNING ACTIVITY 6.1

Information and Decision-Making

Suppose that you are in the market for a new apartment. (Your instructor may give you a different decision-making problem.) How would you go about making the decision of which new apartment to choose? What information would you gather? What would be the consequences of making a bad decision? Be prepared to share your answers in class.

Interestingly, while decision-making skills are important at all levels of business, they are increasingly important as one moves up the "organizational ladder." Why is this? First, decisions made by higher-level managers tend to be more complicated. The more complex a decision, the more skilled one must be to make effective decisions. Second, higher-level decisions typically have greater impacts. Consider this example. The store manager for a fast-food restaurant makes decisions such as hiring and scheduling hourly workers and ordering products. A district manager may decide about local marketing campaigns and choose suppliers. A corporate executive might make decisions about national marketing campaigns and whether to introduce new products. Deciding whom to schedule for the morning shift is much easier than deciding whether to roll out a new product. In addition, a bad decision on scheduling has relatively minor consequences, while a bad product decision can cost millions of dollars.

Let us be quite clear about the importance of decision-making. Virtually everything in this book is related, to some degree, to making better decisions. All the systems we discuss, all the tools you will learn about, and all the information skills

LEARNING ACTIVITY 6.2

What Kind of Information Do I Need?

You will face many different types of decisions throughout your business career. Some decisions are well defined and have clear alternatives and information requirements. Others are much less clear. For each decision, answer the following questions:

- What information would you need to make the decision?
- How confident are you that you have accurately identified the needs? Why do you feel this way?

Decisions

1. You are the manager of a fast-food restaurant. How many hamburger buns should you order for tomorrow?
2. You are the manager of a retail store and you need a new shift manager. Which employee should you promote?
3. You are an executive for a technology company. Should your company develop a new wearable technology to compete with devices such as Apple Watch?

you will gain are directed toward helping you better understand and use information so that you can make better decisions. To paraphrase Herbert Simon, Nobel Prize–winning economist and technologist, nothing is more important for the well-being of society than that decision-making and problem-solving be well done.

Using Information for Decision-Making

Information and decision-making are inextricably linked; many of our information needs are related to decision-making, and making good decisions requires information. Insufficient, inaccurate, or untimely information has detrimental impacts on decision-making. You need good information to understand the objectives of a decision, the constraints that limit the number of alternatives, and the alternatives themselves. You also need information to forecast the potential outcomes of each alternative. Finally, information provides the means for comparing and selecting among the alternatives. If you have inadequate information related to any of these facets, your ability to make the proper decision is limited.

Types of Decisions

As a business professional, you will face many different types of decisions. Information systems can help you with these decisions. Before getting into how information systems can help you, it may be helpful to understand how information needs vary for different types of decisions. Herbert Simon (1960) developed a widely used view of decisions as falling along a continuum from highly structured to unstructured. (Simon called these *programmed* and *nonprogrammed,* but others have used *structured/unstructured,* which we prefer.) **Structured decisions** are routine and repetitive and often have well-defined procedures for dealing with them. **Unstructured decisions,** in contrast, are novel and do not have agreed-upon procedures for making them. **Semi-structured decisions** are somewhere in between the two extremes; they have elements that are structured and elements that are unstructured. For structured decisions, we know what information we need and how to use that information. For unstructured decisions, we are not sure exactly what information we need, nor do we know exactly how to use the information. This is important to grasp, as it helps us understand how different types of information systems can help with decision-making.

A complementary view comes from Anthony (1965), who developed a framework of managerial activity that is made up of three categories: strategic planning, managerial control, and operational control. *Strategic planning* entails choosing the organization's objectives and deciding how to achieve them; predicting the future of the organization and its environment; and using creative, nonroutine thinking. *Managerial control* involves ensuring the efficient, effective use of resources in achieving the objectives laid out in strategic planning. *Operational control* makes sure that the tasks of the organization are being conducted efficiently and effectively. These tasks typically are routine and well defined.

BUSINESS EXAMPLE BOX 6.1

Why Do Managers Make Bad Decisions?

If decision-making skills are so important to business professionals, why are so many bad decisions made? There are a variety of reasons, including the following:

- Poor decision-making skills
- Time pressures (perception that there is not enough time for systematic decision-making)
- Relying too much on intuition rather than engaging in systematic decision-making
- Being overconfident in decision-making skills, intelligence, or knowledge of the decision context
- Going with the group (groupthink)
- Addressing the wrong objective

Decision-making biases also play a role in making poor decisions. A *decision-making bias* is a cognitive tendency that causes the decision maker to make an incorrect decision. Here are a few of the more common decision-making biases:

- **Negativity bias:** Giving more weight to negative than to positive experiences or information.
- **Confirmation:** Searching for information that supports preconceptions.
- **Loss aversion:** Making losses more important than equal gains.
- **Bandwagon:** Tending to do or believe something because many others do or believe the same thing.
- **Gambler's fallacy:** Believing that random events are influenced by previous random events.

Being aware of these biases is the first step in overcoming them.

Gorry and Scott-Morton (1971) combined Simon's framework with Anthony's framework. This resulted in the framework shown in Table 6.1. Examples of relevant information systems are shown in each row/column intersection. As you can see, information systems are useful at all levels of managerial activity and for all types of organizational decisions.

Let us look at this from the perspective of information needs. Operational control has narrow, well-defined information needs that can be anticipated. Furthermore, the information often comes from internal sources or from external sources that are easily accessed. In contrast, strategic planning's information needs often are broad and poorly defined and come from both internal and external sources.

Operational, structured decisions often do not require any human intervention. Transaction processing systems (TPSs) can deal with these decisions using predetermined rules and processes. There is little or no judgment involved. For example, an order-processing program can "decide" to allow a customer to place an order if the customer's credit card company accepts the charges but will block the transaction if the charge is denied.

While some managerial decisions are structured, many are semistructured. Semistructured decisions can be supported through systems that allow flexible information

TABLE 6.1 Gorry and Scott-Morton's Information Systems Framework			
	Operational Control	Managerial Control	Strategic Planning
Structured	Accounting systems (accounts payable/receivable), order processing, inventory control, order processing	Human resources reporting, short-term forecasting	Investment analysis, distribution system analysis
Semistructured	Production planning	Budget variance analysis	Compensation planning
Unstructured	Cash management, project management	Budget preparation, sales planning	New product planning, social responsibility planning
Sources: Adapted from Gorry and Scott-Morton (1971) and Rainer and Turban (2008).			

retrieval. We can anticipate some information needs but cannot create a program capable of making the decision without relying on human judgment. Since we can anticipate some information needs, information systems often have predefined reports that managers can run to retrieve necessary information. However, it is also important to allow for more flexibility through ad hoc querying, which lets managers retrieve information on the fly. These decisions also require tools that allow for flexibility in analyzing the information once it is retrieved.

Strategic planning often involves unstructured decisions. It is very difficult to anticipate the information needs of managers involved in strategic planning. They may need information ranging from simple, prebuilt reports from internal sources to economists' projections on the future of the economy. In addition, it is hard to determine what the managers will do with the information once they have it.

It is important to understand that as the amount of structure in a decision decreases, the need for flexibility increases. Very structured decisions require little flexibility in terms of information systems. In contrast, the highly unstructured decisions typical of strategic planning require information systems that offer great flexibility.

A Decision-Making Process

Many different decision-making processes exist, each with its own set of proponents. There are many similarities across the various processes. Most seem to fit well with Simon's classic intelligence/design/choice model, so we start this section with a brief discussion of that model and then provide a more detailed process that you can use to make better decisions.

The intelligence phase of Simon's model includes those activities that alert decision makers that there is a need to change the current state. Intelligence, in this context, involves sensing conditions and predicting what conditions will require action. The design phase involves developing alternative approaches to bringing about the required change or achieving the required goals. The most effective and efficient alternative is selected in the choice phase. Some believe that it is worthwhile

TABLE 6.2 Information Systems and Decisions Stages	
Phase	**Information System**
Intelligence	Data analysis (spreadsheets, statistical software, visualization) Communication systems Data retrieval and document management systems
Design	Group support systems Communication systems Document management systems Data analysis systems (spreadsheets, statistical software, visualization)
Choice	Data analysis systems
Implementation	Project management systems Data analysis systems Communication systems

to add a fourth phase, implementation, in which the activities in the chosen alternative are carried out. Table 6.2 provides some examples of information systems that can be used at each stage of Simon's model.

While we like Simon's model very much, it may be worthwhile to expand his model to provide more actionable details.

Identify and Clearly Define the Problem

In many ways, this is the most critical phase in the decision-making process. If you do not clearly define the problem to be solved, you may spend considerable time and energy in solving the wrong problem. A good problem definition should produce a clear, concise problem statement that describes both the initial state (where you are now) and desired state (where you want to be).

One common issue with problem definition is confusing the core problem with symptoms of the problem. For example, a sales manager might notice that sales from new representatives are decreasing. This could be a symptom of poor training, poor selection of new representatives, or a number of other underlying problems. There are techniques you can use to get at the core problem. The "5 Whys" is a popular example that is easy to understand and apply. To use the technique, simply start with the apparent problem or issue, such as "Our sales are declining." Then ask yourself (or your group) "Why?" and keep asking "Why?" until you uncover the root cause of the problem. By the way, there is nothing magical about the "5" in 5 Whys. It is fine to ask more than five questions to get at the root cause. One criticism of the 5 Whys is that it is sometimes too simple to uncover causes in some situations. Also, the technique does not allow for identifying multiple root causes. Despite its limitations, the 5 Whys remains an effective and popular technique for uncovering root causes.

Having a well-defined, concise problem statement has other benefits as well. One important benefit is that it allows all parties to agree on the core problem. This is very helpful in later phases. In addition, such a statement enables more effective communication of the problem to various stakeholders.

Determine Requirements and Goals

Once you have your problem well defined, you should define the decision's requirements. **Requirements** are the conditions that any acceptable solution must provide. The key word here is *must*. Most decisions have conditions that must be met and conditions that are desirable but not required. One major benefit of having well-defined requirements is that they often help you eliminate infeasible alternatives early in the decision-making process. For example, if you have an absolute budget for a new automobile, you can quickly eliminate many possibilities on the basis of cost.

In contrast to requirements, **decision goals** go beyond the minimum, essential requirements. One way to distinguish between requirements and goals is to think of requirements as "must haves" and goals as "like to haves." Interestingly, goals often conflict with one another. Consider the example of deciding on a new car. Goals might include low cost, excellent fuel efficiency, and high performance. While these are all reasonable goals, fuel efficiency and performance may be in conflict. Another interesting aspect of goals is that they sometimes suggest new requirements. Also keep in mind that as you analyze your decision, you may realize that an original requirement is actually not required and is therefore a goal. For example, when searching for an apartment, you may originally consider having two bathrooms a requirement. However, if you would choose a one-bath apartment that was fantastic in every other aspect, the "two baths" requirement is really a goal.

Identify Alternatives

Once you understand your problem's requirements and goals, it is time to identify **alternatives**. An alternative is a method for transforming the current condition into the desired state. An alternative often goes through two stages: generation and refinement. Brainstorming is often used to generate alternatives, but there are other methods that can be effective. For example, the *nominal group technique* is a structured method for reaching consensus within a group while encouraging input from all group members. While the technique was developed to help groups reach consensus, it can also be used to generate alternative solutions to a problem. Regardless of the method used to identify alternatives, it is important to provide clear descriptions of each alternative. These descriptions should plainly show how the alternative solves the problem and how it is different from other alternatives.

Define the Criteria

Decision criteria are objective measures of the requirements and goals that help you discriminate among the alternatives. Criteria are used to determine how well each alternative meets a particular goal or requirement. It is important that each criterion be independent of other criteria. (It should stand alone.) Also, each goal and requirement should be represented by at least one criterion. You should also avoid redundant criteria. Finally, you should try to use relatively few criteria (keeping in mind that each requirement and goal must be represented by at least one criterion) to keep things manageable.

Requirements, Goals, and Criteria

Suppose that you are in the market for a new vehicle. Establish at least five requirements and goals (five combined). For each requirement and goal, determine one or more criteria. Be prepared to share your answers.

Select a Decision-Making Technique/Tool

There are many decision-making tools that can be applied to various decision-making situations. While it is beyond our scope here to give a comprehensive list, we offer several that are easy to understand and apply while also being quite useful.

The techniques vary in their complexity and level of subjectivity. Generally, you want to use the simplest appropriate decision-making technique. Be sure to also consider the importance of the decision. More important decisions usually benefit from more complex, sophisticated techniques.

Pros/Cons Analysis

Pros/cons analysis is conceptually simple and requires no mathematical skills. You simply list the advantages and disadvantages of each alternative and then choose the alternative with the strongest advantages and weakest disadvantages. This technique is good for simple decisions that have a small number of alternatives and few discriminating criteria. Where this method gets tricky is in figuring out what "strongest" and "weakest" mean operationally. Also, this method is highly subjective. Despite its weaknesses, a pros/cons analysis is often quite helpful in choosing among alternative courses of action.

Paired Comparisons

The paired comparisons technique, as the name implies, requires you to evaluate each alternative against all other alternatives. There are various ways you can carry out this analysis. In its simplest form, it is kind of like a single-elimination sports tournament. Start by picking one pair of alternatives. Compare the two using the criteria you established earlier. The losing alternative is out of the "tournament." Pair the winner against the next alternative. Keep doing this until only one winning alternative remains.

This method, while simple, is inferior to the more sophisticated variation of paired comparisons. In the more complex variant, you build a matrix, such as the one shown in Figure 6.1. The top row of the matrix is partially grayed out because it represents duplicate comparisons. The diagonal is grayed out because there is no need to compare an alternative to itself. Each remaining cell indicates which of the two alternatives is better. (This is easier if you assign a letter to each alternative.) Once all the comparisons are done, choose the alternative with the most "wins." In our example, you would choose alternative D.

	A	B	C	D
A				
B	A			
C	A	C		
D	D	D	D	

Figure 6.1 Paired Comparisons Matrix

You can also indicate the strength of the "win" by entering a number following the better alternative. Use a scale such as 0 (no difference) to 5 (major difference). Add up all the numbers for each alternative and convert them to a percentage. Choose the alternative with the highest percentage.

Decision Matrix

The decision matrix technique is widely used in business, especially for choosing among projects or products. The decision matrix uses a grid that lists all the alternatives in the first column and all the criteria in the first row, as shown in Table 6.3. Fill in the grid by rating each alternative on the criteria using a scale to indicate how well the alternative scores on the criterion. In our example, we use a 10-point scale, with 10 indicating that the alternative satisfies the criterion very well. Add up the scores for each alternative; choose the one with the highest total score.

The above example considers all criteria to be of equal importance. If this is not the case, you can use a weighted version of the decision matrix. To do this, assign each criterion a weight relative to its importance. One way to do this is to distribute 100% across the criteria, as shown in Table 6.4. Once you have the weights and the ratings for each alternative, multiply each score by that criterion's weight. The sum of

TABLE 6.3 Unweighted Decision Matrix						
	Cost	Quality	Experience	Flexibility	Complete	Score
Proposal 1	6	8	4	7	8	33
Proposal 2	8	6	5	8	8	35
Proposal 3	4	4	6	6	9	29
Proposal 4	1	8	7	6	9	31

TABLE 6.4 Weighted Decision Matrix						
	Cost	Quality	Experience	Flexibility	Complete	Score
Proposal 1	6	8	4	7	8	7.00
Proposal 2	8	6	5	8	8	7.10
Proposal 3	4	4	6	6	9	5.65
Proposal 4	1	8	7	6	9	6.20
Weight	0.25	0.30	0.10	0.10	0.25	

these is the overall score for the alternative. Complete this operation for each alternative and choose the one with the highest sum.

Evaluate the Alternatives Using the Criteria
Once you have chosen a decision-making tool, apply it to evaluate each of the alternatives. It is very important that you use your criteria when assessing the alternatives. This may seem obvious, but sometimes the criteria get lost along the way and are not used effectively.

Check That the Solution Solves the Problem
It is worthwhile to validate the chosen alternative to make sure that it actually solves the original problem. Be sure that the alternative will bring about the desired state, meets all requirements, and best achieves the goals.

It is important to understand that the entire process is dependent on the problem definition. If the problem is poorly defined, you are likely to develop inappropriate requirements and goals, which leads to incorrect criteria. If you have the wrong criteria, your alternative evaluation will be faulty, and you are likely to make the wrong choice. So each step in the process is dependent on the steps that precede it. Because of this, it is important to ensure that each step is carried out correctly.

Information Retrieval and Analysis Tools

Now that you understand more about the relationship between decision structure and information needs, we can discuss how information systems help with decision-making. In this section, we discuss two categories of tools: information retrieval and information analysis tools. (There are other information and communication tools that are helpful · for decision-making, such as email and knowledge management systems, but we discuss them in Section III.) It is worth noting that the tools we discuss here can be considered components of a decision support system, which we cover in Chapter 14.

When you analyze information, the information has to come from somewhere. Often this information comes from an organization's databases, so you need tools to help you retrieve the information you need. Information may also be stored in organizational documents. Information retrieval tools include database management systems, reporting tools, and document management tools. As we discussed in Chapter 5, one element of database management systems is data retrieval, which is called *querying*. Creating complex database queries requires specialized knowledge, so many managers use reporting tools such as Business Objects or Crystal Reports to generate reports to support their decision-making. These **reporting tools** typically allow users to create reports visually rather than requiring special commands. This allows managers to quickly generate reports as they refine their information needs. Figure 6.2 shows a Business Objects screen. The user simply drags data objects from the left-hand pane to the Result Objects area to create a query, which serves as the basis of the desired report.

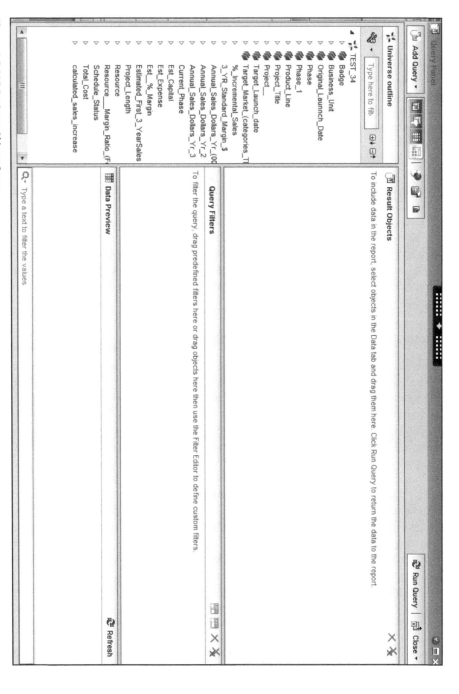

Figure 6.2 SAP Business Objects Screen

LEARNING ACTIVITY 6.4

"What If" I Get a 75% on the Final?

"What grade will I get if I score a 75% on the final?" You have probably asked yourself a similar question. While you can certainly figure this out without a computer, a properly designed spreadsheet can make this much easier. Create a spreadsheet that calculates your overall score in the course using the weightings for various factors as indicated by your instructor. Bring the spreadsheet to class, either electronically on your device or with knowledge of the formulas you used. Be prepared to discuss.

Managers sometimes need information that is stored in documents such as memos and policy documents. **Document management systems** include functions that help users locate and retrieve such documents. There are multiple aspects to a document management system. Some help with the creation of the document, either directly or through integration with other systems. There also must be some means to capture the document. (Place the document into the document management system.) Not all documents originate from a computerized system. Nondigital documents must be scanned so that they can be managed by the document management system. Once the document is in the system, it needs to be indexed to facilitate faster and easier retrieval.

In addition, document management systems help with versioning (keeping track of various versions of a document). Some document management systems help workers share documents. The system may also help with distributing documents properly.

Spreadsheets are a type of decision support system. (We will discuss decision support systems more in Chapter 14.) The electronic spreadsheet played an important role in introducing the use of personal computers to businesses. For an interesting history of the electronic spreadsheet, see http://dssresources.com/history/sshistory.html.

In terms of decision support, spreadsheets provide two very important functions: "what-if" analysis and goal-seek analysis. **"What-if" analysis**, which is a form of sensitivity analysis, involves seeing how changes in one or more input variables impact the value of one or more outcome variables. For example, if you are purchasing a car, you can see the impact of changing the interest rate of your loan on the monthly payment, as shown in Figure 6.3. You can change any of the input variables to see a range of possible outcomes.

Some electronic spreadsheet software (such as Microsoft Excel) offers ways to quickly generate a number of scenarios so that you can see the impact of various changes on outcome variables. Excel's Scenario Manager is an example. Figure 6.4 shows a Scenario Manager summary for our car payment calculator. There are three scenarios showing the monthly payment, total payments, and total paid for three payment periods: 36 months, 48 months, and 60 months.

Goal seeking is the opposite of what-if analysis. In a what-if analysis, you specify changes to the input values to see the impact on the outcome variables. When

◢	A	B	C	D	E	F	G
1							
2	Purchase Price	20,000.00		Interest Rate	0.06		
3	Down Payment	2,000.00		Period (Months)	48		
4	Trade in	2,000.00					
5	Amount Financed	16,000.00					
6							
7	Monthly Payments	375.76					
8	Total Payments	18,036.50					
9	Total Paid	22,036.50					
10							

Figure 6.3 Spreadsheet Example

◢	A	B	C	D	E	F	G	H	I
1									
2		Scenario Summary							
3				Current Values:	36 Months	48 months	60 months		
5		Changing Cells:							
6			Months	48	36	48	60		
7		Result Cells:							
8			Payment	375.76	486.75	375.76	309.32		
9			TotalPayments	18,036.50	17,523.04	18,036.50	18,559.49		
10			TotalPaid	22,036.50	21,523.04	22,036.50	22,559.49		
11		Notes: Current Values column represents values of changing cells at							
12		time Scenario Summary Report was created. Changing cells for each							
13		scenario are highlighted in gray.							
14									

Figure 6.4 Scenario Manager Summary

performing a **goal-seek analysis,** you specify the value of the outcome variable you seek (the goal), and the spreadsheet software determines the value of a particular input that will produce the desired output. Suppose that you have set a budget of $400 per month for your new car, you have received a price quote of $25,000 for the model you want to buy, and your current car is worth $2,000 as a trade-in. You could use goal-seek analysis to determine the down payment you would have to provide to meet your monthly payment goal. The "before" (original) and "after" for this example are shown in Figure 6.5. (Note that we are only showing part of the spreadsheet.) As you can see, you would need almost $6,000 for a down payment to achieve your payment goal.

The reason that spreadsheets can provide such useful tools is that in building the spreadsheet, instead of putting in numbers for some of the cells, you put in relationships in the form of formulas and functions. Figure 6.6 shows the formulas and functions for our car-buying spreadsheet. Because the spreadsheet is built around the formulas and functions, changing one of the input variables quickly shows us the impacts on all related values. If you want to learn more about using spreadsheets, please see Appendix F.

Original				After Goal-Seek Analysis			
	A	B	C		A	B	C
1				1			
2	Purchase price	25,000.00		2	Purchase price	25,000.00	
3	Down payment	2,000.00		3	Down payment	5,967.87	
4	Trade in	2,000.00		4	Trade in	2,000.00	
5	Amount financed	21,000.00		5	Amount financed	17,032.13	
6				6			
7	Monthly payment	493.19		7	Monthly payment	400.00	
8	Total payments	23,672.91		8	Total payments	19,200.00	
9	Total paid	27,672.91		9	Total paid	27,167.87	
10				10			

Figure 6.5 Results of Goal-Seek

	A	B	C	D	E
1					
2	Purchase price	25000		Interest rate	0.06
3	Down payment	2000		Period (months)	48
4	Trade in	2000			
5	Amount financed	=B2-SUM(B3:B4)			
6					
7	Monthly payment	=PMT(E2/12,E3,B5)*-1			
8	Total payments	=B7*E3			
9	Total paid	=B8+SUM(B3:B4)			
10					

Figure 6.6 Spreadsheet Showing Formulas

While spreadsheets can perform many types of statistical analysis, they typically are unable to apply more sophisticated statistical methods. When more advanced techniques are needed, you must turn to dedicated statistical software, such as SPSS, Stata, R, and SAS (among others). These applications have the capability to analyze a large number of data points quickly and accurately. Some applications specialize in particular families of statistical techniques. For example, SmartPLS (http://www.smartpls.de) specializes in a technique called *partial least squares*. Others have broad capabilities ranging from simple descriptive statistics to very complex, sophisticated statistical methods, such as time-series analysis for economic forecasting. Many include programming languages that can be used to develop custom algorithms. These applications vary widely in their complexity. Some are intended only for expert users with in-depth knowledge of statistics and the program itself. Others are specifically intended for more casual users.

There are also software tools that help you visualize data. While most statistical software applications have some **data visualization** capabilities, there are also programs that are intended specifically for data visualization. Visualizing data can help you uncover trends and relationships in data that might be less apparent when viewing the data in tables. Figure 6.7 shows a variety of graphs of sales based on customer and location. These charts were produced using Tableau, a popular visualization

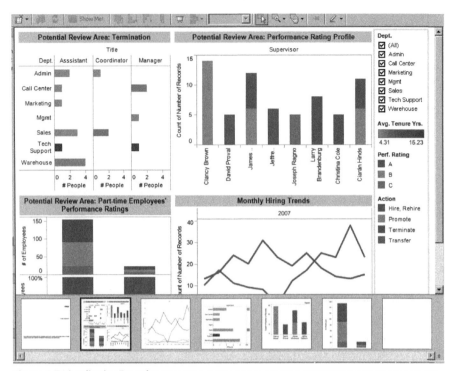

Figure 6.7 Visualization Example Source: https://public.tableau.com/views/Hiring-Terminating-Trends-JoshLock hartTableauAssignment/StaffingTrends?:embed=y&:showVizHome=no&:display_count=y&:display_static_image=y&:boot strapWhenNotified=true.

application. It is worth noting that spreadsheet software can produce charts but typically is not as powerful or flexible as dedicated visualization software.

In Chapter 5, we discussed how Big Data and Internet of Things devices, such as fitness trackers, capture a significant amount of information. This information is not easily used by decision makers until it is presented in usable ways. In this chapter, we discussed how spreadsheets and visualization tools can help make data usable. The same is true for Big Data and other data captured through Internet of Things devices. Utilizing these tools allows for vast amounts of data to be analyzed for informed decision-making.

Chapter Summary

The focus of this chapter is using information for decision-making. We discussed a variety of tools that can be used to help retrieve, analyze, and understand information when making decisions. We also presented a systematic process for making decisions. Finally, we provided you with several techniques that are helpful in decision-making.

STATS BOX 6.1:

The Cost of Spreadsheet Errors

Spreadsheets are widely used analytical tools. Unfortunately, electronic spreadsheets often contain errors that can prove costly. F1F9, a large financial modeling and training firm, estimates that almost 90% of corporate spreadsheets contain meaningful errors. These errors could cost as much as $10 billion annually. Here are a few examples of the costs of spreadsheet errors:

- Eastman Kodak Co. had to restate its income statement in 2005 because of several errors. One of the most severe was an $11 million error in an employee's severance pay package. The error was caused when too many zeros were added to the employee's severance accrual.
- A spreadsheet error caused the London 2012 Olympic organizing committee to oversell tickets to synchronized swimming events by 10,000 tickets. The error was caused when a staff member entered 20,000 rather than 10,000 into a spreadsheet.
- A cell formatting misunderstanding led to the United Kingdom's intelligence agency, MI5, bugging more than 100 wrong phones.
- In 2013, the "London Whale" disaster led to JPMorgan Chase suffering $6 billion in trading losses and was due in part to an Excel spreadsheet showing underestimated risks from various investments.
- A copy and paste error in a spreadsheet cost TransAlta, a Canadian power generator, to lose $24 million. This happened when they accidentally bought power from the U.S. power grid at a cost higher than they should have.

You can learn more about these and other costly spreadsheet errors at http://www.eusprig.org/stories.htm and https://incisive.com/spreadsheet-error-horror-stories/.

Here are the main points discussed in the chapter:

- Good decision-making requires good information and good decision-making skills, processes, and tools.
- Decision-making skills are critical to a successful business career. These skills are highly sought after by business employers.
- Decision-making skills become increasingly important as one moves to higher levels of responsibility in a business. This is because decisions made at higher levels are more complex and have greater impacts.
- Making good decisions requires good information. Insufficient, inaccurate, or untimely information hurts decision-making.
- Information impacts the following with respect to decision-making: the constraints that limit the number of alternatives, the alternatives themselves, the forecast of the potential outcomes from each alternative, and the means for comparing and selecting among the alternatives.
- Structured decisions are routine and repetitive and often have well-defined procedures for making them. Unstructured decisions are novel;

we do not know exactly how to go about making them, what information is needed, or how to use that information. Semistructured decisions have elements of both structured and unstructured decisions.

- Decision-making can be done by completing the following steps:
  1. Identify and clearly define the problem.
  2. Define the requirements and goals of the decision.
  3. Identify alternatives.
  4. Define the decision criteria.
  5. Select the appropriate decision-making tools.
  6. Evaluate the alternatives using the criteria.
  7. Check that the solution solves the problem.
- Many information analysis tools exist, including tools for information retrieval, information analysis, knowledge management, and communication.
- Information retrieval tools include database management systems, reporting tools, and document management tools.
- Information analysis tools include electronic spreadsheets, statistical software, and data visualization tools.

Review Questions

1. Explain why decision-making skills are more important at higher levels of an organization.
2. Compare and contrast structured, semistructured, and unstructured decisions.
3. Explain the relationship between decision type (structured, semistructured, unstructured) and managerial activity (strategic planning, managerial control, operational control).
4. Name and briefly describe the phases of Simon's model of decision-making.
5. For each phase of Simon's model of decision-making, give an example of an information system that supports the phase.
6. Name the phases in the detailed decision-making process described in the chapter.
7. Contrast requirements and goals in the context of decision-making.
8. Name and briefly describe the decision-making tools discussed in the chapter.
9. Name and briefly describe the information retrieval tools discussed in the chapter.
10. Compare and contrast the "what-if" and "goal-seek" analyses.

Reflection Questions

1. What is the most important thing you learned in this chapter? Why is it important?
2. What topics are unclear? What about them is unclear?
3. What relationships do you see between what you learned in this chapter and what you have learned in earlier chapters?
4. Think about a difficult or important decision you have had to make. What made the decision difficult? How could the material in this chapter have helped you make that decision?
5. Think about a time when you made a poor decision. What could you have done that would have led to a better decision?
6. Briefly discuss how a reporting tool can be used to help you make semistructured and unstructured decisions.
7. Why are document management systems useful when making unstructured decisions?
8. How can following a disciplined decision-making process (such as the one described in this chapter) help you justify a decision?
9. How do you decide how much time and effort to put into making a decision?
10. Suppose that you have several job offers from which to choose. Describe how you would go about deciding which one to accept.

Additional Learning Activities

6.A1. Suppose that you have three decisions to make.
   a. What elective class to take in your final semester
   b. Which of several postgraduation job offers to accept
   c. Where to go for your graduation trip
   Rank each decision according to how much time and effort you would put into making the decision. Explain your rating. Be prepared to share your answers in class.
6.A2. Suppose that you have three job offers from which to choose.
   a. Determine the requirements and goals for deciding which offer to accept. (Have at least five total requirements and goals.)
   b. Determine the criterion you would use to measure each requirement/goal.
   c. Which decision-making tool would you use to make this decision? Why?
6.A3. Suppose that you are a marketing manager who needs to decide among several different advertising outlets.
   a. Determine the requirements and goals for deciding which outlet to use. (Have at least five total requirements and goals.)

b. Determine the criterion you would use to measure each requirement/goal.

c. Which decision-making tool would you use to make this decision? Why?

6.A4. Go to the course website, https://www.prospectpressvt.com/textbooks/belanger-information-systems-for-business-4-0, and locate the Car Buying spreadsheet in the Student Resources. Use the spreadsheet to determine the monthly payment for two different interest rates for two different periods (for example, 6% and 8% for 36 and 60 months). For each combination, report the monthly payment, total payments, and interest paid. Briefly describe how you could use this spreadsheet when car shopping.

6.A5. Go to the course website, https://www.prospectpressvt.com/textbooks/belanger-information-systems-for-business-4-0, and locate the Monthly Commission spreadsheet. Determine how many units you would have to sell to earn $3,000 in commission payments. How did you go about solving this problem? How long did it take you? Be prepared to share your experience in class.

6.A6. Go to the course website, https://www.prospectpressvt.com/textbooks/belanger-information-systems-for-business-4-0, and locate the What-If spreadsheet. Follow the instructions in the spreadsheet.

6.A7. Go to http://simile-widgets.org/ and select one of the links on the right (Exhibit, Timeline, Timeplot, or Runway). Look at the visualization it created. Then click on one of the More Live Examples underneath the visualization that is on that page. Prepare a paragraph that discusses the following:

1. Which dataset did you choose?

2. What story did you want the visualization to tell?

3. How effective was the visualization at telling that story?

4. How effective was the visualization at helping you to better understand the data?

Be prepared to discuss your experience.

6.A8. Pick four competing elective classes that you might take next semester. (You can only take one of the four.) List the classes, then do the following:

a. Prepare a paired comparison matrix to show which class you would take.

b. Prepare an unweighted decision matrix with at least three criteria (such as time of day or your interest in the class).

c. Prepare a weighted decision matrix using the criteria from (b).

Which of the three methods was the most effective? Why?

References

Anthony, R. N. 1965. *Planning and Control Systems: Framework for Analysis.* Cambridge, MA: Harvard University Graduate School of Business Administration.

Gorry, A., and M. Scott-Morton. 1971. "A Framework for Information Systems." *Sloan Management Review* 13(1): 56–79.

Rainer, R. K., and E. Turban. 2008. *Introduction to Information Systems: Supporting and Transforming Business.* 2nd ed. New York: John Wiley & Sons.

Simon, H. A. 1960. *The New Science of Management Decision.* New York: Harper & Row.

Glossary

Data visualization: A visual representation of data with the goal of clearly communicating or better understanding the meaning of the data.

Decision: A choice among alternatives.

Decision alternative: A method for transforming the current condition into the desired state.

Decision criteria: Objective measures of decision requirements and goals that discriminate among the alternatives.

Decision goal: Desired decision solution requirements that go beyond the minimum, essential requirements.

Document management system: A system that assists with managing, locating, retrieving, and tracking documents.

Goal-seek analysis: An analysis that determines the value of a particular input variable that will produce the desired output (the goal).

Reporting tools: Information systems tools that allow users to create reports without knowing special commands.

Requirement (decision): A condition that any acceptable solution must provide.

Semistructured decision: A decision for which some elements are structured and others are unstructured.

Structured decision: A decision that is routine and repetitive and often has well-defined procedures for making it.

Unstructured decision: A decision that is novel and therefore has no agreed-upon, well-understood procedure for making it.

"What-if" analysis: An analysis that involves seeing how changes in one or more input variables impact the value of one or more outcome variables. Also known as *sensitivity analysis*.

Chapter 7

Transmitting Information

Learning Objectives

By reading and completing the activities in this chapter, you will be able to:

- Discuss key components and characteristics of networks
- Explain the different types of networks
- Explain what the Internet is and what it is used for
- Identify the main networking architectures and infrastructures
- Describe various Web 2.0 and Web 3.0 concepts

Chapter Outline

Introduction and Definitions
Focusing Story: The Rise of Mobile Wi-Fi: Hotspots
    Learning Activity 7.1: How Does That Car Drive Itself?
Types of Networks
The Internet
    Learning Activity 7.2: Broadband in Developed Countries
    Learning Activity 7.3: Who Is Using the Internet?
    Learning Activity 7.4: A Future Internet? Internet2 and Business
Networking Architectures
    Learning Activity 7.5: Architectures and Principles
Communicating Information in Modern Organizations
    Learning Activity 7.6: Web 3.0 versus the Internet of Things

Introduction and Definitions

Networks

To begin our discussion of **networks,** we should first define what a computer network is. Basically, it is a collection of interconnected devices that allow users and systems to communicate and share resources. The devices can be computers, servers, routers, or a wide variety of other telecommunication devices.

Network Components and Characteristics

Setting up a network requires a variety of devices, and depending on which devices are used, a different type of network is established. For example, the wireless networks discussed in the focusing story require a router to set up the network and

The Rise of Mobile Wi-Fi: Hotspots

The interconnected world as most students know it—with easy and often free access to the Internet from home, school, the coffee shop, the airport, and even on the train—is actually fairly recent. We used to need cables (wires) to connect to other systems and the Internet. Really.

But now that we have tasted the freedom of the wireless world, people want more and more. However, there are places in this country that lack the access many in a university environment may take for granted. That's where hotspots come in. Many people utilize a device the size of a credit card (but thicker) to access an Internet connection anywhere. Many others utilize their smartphones as a mobile hotspot to take advantage of the data that they are already paying for on their wireless smartphone plan (unless you are lucky enough to have an unlimited plan!). You are likely familiar with public hotspots at airports, coffee shops, and restaurants. There are also hotspots in hospitals, campgrounds, train stations, and even supermarkets. Those are considered public hotspots, while private (also called tethered) hotspots usually come from the use of a smartphone or tablet with a cellular connection. Most phone-based hotspots, however, have a limited usage of data at full speed (after which the hotspot runs at low data speed). Most smartphone technologies allow 5 to 10 connections to a hotspot. The main disadvantage of a smartphone-based hotspot compared with a dedicated device is that the hotspot tends to use more of your phone's battery (Nadel, 2018).

Hotspots have huge benefits in enabling business and education to occur in previously impractical locations. For example, a friend of one of the authors runs a side business creating webpages from his farm property. Prior to hotspots, his best option was to utilize dial-up or satellite Internet technology. At his office job, he connected via a high-speed Internet connection, and his clients expected that he could do the same as part of his business services. As it became obvious that dial-up and satellite Internet were not going to cut it for him to continue his side business, he decided to try a hotspot connection. This proved to be the perfect solution. Although it was much more expensive than a cable or DSL connection, it ultimately allowed him to keep his business running.

In education, hotspots play an even more important role. For example, universities are increasingly using distance learning, which became a necessity during the COVID-19 pandemic of 2020. Usually, distance education capitalizes on the broadband connections to the Internet that most people have access to. However, students in rural areas, such as Native American tribal reservations and farm country, can have a hard time participating in the requirements of distance education. To address these concerns, universities have encouraged students to try utilizing hotspot connections, which has allowed many students to participate in classes and get the level of education they expect from a public university. Importantly, some public schools closed during the pandemic started using buses to offer Wi-Fi access to students in rural areas (Hui 2020).

These examples show how utilizing hotspots opens doors in ways that were before impractical, if not impossible. Think about all the other avenues this opens up.

Focusing Questions

1. What potential challenges exist for using hotspots for business?
2. What other creative business uses can you think of for hotspots?
3. What do you think could be the next evolution in wireless networks?

LEARNING ACTIVITY 7.1

How Does That Car Drive Itself?

For this learning activity, watch the following two YouTube videos. While you do, make a list of all the different pieces of information and technology that are required for a car to drive itself. After you have done this, find out where that information comes from and how it interacts with other information for a car to successfully drive itself. Be prepared to discuss your list in class.

- https://www.youtube.com/watch?v=CqSDWoAhvLU
- https://www.youtube.com/watch?v=TsaES–OTzM

devices that have wireless network adapter cards, like the Apple iPad or most laptops, to connect to it.

A **router** is an intelligent device that controls the flow of transmissions in and out of a network. It takes its name from its main purpose, which is routing traffic to appropriate devices on the network (by using a routing table that contains address information for the devices on the network) or to other networks. Routers oftentimes include connections for Ethernet cables while also allowing wireless signals that many devices in businesses and homes use. See Table 7.1 for a list of common router manufacturers and links to their current product lines.

We have been talking about devices having specific addresses. How is that done? It is the result of each device having a **network interface card (NIC, or network adapter)** that provides physical access to a device because it has a unique ID written on a chip that is mounted on the card. Wireless NICs, just like wired NICs, also have their own unique addresses. The network adapter for wired networks is where you plug in an Ethernet cable on your computer.

Finally, basic networks can also make use of **repeaters** to retransmit a data signal that they receive after eliminating noise in the signal and regenerating it for strength. These are often used when the signal from a wireless router weakens and is not usable in certain parts of a building or property that are some distance away from where the router is located.

TABLE 7.1 Examples of Router Manufacturers

Manufacturer	Link to Sample Product
ASUS	http://www.asus.com/us/Networking/Wireless-Routers-Products/
D-Link	https://us.dlink.com/en/consumer/wifi-routers
Linksys	http://www.linksys.com/us/c/wireless-routers/
NETGEAR	https://www.netgear.com/home/products/networking/wifi-routers/
TP-Link	https://www.tp-link.com/us/home-networking/wifi-router/
TRENDnet	https://www.trendnet.com/products/wifi/routers

Types of Networks

Wired versus Wireless Networks

There are many ways to categorize networks. First, networks can be wired or wireless. Wired networks make use of physical cables, which can be copper wires (like the lines to your phones at home or the Ethernet cable we talked about in the previous section), coaxial cables (like the ones used for cable television), or fiber-optic cables (which are made of glass fiber and transmit using light signals) (see Figure 7.1). The possible speed of transmission increases from wires to coaxial cables to fiber-optic cables, with fiber-optic cables allowing speeds hundreds of times faster than coaxial cables and thousands of times faster than basic wires.

There are also quite a few wireless networks besides the Wi-Fi you are used to. All of them make use of frequencies to transmit data (and voice). For example, some transmissions are done using microwave transmitters that must be placed no more than 30 miles apart, since they must "see" each other to transmit. The satellites used for data transmissions are typically 22,000 miles above the equator in space. (They are called *geosynchronous satellites* because they keep a position relative to the Earth.) Those are the same systems that send you satellite radio signals, GPS coordinates, and TV signals from remote locations. Infrared communication can also be used to create a wireless network for devices in close proximity. For example, wireless mice, keyboards, and other such devices often use infrared signals to connect to one another. Radio signals are also used for connecting devices. Think of your cordless phone or garage door opener. Cellular networks also use radio communication technologies to offer another wireless alternative. As you know, cell phones connect to a local antenna that relays calls from one area to the next. Finally, Bluetooth is another standard that uses short-wavelength radio transmissions to connect devices (like we discussed for infrared technology) such as wireless mice, keyboards, or headphones.

With all the wireless technology options, what is a wireless, or Wi-Fi, network? First of all, Wi-Fi is a name owned by the Wi-Fi Alliance. Today it has become standard terminology to refer to a wireless network that you can access with one of your wireless devices, such as a laptop or smartphone. Wi-Fi networks also make use of radio technology, but in a way different from some of the technologies discussed above. A Wi-Fi network consists of a wireless access point (WAP, or hotspot) that

Figure 7.1 Types of Cables—from left to right: coaxial, Ethernet, fiber optic Sources: istock.com/ piotr_malczyk, istock.com/hatchapong, and istock.com/zentilia

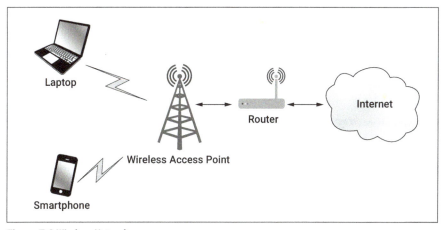

Figure 7.2 Wireless Network

is itself connected to a wired network (see Figure 7.2). The WAP sends the transmissions it receives from wireless devices to a wired device connected to the Internet.

Broadband Networks

Another way to classify networks is by whether they are baseband or broadband; these classifications refer to the speed and frequency of the bandwidth of the network. Today, the networks you use to connect to the Internet are likely to be broadband networks. They are faster, with download speeds (transfers from the servers to your computer) in megabits per second, although most broadband networks today transfer at rates in gigabits per second. The types of networks that are classified as broadband include cable- and fiber-based networks, as well as some about which you may not have heard, like integrated services digital network (ISDN) and T-lines. We discuss these networks in greater depth in Appendix H. Wireless networks are also now called mobile broadband with the advent of 4G (fourth generation) and 5G mobile services. 4G emerged in 2006 as a high-speed mobile network, and a few 5G networks became operational in 2018.

Network Coverage

LAN vs WAN

Networks can also be classified as a function of the area of coverage or geography of the network. The two most significant ones are the local area network (LAN) and the wide area network (WAN), but other variations exist. See Table 7.2 for a list of different terms you might hear when people discuss types of networks.

Local area networks (LANs) connect devices in a limited geographical area (a local area of usually fewer than five kilometers). Examples include homes, departments, a school, or a floor of a large office building. Because of the limited geography, LANs can offer very good transmission speeds, especially for wired LANs. To be part of the LAN, a cable must connect the NIC of the computer or device with the

Ethernet

router/bridge/switch of the LAN's **hub** (for wired LANs). The most popular LAN technology today is Ethernet, which is represented in Figure 7.3, although some alternatives do exist.

Wide area networks (WANs) connect devices over a large geographic area, which can span any distance you can imagine, from citywide to continentwide to worldwide. Unlike LANs, WANs often make use of various communication media

TABLE 7.2 Network Types (Geographical Classification)		
Type of Network	**Description**	**Notes**
Local area network (LAN)	A network that connects devices in a limited geographical area (a local area of usually fewer than five kilometers).	Most organizations use many LANs interconnected via backbone networks (see below).
Wide area network (WAN)	A network that connects devices over a large geographic area, which can span any distances you can imagine, from citywide to continentwide to worldwide.	These networks can be implemented via private networks or public transmission systems, such as the Internet.
Personal area network (PAN)	A network connecting personal devices to a personal computer (i.e., mouse, microphone, printer, etc.,) over a very short distance.	Often uses USB connections (wired) or Bluetooth or infrared technologies (wireless).
Home area network (HAN)	A network used within a home office, allowing PCs to share devices such as printers, routers, and scanners.	Sometimes referred to as an office area network (OAN).
Backbone network (BBN)	A network that serves to interconnect other networks (like LANs) or network segments (subnetworks).	Often used in large buildings or areas that have multiple LANs that need to communicate with each other.

Figure 7.3 An Ethernet Local Area Network

(like phone lines, coaxial cables, and satellite) as well as several different providers (phone company and cable company), even for one given network.

The Internet

One network we have yet to discuss is the Internet. The **Internet** is a publicly accessible worldwide network of networks. One way to explain how the Internet works is to think of a road system (Van Slyke and Bélanger 2003). In many countries, there are interstate highway systems that link major cities together. These interstate highways

---

LEARNING ACTIVITY 7.2

Broadband in Developed Countries

The percentage of the population in a country that has access to broadband networks can have a significant impact on the use of applications such as electronic business, electronic government, and other network-dependent applications. According to the Organisation for Economic Co-operation and Development (OECD) data shown below, as of June 2019, Japan has more than 176.6 mobile broadband subscriptions per 100 people. In other words, in a group of 100 Japanese citizens, there would be more than 176 mobile broadband connections to the Internet. As you can see, this is not the same for all countries in the developed world.

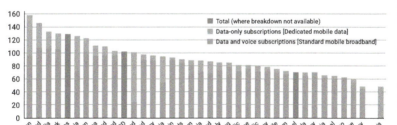

1.2.2. OECD Mobile broadband subscriptions per 100 inhabitants, by technology, June 2017

For this learning activity, identify and be prepared to discuss the following:

- Why do you think some industrialized countries have more mobile broadband subscriptions than others?
- What new opportunities emerge when essentially an entire country is connected to the Internet with mobile broadband connections?
- What different opportunities exist in countries where the majority of people do not have access to the Internet via mobile broadband connections?

Source: Based on data from the OECD: OECD Broadband Portal, http://www.oecd.org/sti/broadband/broadband -statistics/.

LEARNING ACTIVITY 7.3

Who Is Using the Internet?

The Internet is truly a global phenomenon, with connectivity in almost every country in the world. In 2019, the International Telecommunications Union (ITU) reported that more than 4.1 billion individuals used the Internet around the world. This includes 82.5% of Europeans, 77.2% of individuals in the Americas, 48.4% in the Asia and Pacific region, and 28.2% in Africa. This represents a wide variety of individuals with different languages, cultures, and applications. For this activity, find statistics about Internet users in terms of gender, age, culture, language, or other characteristics that your instructor will provide you. Be prepared to discuss the following in class:

1. How is usage of the Internet likely different around the world? Why would there be such differences (or why are there no differences)?
2. What is the likely future of the Internet in terms of diversity? How will that affect businesses? Governments? Societies?

Source: *Measuring Digital Development: Facts and Figures 2019.* ITU Publications. https://www.itu.int/en/ITU-D/Statistics/Documents/facts/FactsFigures2019.pdf.

intersect one another to form a complex road system. However, not every place in the country is accessible directly from an interstate highway. Smaller roads branch off from the highways to allow travelers to reach virtually any area of a country. The Internet works in a similar way. Main connections, called the *Internet backbones*, carry the bulk of the data traffic on the Internet. Networks from major Internet service providers (ISPs) make up the backbone. These networks extend around the globe in a way similar to how interstate highway systems extend across many countries.

For the Internet to allow global traffic, routers are used to interconnect the various networks. Every host or computer that is a full participant (permanently connected) on the Internet has a unique address called an *IP (Internet Protocol)* address. Appendix H includes more details on the Internet's protocols and addressing, but it is important to note that the IP addressing scheme most students have seen (such as 128.192.68.1) is outdated because it has run out of addresses to allocate to businesses, organizations, and countries. It is called **IPv4** (for version 4). The new IP addressing scheme is called IPv6. In simple terms, **IPv6** expands the IP address size from the current 4 bytes (the four sections of the address above) to 16 bytes (such as 2002:1AC4:DB32:EF10). In fact, on Thursday, February 3, 2011, the **Internet Assigned Numbers Authority (IANA)**, the organization responsible for assigning IP addresses, assigned the last blocks of IPv4 addresses. With IoT devices, which we previously discussed, you can imagine that the number of IP addresses needed for everything to be connected is growing exponentially. It would not have been possible to implement IoT to the scale we have today with the addresses available in IPv4.

The Internet started as a project by the United States Defense Advanced Research Projects Agency (DARPA) in collaboration with universities. The Internet then

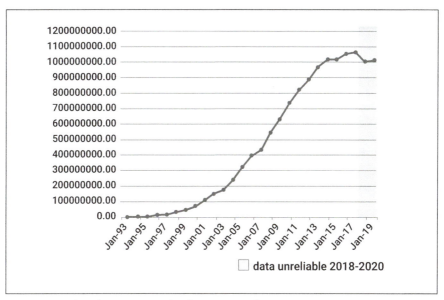

Figure 7.4 Number of Hosts Connected to the Internet as of January 2019 Source: Based on data from the Internet Systems Consortium, ISC Domain Survey, https://www.isc.org/survey/.

interoperability

grew by absorbing various networks worldwide. Today, the Internet has grown so much that there are more than one billion host computers connected, based on data collected by the Internet Systems Consortium, which is the developer of the open source infrastructure software BIND DNS (see Figure 7.4). One of the core characteristics of the Internet that has contributed to this growth is interoperability, which is the ability of heterogeneous (different) systems to communicate with one another. For example, systems based on different hardware platforms and running different software can communicate together on the Internet if they use the Internet's protocols (described in Appendix H).

Web = application
↳ Internet = network

Internet Applications

The Internet is a network, not an application. When asked what the Internet is, students often describe it as a place where there is all this information you can access. But this definition actually refers to the **World Wide Web (the Web)**. They are *not the same*! The Web is one application using the Internet as its carrying network. Examples of other applications that make use of the Internet include email, file transfer, instant messaging, Internet telephony, and desktop videoconferencing. Table 7.3 briefly describes some of these other traditional Internet-based applications. Later in the chapter, we will discuss more recent applications, labeled as Web 2.0 applications.

The Web allows users to access an amazing variety of resources using the infrastructure of the Internet. The Web is the result of work performed by Tim Berners-Lee at

TABLE 7.3 Traditional Internet-Based Applications	
Application	**Description**
World Wide Web (the Web)	Graphical interface to worldwide resources.
Electronic mail (email)	Allows users to send and receive messages through computer networks.
Instant messaging	Allows multiple users to communicate synchronously by sending and receiving short text messages online.
Voice over IP (Internet telephony)	Allows voice data to be sent over an IP-based network, such as the Internet.
Desktop videoconferencing	Allows individuals in different locations to communicate via voice and video on personal computers.
Peer-to-peer file sharing	Allows file sharing between specific individuals or systems across the Internet.
Online application sharing	Allows users to share documents, calendars, or other applications using websites.
File Transfer Protocol (FTP)	Allows users to move files back and forth between nodes on the network. FTP is *anonymous* when files are loaded on such public sites.

CERN (the European Organization for Nuclear Research), who proposed the hypertext transfer protocol (HTTP) to interconnect files on the Internet. That is why many Web addresses start with "http," indicating that the page is located on a server that supports the hypertext transfer protocol. You are also certainly familiar with hyperlinks, the (often) underlined blue links that allow you to navigate back and forth among pages or to a specific section of a Web page. When you click on a link, your browser sends a request to the Web server that houses the requested page. The server then sends the document over the Internet to your browser. Figure 7.5 illustrates how the Web application makes use of the Internet.

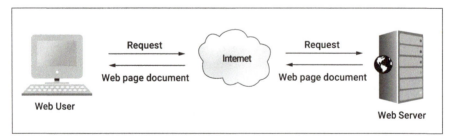

Figure 7.5 The Internet and the Web

BUSINESS EXAMPLE BOX 7.1

Internet Application of the Day: Distance Learning

Ever heard of the University of Phoenix or DeVry University? You might have seen ads about the completely online degree programs available at these and many online universities where one can get an associate, bachelor's, master's, or even PhD degree. Online leaning is not new, but it has grown substantially because it offers a cost-efficient way to deliver education and allows institutions to have a broader reach to individuals who would otherwise not be able to attend school. For a long time, these degrees were thought of as "not as good" as regular programs with face-to-face instruction. Then, distance learning became more mainstream, and most universities offered some degrees or at least some of their courses online only. By 2015, more than one-quarter of students were taking online courses (28%), with public institutions having 72.7% of undergraduate students taking online courses (Online Learning Consortium 2015).

And then the COVID-19 pandemic occurred in 2020, leading to a significant change in the world of education forever, with distance learning becoming the main approach to education during the pandemic. Thanks to the Internet, universities worldwide transitioned to online learning in a matter of weeks. For younger children, more than 1.2 billion of them worldwide (in 186 countries) were suddenly out of classrooms. However, many children around the world still live in areas where the Internet is not widely available or affordable. Even in Western countries like the United States, close to 3 million students do not have Internet access at home (Al-Arshani 2020). This led many school districts to add Wi-Fi hotspots, like those discussed in the focusing story, so that students could access their online courses even if they had no Internet at home. In some areas of the United States, hotspots were created on school buses that were either parked in school parking lots or traveled to where the students lived.

Besides access to the Internet, many challenges exist for successfully using distance learning. One of them is maintaining interaction with the instructor and other students. There are also challenges handling technical issues with the use of the various software platforms, such as video conferencing for live class sessions, online office hours, and group work using shared folders and files. Importantly, students need to develop their own computer skills, learn proper time management, and be self-motivated to complete all of their work on time.

Online learning, facilitated by the rise of the Internet, had already changed the availability of education, providing a broader reach to degrees for individuals with limited availability (e.g., working full time) or unable to physically attend class because of geographical constraints (e.g., living in remote areas) or physical disabilities. Online learning has now also enabled students of all ages to continue to learn when going to a face-to-face class was no longer an option because of the pandemic.

Sources: Kumar, S. 2015. "5 Common Problems Faced by Students in eLearning and How to Overcome Them." *eLearning Industry*. https://elearningindustry.com/5-common-problems-faced-by-students-in-elearning-overcome; Li, C., and F. Lalani. 2020. "The COVID-19 Pandemic Has Changed Education Forever. This Is How." *World Economic Forum*, April 29. https://www.weforum.org/agenda/2020/04/coronavirus-education-global-covid19-online-digital-learning/; Online Learning Consortium. 2015. "Online Report Card." https://onlinelearningconsortium.org/read/online-report-card-tracking-online-education-united-states-2015/; Al-Arshani, S. 2020. "School Districts across the Country Are Using School Buses to Deliver WiFi to Students Who Lack Access." *Insider*, Apr 1. https://www.insider.com/wifi-buses-being-used-across-country-to-give-kids-internet-2020-3.

Internet versus Intranet

Most organizations today, including businesses, schools, and governments, use an **intranet**. It sounds similar to "Internet" because it is the use of Internet technologies and related applications inside an organization. This means that security controls are in place to ensure that only individuals inside the organization have access to the applications on the intranet.

The Internet of Things

We discussed the **Internet of Things (IoT) in Chapter 1 and when we introduced Section II of the book**. An important aspect of IoT is that all of these millions of devices, large or small, need to have a unique IP address so that they can capture, transmit, and receive data. This is why IPv6 is needed for IoT to really function well. This also means that the IoT devices are constantly reachable, and hence the security issues arise that we are going to discuss in the next chapter.

IoT has already made significant changes to health care, automobiles, and the home. For example, someone with heart issues can have an internal monitor connected to the Internet that keeps doctors informed of his or her heart status in real time. In the automobile industry, cars are parking and driving themselves. This is all done using the interconnectivity of various devices in the car (as you likely learned in Learning Activity 7.1). IoT in the home includes smart thermostats that regulate the temperature of the house and locks that control access to the home remotely. Nest is an example of a thermostat that learns when you are home or away and the desired temperature for your house and then automatically sets itself based on your patterns. It also allows users to control the temperature of the house from anywhere they are if they have an Internet connection. Your smart refrigerator can tell you when you have outdated food items. You can see who is at your door from wherever you are, offering new home security possibilities. You can also track where your family or your friends are in real time.

As more devices connect to the Internet, more possibilities emerge for changing the way we live our lives. Think of your fitness trackers that can encourage you to exercise or apps on your smartphone to lock/unlock or start your car remotely. How does the app communicate with the car? Via the Internet. That is just the tip of the iceberg for IoT. IoT Analytics suggests that in mid-2018, more than 7 billion IoT devices were already connected to the Internet, with a prediction of 34.2 billion devices by 2025 (Lueth 2018).

Networking Architectures

Architecture
An architecture is a layout or blueprint for how devices are supposed to work together. **Networking architectures** can then be implemented as specific **networking infrastructures**. The infrastructure includes the actual hardware, software, and networking components that support the processing and transfer of information. In this section, we focus on the higher level picture, the networking architectures, which each have specific advantages and disadvantages. Some architectures have been in existence for a long time (e.g., historical architectures). These include the mainframe, client/server computing, and peer-to-peer networks. Others are more recent (e.g., modern architectures), such as cloud computing and virtualization. We describe these in this section but first explain key principles that should be considered in selecting an architecture.

Architectural Principles
When deciding on an architecture, managers must ensure first and foremost that the architecture will support the business objectives of the organization (ensuring information technology [IT] and business alignment, as discussed in Chapter 4). Then managers must also consider certain architectural principles, such as the following:

- *Ease of implementation:* Some architectures are easier to implement than others. For example, the SaaS (software as a service) architecture we describe later in this chapter requires limited effort by the organization wishing to use it.
- *Flexibility/interoperability:* In some situations, it is important to be able to make changes to how devices are interconnected or which devices are connected to which ones. Some architectures allow for more flexibility concerning devices that are connectable.
- *Control:* Organizations often want to maintain a certain degree of control over which devices can connect and how they are connected to the network. In the days of the mainframe, where everything was connected to one large central computer, organizations had complete control. Today, they must evaluate how much control they need as they make decisions on which architecture to use.

- *Scalability:* When the number of users or devices increases or decreases substantially, it is important for organizations to be able to grow or reduce the size of the network accordingly. This is the concept of **scalability**, and it is a core consideration in selecting an architecture. In fact, the client/server architecture described below overtook the old mainframe architecture we just described partly because it offered substantial scalability.
- *Maintainability:* Once an architecture has been implemented, it must be maintained by installing software patches, upgrading equipment, and replacing equipment when it fails. Maintenance is a task that requires responses 24 hours per day, 7 days per week. One principle that may affect organizations' choices of architecture is how to minimize having to perform this maintenance themselves.
- *Security and reliability:* All technology decisions should take into consideration security, including reliability of the systems, and architectures are no exception. For example, most experts consider a peer-to-peer architecture, where files are shared, as one of the least-secure options for networking. We will discuss security in depth in Chapter 8.

Client/Server Architecture

In the **client/server architecture**, processing and storage tasks are shared and distributed between two types of network systems: clients and servers. Clients are processes that request services from servers. Servers provide services to clients by responding to their requests. Most computers can perform client or server services, even if we mostly think of clients as the personal computers we use to access larger machines called servers.

An example can help clarify the concept of sharing the workload among clients and servers (see Figure 7.6). When you search the Web for information, you are using a browser on your personal computer. The browser is acting as a client that makes requests to a Web server when you click on a hyperlink, asking for that page to be sent to you. The Web server receives the request and sends the document back to your browser. Now assume that you need information about prices in that document. The Web server may send a request to a database server for the information, and then it sends it along with the other document to your browser. The browser then displays the requested document. In this case, the client (browser) requests services but also performs the task of displaying the results (presentation task). The Web server receives the requests, makes its own request to the database server (data task), and then packages it to send it back to you (processing or logic task). For servers and clients to communicate, even when they are on different technology platforms, they require some software that can understand each technology's specific formats and communication protocols and translate them into the other platform's formats and protocols, enabling interoperability. This is the role of middleware.

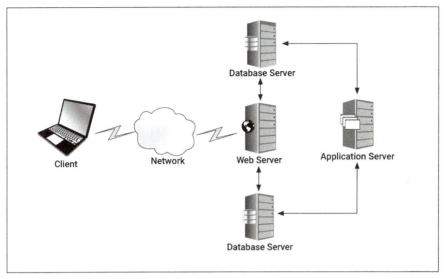

Figure 7.6 Client/Server Computing Architecture

Peer-to-Peer Architecture

The **peer-to-peer architecture** is used substantially on the Internet, but not as much in organizations. In a peer-to-peer network, all systems are equal (acting as both clients and servers), sharing their resources with one another. The sharing of resources can include processing power, disk storage, and network bandwidth. Some students might be familiar with music-file-sharing systems or movie-file-sharing systems, which are good example of peer-to-peer networking.

Wireless Architecture

In the wired versus wireless networks section, we discussed how wireless networks are set up. Organizations must evaluate several decisions regarding what to support and whom to allow on the network when they choose to implement a wireless architecture. Think of your university's wireless network. Who should have access to it? If everyone, is it your tuition fees that pay for the network? Which devices should be allowed on it? What level of security should be provided? All these issues have to be discussed and decided upon by the network architects of your university before they make the wireless network available.

One important decision organizations have to make when considering wireless architectures they want to support is what policies to implement with respect to employees bringing their own wireless devices and using them for work and/or personal purposes. This is referred to as **BYOD (bring your own device)** policies. These devices include smartphones, tablets, and even personal laptops, which some employees prefer to use over office computers. Because these devices are not controlled by the organization, some companies are concerned about the possible security issues.

They are also concerned about employees using the devices for purposes other than work during work time. Yet at the same time, most organizations recognize the increased productivity that has resulted from employees using their devices anytime, anywhere. Additional Learning Activity 7.A1 at the end of the chapter proposes a case study that explores BYOD use in classrooms.

Service-Oriented Architecture

A more recent architecture that has gained substantial interest in the business world is **service-oriented architecture (SOA)**. It is a model or a set of design principles of how to take data from heterogeneous systems and create reusable services. The services can be used in different systems with different technology platforms. Think of an invoice as a service. The accounts payable department probably has a system that generates invoices. The shipping department, however, also has a system that handles invoices for products that are ready to ship. The two departments can make use of the same service, an invoice, with each of their own applications, independent of where and on which platform their applications are. Figure 7.7 shows how SOA works.

As with all architectures, it is important to realize that SOA is not a product itself but an architecture or, in this case, a way to design how devices are interconnected. Note also that the layer that allows SOA to work is similar to middleware, which we

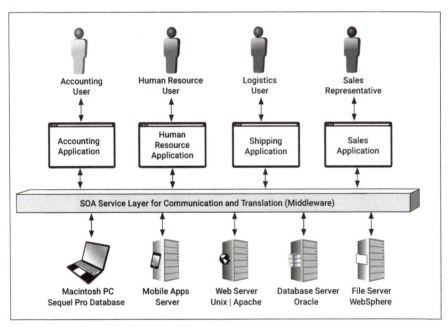

Figure 7.7 Sample Service-Oriented Architecture (SOA) Layout

discussed previously. Because SOA is a vague architecture and requires some coding (using the middleware), it is often too broadly defined by organizations. This can lead to poor implementation, where reusability is not really achieved and the potential gains are not realized. While it seems easy when looking at Figure 7.7 to connect the various heterogeneous systems, it takes time to make sure that all the different data fields and formats can be "understood" and translated by the middleware. Furthermore, in some organizations, those who "own" the data may not agree to let various systems access their data and services.

Cloud Computing

One of the most popular architectures that has appeared in the past decade is cloud computing. In **cloud computing**, an organization acquires or rents computing resources from online providers instead of having its own locally managed hardware and software (see Figure 7.8). Using the Internet, the organization is able to increase or decrease its computing resources when the need arises or decreases. What can be acquired via the cloud? Virtually any computing resource, from hardware for storage and processing, to backup services, to specific applications. Clearly, the cloud includes the SaaS we discuss later in this chapter, and much more. The main providers of cloud services are large organizations with resources to spare, such as Google, IBM, AT&T, Microsoft Azure, and Amazon Web Services (AWS), although specialized providers also exist. You are likely familiar with using the cloud to store documents if you have used Dropbox, Google Drive, iCloud, or one of the many other similar services available to you today. Importantly, the cloud architecture is a major facilitator of IoT, since the devices themselves do not have enough capacity to store all of the data they collect. In some ways, IoT devices generate massive amounts of data, and cloud computing provides the means to transfer and store the data.

The four main forms of cloud computing are IaaS, PaaS, SaaS, and BPaaS (business process as a service). The interaction among them is displayed in Figure 7.9.

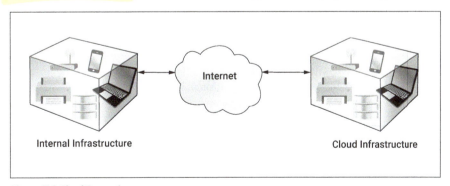

Internet

Internal Infrastructure

Cloud Infrastructure

Figure 7.8 Cloud Computing

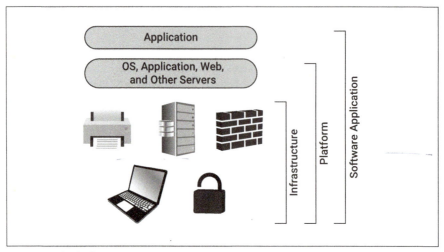

Figure 7.9 Cloud Infrastructure Services

Infrastructure as a Service (IaaS)

The availability of cloud computing has provided the ability to offer infrastructure as a service (IaaS), which is an Internet-based computer operating system that is maintained and updated by a third party at a remote location. Utilizing a system such as this allows businesses to pass the running of their computer systems to another company so they can focus on their business competencies. Companies that offer these services usually utilize virtualization, which is discussed below, to provide many different operating systems on one server. It also allows for the dynamic scaling of infrastructure when demand increases in areas such as online storage and backup. Some examples of companies providing IaaS are Amazon.com and Symantec.

Platform as a Service (PaaS)

Another service provided by the ubiquity of Internet connections is known as platform as a service (PaaS), which provides a cloud-based platform for running websites and databases and executing programming languages. In the current Internet era, it is important for systems to be operational 24 hours per day and 365 days per year. If a company wants to keep its system up at this level, it needs to invest not only in staff whose job it will be to provide this level of uptime but also in backup systems that provide redundancy when electronic systems inevitably need maintenance or even fail. Third parties that provide these services are able to gain efficiencies of scale by spreading the cost of supporting a system like this over many customers, thus reducing the total number of systems needed (though virtualization) as well as the personnel necessary to run it. Examples of PaaS offerings include Google's App Engine and Microsoft's Azure.

Software as a Service (SaaS)

Another architecture or model of computing is **software as a service (SaaS)**, which is basically the acquisition (or rental) of software via a subscription model (see Figure 7.10). SaaS, often called "software on demand," allows organizations to rent software from a provider, frequently using the Internet to access this software application. SaaS software tends to be mostly straightforward applications that businesses need but that do not provide specific competitive advantages. For example, the most popular SaaS applications include customer relationship management software, sales force automation, human resources management software, and desktop applications (like word processing). One form of SaaS that you are likely familiar with is Google Docs. Others include Web meeting technologies such as Cisco's WebEx, customer relationship management technologies such as Salesforce.com, and enterprise resource planning software such as SAP Business ByDesign.

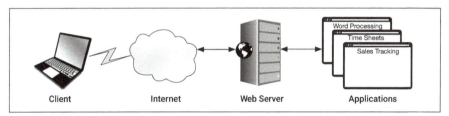

Figure 7.10 Software as a Service

Business Process as a Service (BPaaS)

A newly emerging delivered service is **business process as a service (BPaaS)**, which leverages the SaaS, IaaS, and PaaS already used by the company. BPaaS provides companies with a way to provide services in a streamlined fashion that adapts and grows as the businesses' needs to provide services change and expand. BPaaS infrastructures are implemented in such a way as to allow multiple languages to be displayed to the end user and to scale very rapidly as business grows. These business processes include any processes that a company may provide, including email, pack-

---

LEARNING ACTIVITY 7.5

Architecture and Principles

The purpose of this activity is to further explore one of the networking architectures discussed in this chapter. More specifically, the goal is to map each architectural principle to one of the networking architectures that your instructor will assign to you. Perform necessary research as a group and clearly highlight how the given architecture ranks on each architectural principle (high, medium, low), and make sure to describe why. For example, you could state that the client/server architecture ranks high on scalability because clients and servers can be added to the network easily and as needed. Prepare a presentation for the class.

age shipping, and customer credit management. One of the largest providers of BPaaS is ADP, a payroll processing company. By utilizing ADP's payroll processing BPaaS, companies can focus on the products and services they are in business to offer and let ADP handle their payroll processing needs.

Virtualization

Another recent trend in computing architectures is virtualization. As can be seen in Figure 7.11, **virtualization** allows one physical device, such as a server or computer, to operate as if it were several machines. Virtualization software creates virtual environments called *virtual machines* on the hardware platform. Each virtual machine can then run its own operating system and applications. If you own an Apple Mac computer, its operating system allows you to perform virtualization so you can run a Windows operating system on your Mac.

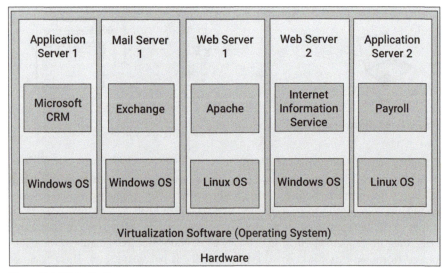

Figure 7.11 Virtualization

Communicating Information in Modern Organizations

The previous section briefly reviewed some of the most used or most recent networking architectures. How do we make use of these architectures? Well, they serve as the communication backbone for all the applications that organizations employ to perform tasks. These tasks are called *business processes*, and we will dedicate Chapter 11 to discussing them more in detail. In this section, however, we want to continue our discussion about the communication applications running on networks with a review of the more recent Web 2.0 applications.

Web 2.0

Web 2.0 refers to the second generation of applications on the Internet. The main difference between the initial set of applications, like email or Web surfing, and Web 2.0 applications is the addition of interactivity, where the user becomes a participant in the interaction. For example, through **wikis** like Wikipedia (http://www.wikipedia.org), users can participate in the design of the site by adding their own definitions and corrections to the information. The set of technologies referred to as Web 2.0 includes wikis, social networking, blogs, mashups, Twitter, and much more. Table 7.4 presents a list of some popular Web 2.0 technologies and their descriptions.

The Power of Wikinomics (Crowdsourcing)

You have heard of wikis and Wikipedia, but what if you could take advantage of everyone's voluntary contributions to knowledge for business purposes? This is referred to by some as **crowdsourcing** or **mass collaboration**, where individuals voluntarily come together to solve a problem. Others have given it the name **wikinomics**. More specifically, in a book titled *Wikinomics: How Mass Collaboration Changes Everything* (2007), Don Tapscott and Anthony Williams give the example of Goldcorp.

Goldcorp, a Canadian mining company with a 50-year-old mine considered to be tapped out, needed to clearly identify new deposits, but in-house geological experts did not seem to be able to find them. The company's CEO, Rob McEwen, then took a very unusual step by publishing all proprietary geological data on the Web (mining is usually a very secretive industry). He challenged "Web prospectors" in a contest to find new mining deposits and offered $575,000 in prize money. Approximately 1,000 individuals from 50 countries examined the available information, and 110 target locations were submitted to Goldcorp; 50% of them were previously unknown to the company. The amazing result is that 80% of the newly identified deposits yielded substantial quantities of gold.

Another example comes from the Boston Marathon bombing. On April 15, 2013, during the Boston Marathon, a bomb was detonated near the finish line, killing three people and injuring an estimated 264 others. Luckily for the investigators, there are commercial cameras all around the area where the bombing took place. However, there were holes, or blind spots, in what these cameras recorded. This is where the crowdsourcing came in. Police were able to tap into the approximately 500,000 people watching the marathon and the videos they were recording. These people all provided digital recordings, which helped with the investigation. Unfortunately, crowdsourcing did not work as well for the investigation. A second group of people thought they would help the investigators review the vast number of videos they had. This resulted in an innocent person being suspected of committing the crime based on an overzealous ambition to help find the person who set off the bomb (Wadhwa 2013).

There are many other examples of crowdsourcing, such as McDonald's creation of a "My Burger" campaign in Germany, where customers created their own burgers and the top five creations were actually released in McDonald's restaurants. This campaign brought new ideas, new clients, and lots of "free" advertising (Kearns 2015). Other crowdsourcing campaigns you might be familiar with include Starbuck's "White Cup" (design the Starbucks cup), Waze (share traffic information), Lay's Chips' "Do Us a Flavor" (create new flavors for chips), or Greenpeace's "Save the Arctic" (against oil drilling) campaigns (Kearns 2015).

TABLE 7.4 Popular Web 2.0 Technologies		
Web 2.0 Technology	**Description**	**Examples**
Wikis	Website that allows individuals to jointly create and edit Web pages about a chosen topic.	• Wikipedia: http://www.wikipedia.org • Wiki Quotes: https://www.wikiquote.org/
Instant messaging and presence awareness	Allows several individuals to communicate via real-time text-based messages. **Presence awareness** indicates when colleagues are currently online and connected.	• Discord: https://discordapp.com/
Collaborative content	Allows several individuals to contribute to and share stored data and documents, such as text documents, movies, pictures, calendars, and more. The collaborative content systems control user access and versioning, preventing two users from changing a document simultaneously.	• Doodle (calendar): http://www.doodle.com • Google Docs (documents): http://docs.google.com
Web conferencing (voice, video, and data)	Allows individuals to conduct live meetings via the Internet, including training sessions or sales presentations. All participants use their own computers to connect to the meeting and can use voice (voice over IP) and/or video, as well as send files, use instant messaging, or even share their personal desktop with others.	• Skype: http://www.skype.com • Google+ Hangouts: https://hangouts.google.com/
Social networking	Allows individuals to participate in a community of users (individuals or organizations) where they are connected with one another via links (e.g., friendship or business relationships).	• Facebook: http://www.facebook.com • LinkedIn (business social network): http://www.linkedin.com
Blogs	Allows individuals to write commentaries or opinions on anything they want on websites that others can read. Usually, blogs are updated regularly and allow others to leave comments. The name comes from "Web log."	• Create your own blog: http://www.blogger.com; http://www.wordpress.com
Twitter (microblogging)	Allows individuals to write tweets, which are short text messages, on their user page for others to "follow." Initially up to 140 characters, this was increased to 280 in late 2017. Twitter is very popular among celebrities for sharing their day-to-day activities with fans.	• Twitter: http://www.twitter.com
RSS (Really Simple Syndication)	Allows individuals to receive frequently updated information like blogs or news headlines directly on their personal computers.	• Search your university's news pages for where you can sign up to receive news as an RSS feed. • CNN RSS Feeds: https://www.cnn.com/services/rss/

As you can see from Table 7.4, there are many interesting Web 2.0 technologies that have changed the way we interact with organizations and with one another. Individuals are increasingly connected via technologies, with many advantages and disadvantages.

Web 3.0

Web 2.0 technologies are all around us today, but a new technology is emerging into the marketplace: **Web 3.0.** If Web 2.0 introduced interactivity and user participation, what is Web 3.0 bringing to us? The semantic Web. This is when computer systems start to understand the meaning (semantics) of information and data. In other words, the systems will consider the context of the information to give meaning to other information or data connected to it. For example, when you search the Web today, you have to identify which keywords are likely to give you the information you need. In the semantic Web, the systems will understand what you need and give you exactly that information. Searches will become much more intelligent and results much more personalized. For example, Google implements this when you search for restaurants on your smartphone and it suggests local restaurants you might enjoy based on such information as your location, prior searches, and restaurants you have previously clicked on.

STATS BOX 7.2:

Diffusion of Innovations Timeline

Social networking is everywhere, and there is no doubt about its popularity. What is quite interesting is that it achieved this status very rapidly compared to other major innovations. The table below compares how long it took various technologies to achieve 50 million users. Of course, in the days of the telephone, there were fewer people, so this is not a scientific analysis. Yet it does tell something about today's interconnected world.

Innovation	Year Introduced	Year Reached 50 Million Units	Years It Took
Telephone	1876	1926	50
Television	1928	1950	22
Cellular phone	1983	1995	12
Internet	1990	1997	7
Facebook	2004	2007	3
WeChat	2011	2012	1
Pokémon Go	2016	2016	19 days

Sources: Desjardins, J. 2018. "How Long Does It Take to Hit 50 Million Users?" Visual Capitalist, June 8. https://www.visualcapitalist.com/how-long-does-it-take-to-hit-50-million-users/.

LEARNING ACTIVITY 7.6

Web 3.0 versus the Internet of Things

You may have noticed that Web 3.0 and the Internet of Things are similar in nature. For this learning activity, identify two ways that Web 3.0 is being used and two ways that the Internet of Things is being used. Identify how these implementations are similar and how they are different. Be sure to pay attention to how the understanding of information and data (semantics) is used with each of these technologies. Be prepared to discuss your lists of similarities and differences with the class.

Chapter Summary

In this chapter, we discussed networking concepts and how information is transmitted. We first defined networks and discussed various types, including the Internet and the Web. We then discussed networking architectures and concluded with Web 2.0 technologies and Web 3.0.

Here are the main points discussed in the chapter:

- A network is a collection of interconnected devices that allows users and systems to communicate and share resources. The network requires connecting devices such as routers, bridges, or switches. Each device on the network requires a network interface card (NIC, or network adapter). Repeaters and hubs can be used to regenerate signals on networks.
- Networks can be classified as wired (using physical connections) or wireless (using airwaves for connections). Wireless networks include Wi-Fi, microwave signals, satellite signals, infrared signals, and radio signals, including cellular networks and Bluetooth. Networks can also be classified as local area networks (LANs), which connect devices in a limited geographical area, or wide area networks (WANs), which connect devices over a large geographic area like a city, a country, or the world. Other networks include personal area networks (PANs), home area networks (HANs), virtual private networks (VPNs), and backbone networks (BBNs).
- The Internet is a publicly accessible worldwide network of networks. It uses routers to interconnect the various networks, and every host or computer that is a full participant (permanently connected) on the Internet has a unique address called an IP (Internet Protocol) address. The Internet is a network, not an application. The many applications that can make use of the Internet include the Web, which is a graphical interface to worldwide resources, as well as electronic mail, instant messaging, voice over IP (Internet telephony), desktop videoconferencing, peer-to-peer file sharing, online application sharing, file transfers (FTP), newsgroups, and many more.

- Network architectures describe how devices are supposed to work together. Every architecture has its advantages and disadvantages in terms of ease of implementation, flexibility and interoperability, control, scalability, and security and reliability. The main networking architectures we discussed include client/server architecture, where processing and storage tasks are shared and distributed between clients and servers; peer-to-peer architecture, where systems are equal in sharing their resources with one another; wireless architecture, which defines how devices are to be connected to the wireless network; service-oriented architecture (SOA), which allows data from heterogeneous systems to be used to create reusable services; cloud computing architecture, which allows an organization to acquire computing resources from online providers instead of having its own locally managed hardware and software; and the virtualization model, where one physical device can operate as if it were several machines.
- Web 2.0 refers to the second generation of applications on the Internet, where the user becomes a participant in the interaction. Web 2.0 technologies include wikis, which allow individuals to jointly create and edit Web pages about a chosen topic; instant messaging and presence awareness, which allow individuals to communicate via real-time text-based messages with presence awareness indicating when colleagues are currently online and connected; collaborative content, which allows several individuals to contribute to and share stored data and documents; Web conferencing, which allows individuals to conduct live meetings via the Internet; social networking, which allows individuals to participate in a community of users connected with each other; blogs, which allow individuals to write commentaries or opinions on anything they want on websites others can read; mashups, which allow users or developers to combine data or applications from several sources to create new ways to view data or new aggregated results; Twitter, which allows individuals to write short text messages of up to 280 characters on their user page for others to "follow"; and RSS, which allows individuals to receive frequently updated information like blogs or news headlines directly on their personal computers.

Review Questions

1. What is a network?
2. What are the key components and characteristics of a network?
3. What is a wireless network, and what are some examples of it?
4. What are the differences between local area networks (LANs) and wide area networks (WANs)?

5. What is the Internet, and what is the difference between the Internet and the World Wide Web?
6. What is the role of network architectures, and what principles should managers take into consideration when evaluating network architectures?
7. Briefly explain each of the networking architectures presented in the chapter.
8. Explain what Web 2.0 means and provide three examples of Web 2.0 technologies.
9. What is Web 3.0? What other terminology is often related to Web 3.0?
10. What is the Internet of Things?

Reflection Questions

1. What is the most important thing you learned in this chapter? Why is it important?
2. What topics are unclear? What about them is unclear?
3. How can the architectures discussed in this chapter be used to create competitive advantages for the organizations we discussed in Chapter 4?
4. Why do many individuals confuse the Internet with the World Wide Web?
5. Why does it make sense for companies to use architectures like cloud computing?
6. What issues could exist when companies use architectures like cloud computing?
7. Do you think we will eventually live in a completely wireless world, or will there always be wired networks? Why?
8. In your opinion, which architectural principle is most important in deciding on a network architecture?
9. Is the Internet likely to continue growing in terms of users, hosts, and connected networks? Why or why not?
10. Identify three precise examples of how a business can leverage IoT for competitive advantage.

Additional Learning Activities

7.A1. This activity requires you to complete a case analysis regarding BYOD (bring your own device) as implemented in classrooms. The case can be accessed online or may be provided by your instructor. Sipior, J. C., J. Bierstaker, Q. Chung, and J. Lee. 2017. "A Bring-Your-Own-Device Case for Use in the Classroom." *Communications of the Association for Information Systems* 41(10). http://aisel.aisnet.org/cais/vol41/iss1/10.

7.A2. This activity requires you to complete a case analysis regarding social networking. The case may be accessed through your university website,

the Ivey Business School business case website (https://www.iveycases
.com), or the Harvard Business School cases website (https://hbsp
.harvard.edu/home/), or it may be provided to you by your instructor.
Aggarwal, R., S. E. Chick, and F. Simon. 2017. "PatientsLikeMe:
Using Social Network Health Data to Improve Patient Care."
INSEAD, January. Product number: IN1312-PDF-ENG.

7.A3. As a group, research Web 3.0 and develop a presentation for fellow
students that can clearly explain what it is and how it is being utilized.

7.A4. For this activity, in a group conduct research on a company that moved
to remote work (also known as telework) during the 2020 COVID-19
pandemic. Discuss the challenges they faced and the benefits they
identified. Explain what role the Internet played and what would have
happened to this company if the pandemic occurred at a time when
the Internet was not widely available. Present your findings to the class
and identify which companies did well and which did not do well with
remote work during the pandemic.

7.A5. Research voice over IP (VoIP) products. Select and download two of
them that either are freeware (available for free) or offer a test version.
Use the two products several times and identify their main advantages
and weaknesses. Prepare a recommendation report for which one
would be most useful for (a) students and (b) a small business. Make
sure to discuss why.

7.A6. Research the advantages of 4G wireless networks and 5G wireless
networks. Discuss whether you believe it is worth upgrading phones
from 4G to 5G.

7.A7. This activity requires you to complete a case analysis regarding cloud
computing. The case can be accessed online or may be provided by your
instructor. Cain, J., M. Levorchick, A. Matuszak, A. Pohlman, and
D. Havelka, 2015. "eLoanDocs: Riding the Tide of Technology
without Wiping Out." *Communications of the Association for Information
Systems* 36(38). http://aisel.aisnet.org/cais/vol36/iss1/38.

7.A8. Reflect on how you used Web 2.0 technologies as schools transitioned
to online learning during the spring semester of 2020 as the COVID-
19 pandemic occurred. What Web 2.0 technologies did you use?
Which worked great and which were challenging? Why?

References

Hui, T. K. 2020. "Pandemic Closed NC Schools. Now Some Buses Will Have
Wi-Fi So Students Can Go Online." *The News & Observer*, May 6. https://www
.newsobserver.com/news/local/education/article242535261.html.

Kearns, K. 2015. "9 Great Examples of Crowdsourcing in the Age of Empowered
Consumers." *Tweak Your Biz*, July 10. https://tweakyourbiz.com/marketing/9
-great-examples-crowdsourcing-age-empowered-consumers.

Lueth, K. L. 2018. "State of the IoT 2018: Number of IoT Devices Now at 7B—Market Accelerating." *IoT Analytics.* https://iot-analytics.com/state-of-the-iot-update-q1-q2-2018-number-of-iot-devices-now-7b/.

Nadel, B. 2020. "How to Use a Smartphone as a Mobile Hotspot." *Computerworld,* July 2. https://www.computerworld.com/article/2499772/how-to-use-a-smartphone-as-a-mobile-hotspot.html.

Tapscott, D., and A. D. Williams. *Wikinomics: How Mass Collaboration Changes Everything.* Penguin. 2008.

Van Slyke, C., and F. Bélanger, 2003. *Electronic Business Technologies.* New York: John Wiley & Sons.

Wadhwa, T. 2013. "Lessons from Crowdsourcing the Boston Bombing Investigation." *Forbes,* April 22. http://www.forbes.com/sites/tarunwadhwa/2013/04/22/lessons-from-crowdsourcing-the-boston-marathon-bombings-investigation/.

Glossary

Backbone network (BBN): A network that serves to interconnect other networks (like LANs) or network segments (subnetworks).

Blog (Web log): A website that allows an individual to write commentaries or opinions on anything for anyone to read.

Bluetooth: A wireless network that uses short-wavelength radio transmissions to connect devices such as wireless mice, keyboards, or headphones.

Business process as a service (BPaaS): Provides companies with a way to provide services in a streamlined fashion that adapts and grows as the businesses' needs to provide services change and expand.

Bring Your Own Device (BYOD): A policy regarding the use of personally owned wireless devices (laptops, tablets, and smartphones) by employees in the workplace.

Cellular networks: Networks that use radio communication over local antennas to relay calls from one area to the next.

Client/server architecture: A computing model where the processing and storage tasks are shared and distributed between clients and servers.

Clients: Processes that request services from servers.

Cloud computing: A computing model where an organization acquires or rents computing resources from online providers instead of having its own locally managed hardware and software.

Collaborative content: A situation in which several individuals contribute to and share stored data and documents.

Crowdsourcing: Mass collaboration of information to solve a problem. Also known as *wikinomics.*

Desktop videoconferencing: A system allowing individuals in different locations to communicate via voice and video on personal computers.

Electronic mail (email): Application used to send and receive messages through computer networks.

File Transfer Protocol (FTP): A system allowing users to move files back and forth between nodes on the network.

Home area network (HAN): A LAN used within a home office, allowing PCs to share devices such as printers, routers, or scanners.

Hub: A form of repeater that has multiple ports to connect many devices.

Instant messaging: A system allowing multiple users to communicate synchronously by sending and receiving short text messages online.

Instant messaging and presence awareness: A system allowing several individuals to communicate via real-time text-based messages with presence awareness indicating when colleagues are currently online and connected.

Internet: The publicly accessible worldwide network of networks.

Internet Assigned Numbers Authority (IANA): The organization responsible for assigning IP addresses.

Internet of Things (IoT): The connection of physical everyday objects to the Internet that allows them to communicate with one another and other computers.

Internet2: A consortium of research and education institutions, industry leaders, and government agencies that operates the Internet2 network.

Internet2 network: A fiber optics–based network used for high-speed transfers among research institutions and for testing and researching networking technology.

Intranet: The use of Internet technologies and related applications inside an organization.

Internet Protocol Version 4 (IPv4): An older IP addressing scheme that uses 4 bytes for addresses (such as 128.192.68.1) and ran out of addresses to allocate in 2011.

Internet Protocol Version 6 (IPv6): A new IP addressing scheme that uses 16 bytes.

Local area network (LAN): A network to connect devices in a limited geographical area (usually fewer than five kilometers).

Mashup: A situation in which users or developers combine data or applications from several sources to create new ways to view data or create new aggregated results.

Mass collaboration: More than one individual gathering information. Also known as *wikinomics* or *crowdsourcing*.

Network: A collection of interconnected devices that allow users and systems to communicate and share resources.

Network architecture: The layout or blueprint for how devices are supposed to work together.

Network infrastructure: The actual hardware, software, and networking components that support the processing and transfer of information.

Network interface card (NIC, or network adapter): An interface that provides physical access to a device because it has a unique ID written on a chip that is mounted on the card.

Online application sharing: A system allowing users to share documents, calendars, or other applications using websites.

Peer-to-peer architecture: A computing model where all systems are equal (acting as both clients and servers), sharing their resources with one another.

Peer-to-peer file sharing: File sharing between specific individuals or systems across the Internet.

Personal area network (PAN): A network connecting personal devices to a personal computer (e.g., mouse, microphone, printer) over a very short distance.

Really Simple Syndication (RSS): A format allowing individuals to receive frequently updated information like blogs or news headlines directly on their personal computers.

Repeater: A device that retransmits a data signal that it receives after eliminating noise in the signal and regenerating it for strength.

Router: An intelligent device that controls the flow of transmissions in and out of a network.

Scalability: The ability to grow or reduce the size of the network as required.

Servers: Processes that provide services to clients by responding to their requests.

Service-oriented architecture (SOA): A computing model or set of design principles for how to take data from heterogeneous systems and create reusable services.

Social networking: Individuals participating in a community of users connected with one another.

Software as a service (SaaS): An architecture or model of computing where the acquisition (or rental) of software is done via a subscription model.

Twitter: A technology allowing individuals to write short text messages of up to 280 characters on their user pages for others to "follow."

Virtualization: A computing model that allows one physical device, such as a server or computer, to operate as if it were several machines.

Voice over IP (Internet telephony): Voice data sent over an IP-based network, such as the Internet.

Web 2.0: The second generation of applications on the Internet, where the user becomes a participant in the interaction.

Web 3.0: Computer systems that understand the meaning (semantics) of information and data, resulting in more intelligent searches and more personalized results. Also known as the *semantic Web*.

Web conferencing: A technology allowing individuals to conduct live meetings via the Internet.

Wide area networks (WANs): A network connecting devices over a large geographic area, such as a city, a country, or the world.

Wi-Fi: The network name owned by the Wi-Fi Alliance; a wireless network that uses radio technology.

Wikinomics: A term to describe when individuals voluntarily come together to solve a problem online. Also known as *crowdsourcing*.

Wikis: Web pages individuals jointly create and edit about a chosen topic.

Wired network: A network that makes use of physical cables (copper wires, coaxial, or fiber-optic cables) for connections.

Wireless network: A network that makes use of frequencies to transmit signals.

World Wide Web (the Web): The graphical interface to worldwide resources available on the Internet.

Chapter 8

Securing Information

Learning Objectives

By reading and completing the activities in this chapter, you will be able to:

- Discuss the relationship between risk management and information security
- Explain the various types of threats to the security of information
- Identify the main goals of information security
- Discuss the different categorizations of security technologies and solutions
- Explain the basic functioning of security technologies and solutions, such as passwords and password managers, two-factor authentication, firewalls, biometrics, encryption, virus protection, and wireless security
- Discuss the main purposes and content of security policies

Chapter Outline

Introduction and Definitions
Focusing Story: My Mac Is More Secured than Your Windows-Based PC!
  Learning Activity 8.1: How Protected Is Your Computer?
Information Security Threats
  Learning Activity 8.2: Detecting Phishing
  Learning Activity 8.3: What Do Hackers Break into the Most?
Security Technologies and Solutions
  Learning Activity 8.4: How Strong Is Your Password?
  Learning Activity 8.5: Biometrics on the PC
  Learning Activity 8.6: Breaking the Encryption
  Learning Activity 8.7: IoT Security
  Learning Activity 8.8: Where Is the Security?

Introduction and Definitions

Information security is an important topic for everyone in today's interconnected world. What is **information security**? It is defined as the set of protections put in place to safeguard information systems and/or data from security threats such as unauthorized access, use, disclosure, disruption, modification, or destruction.

My Mac Is More Secured than Your Windows-Based PC!

We are sure some of you have heard it and probably even shared the rumors with others. If you get a Mac (an Apple computer with a Macintosh operating system), you are much less likely to have security breaches than if you get a regular PC—plus they are really cool. We might agree with the cool part, but what is the status of Apple computer security?

For quite a while, there were few viruses and other such programs (called **malware**) that targeted Mac systems. That resulted in a reputation that Mac systems were substantially safer than Windows-based PCs. Experts do recognize that this is still true—Apple has more control over the operating system and the hardware that it runs on, thus making it more secure—but they also warn that this does not mean there are no security issues! One Mac lover and security researcher named Charlie Miller first hacked an iPhone in 2007 and then a MacBook Air in two minutes at a competition in Vancouver a few months later.

When Apple released the iPhone, it generated new interest in Apple's platforms and might have attracted the attention of many more hackers who had targeted Microsoft's platforms before. As a result, many hackers started to identify security flaws in the iPhone, such as the ability to bypass the phone's PIN by placing it in recovery mode or the ability to extract encrypted data using basic forensic software (i.e., software used by investigators to perform such tasks as logging into a computer or recovering lost or deleted files). This software is available for free online, and a YouTube video even explains how to do this. Of course, Apple released fixes for these security issues as soon as they could. Nevertheless, as a result of these issues, many firms that deal with sensitive data, such as law firms, decided iPhones were too risky to use. New security issues also surfaced with the release of the iPad. Over the years, many ways to hack a Mac have been discovered and exploited. In 2017, Mac users started being victims of **ransomware** when hackers used the "Find My iPhone" feature with a bunch of iCloud usernames and passwords they stole to remote-lock users' devices. In 2020, a security expert was able to hack a Mac when users would double-click on a Microsoft Office file. Other researchers were able to hack into a new Mac when it first connected to an enterprise Wi-Fi network. This vulnerability was shown at the Black Hat security conference.

Sources: Chin, M. 2017. "Some Mac Users Are Getting Hit with Ransomware—Here's What to Do." *Mashable*, September 22. http://mashable.com/2017/09/22/icloud-hack-find-my-mac/#JzWi0I19TmqY; Hooper, L. 2007. "iPhone: Apple's Worm?" *CRN* 1256 (December 10): 16; Blake, A. 2020. "Is Mac Really More Secure than Windows? We Asked the Experts." *Digital Trends,* April 13. https://www.digitaltrends.com/computing/privacy -macos-or-windows-we-asked-the-experts/; Franceschi-Bicchierai, L. 2020. "Ex-NSA Hacker Finds a Way to Hack Mac Users via Microsoft Office." *Vice*, August 5. https://www.vice.com/en_us/article/jgxamy/hacker -finds-a-way-to-hack-mac-users-via-microsoft-office; Newman, L. H. 2020. "Hacking a Brand New Mac Remotely, Right Out of the Box." *Wired*, August 9. https://www.wired.com/story/mac-remote-hack-wifi -enterprise/.

Focusing Questions

1. Why is forensic software available online to everyone?
2. Why are security issues in iPhones and iPads of concern to Mac users?
3. Will a larger market for Apple platforms change this security landscape?
4. What other Apple devices could create new security issues for Mac users?

Risk Management

Before discussing information security, it is important to understand that information security is one aspect of risk management that organizations pursue. No matter what efforts an organization may make to provide the best security possible, or which technologies and tools it might invest in, there will always be security risks involved. **Risk management** is the process of identifying, assessing, and prioritizing the security risks an organization may face. As a result of this process, organizations may decide to accept the risks, try to mitigate or prevent those risks by investing in security protections, or share the security risks with another organization—for example, by buying insurance. Risk management determines where the company is most vulnerable and how likely it is to be affected by a threat. What are vulnerabilities? In information security, vulnerabilities are weaknesses in the organization's systems (including people), networks, and software that can be exploited to conduct an attack or gain unauthorized access to systems and data. Security threats are those persons or events that can potentially negatively affect the organization (e.g., a virus, a cyberattack, etc.). Therefore, risk management is about determining the potential risk that a threat will take advantage of a vulnerability. While weighing these risks, the company looks at the cost of addressing these issues and makes decisions based on both the financial aspect and the acceptable level of security needed to protect different areas in the company.

While security and risk management do go hand in hand, they have a few differences as well. **Security** is focused on protecting the assets a company has from both external and internal threats. External threats such as hackers and viruses threaten the valuable information a company collects while in business. Information security refers to the protection of data and systems from attacks to ensure the company can run smoothly without interruption. Internal threats are important to guard against as well. For example, security involves monitoring the network to make sure that policies are not being breached in a harmful way (deliberately or accidentally) and that employees are only able to access systems they are authorized to. A newer term is **cybersecurity**. Some people equate information security and cybersecurity, while others suggest that cybersecurity is broader, focusing on the protection of everything that is in electronic form, including devices and networks, whereas information security is focused on the data. Risk management has a larger focus than information security or cybersecurity, including analyzing and balancing risks with the resources available to mitigate them. A representation of how a multitude of threats work together to increase risk is presented in Figure 8.1.

In this chapter, the focus is on the steps that can be taken to mitigate or prevent risks. That is the essence of why information security steps are taken. First, various high-level information security concepts are discussed. This is followed by a discussion of various information security threats and, finally, by solutions to these threats. It is important to note that some of the solutions are employed to counteract multiple threats. There is not a direct one-to-one relationship between threats and solutions.

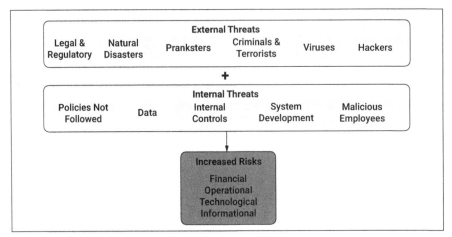

Figure 8.1 Threat Types and Causes That Lead to Increased Risk Exposure

Information Security Concepts

Whether you need to secure your own computer to prevent a virus from destroying your files or an organization needs to secure access to its data and servers to ensure that hackers cannot get information about its clients, the core concepts of security remain the same.

First, security is not only about technology but also about management and people. It is possible to acquire the best-ever security tools, but if employees do not use them or, even worse, decide to bypass the security protections mandated by the organization, the systems and data are vulnerable to attacks or security failures. Top-to-bottom commitment to security in the organization is needed to ensure that security tools are properly used and security policies are followed adequately.

Second, the world of information security is somewhat unfair. How many security weaknesses does the hacker need to break into a system? One. How many security weaknesses does the organization or individual need to fix or protect against? All of them. This is referred to as **asymmetrical security warfare**. One party must do everything to protect itself, while the other party (the attacker) only needs to find one weakness.

Third, as a result of having to protect the computer systems and data from all possible weaknesses, security for individuals, and even more so for organizations, requires what is called *defense in depth*. **Defense in depth** means that there must be multiple layers of security protections in place. For example, if you have very important data on your computer, such as personal financial information, defense in depth would mean having a lock on the computer, a password to access it, a firewall to protect it against breaches from the Internet, and even encryption of the data so they are unreadable without the passcode or password. We will discuss all these technologies in this chapter. Below we explore various levels of security protection in depth.

Security Levels

The concept of defense in depth refers to multiple layers of protection, which can include multiple devices performing similar functions. Similarly, there must exist protection technologies and policies to secure stored information as well as information in transit. Some like to refer to three main levels: (1) protection of the information itself, (2) protection of the computer on which the information is stored, and (3) protection of the network to which the computer is connected. A more fine-grained set of **security levels** is shown in Figure 8.2. In this case, data can be accessed only with proper authorization for the data, the application that uses the data, the host computer on which the data are stored, and the network to which the host is connected.

Figure 8.2 Access Levels for Information Security

An example of securing access to data could be the university grade-entry application that your instructors use. To give you a grade for the semester, the professor needs to have access to the university's network with a user ID and password. To

---

LEARNING ACTIVITY 8.1

How Protected Is Your Computer?

For many individuals, students and nonstudents, security is the job of information technology (IT) specialists and something they do not feel comfortable handling themselves. However, many studies show that the weakest link in security is these same individuals who fail to perform basic steps in securing their computers, data, and networks. Does this describe you? Let's see how secure your computer is by using a vulnerability test software called Nessus from the company Tenable.

1. Visit https://www.tenable.com/products/nessus and Download Nessus Essentials.
2. Register for a free activation code and download the most recent version of the tool for your operating system (like Nessus-X.XX.X-x64 .msi for Windows 7, 8, or 10 if using a 64-bit PC or Nessus-X.XX.X.dmg if using recent MacOS; the X's represent the version number).
3. After installing the tool and inserting the activation code you received by email, it will download additional files and ask you to start a scan. Find your IP address (you can use a search tool and ask for "my IP address"). Use that address as the target, click on the arrow next to Save, and Launch.
4. Click on the results, look at the vulnerabilities tab, and note any vulnerabilities or noncompliant configurations indicated and bring this information to class for discussion. After this, you can click "Remediate All," and it will attempt to fix the identified issues.

change grades, the instructors are required to connect to a specific application on a specific system, which also requires the use of a user ID and password. In addition, only authorized users (instructors) are able to access the grade-entry application. The student data are protected through an additional level of security, since only certain instructors are allowed to change certain grades for certain students. At many universities, an employee ID is required in addition to the user ID and password when accessing specific information about students. Finally, data can be encrypted at the data level, providing yet another level of security. In this case, the instructor (or her computer) will need to have software to handle the encrypted information and make it legible.

Security Goals

Before we discuss each of the security tools, we need to briefly mention that all tools and policies are meant to address one or more core security goals, which are known as CIA: confidentiality, integrity, and availability.

Confidentiality involves making sure that only authorized individuals can access information or data. **Integrity** involves making sure that data are consistent and complete. For example, as a message is transmitted, its content must not be modified unwillingly during the transmission. Finally, **availability** involves ensuring that systems and/or data are available when they are needed. For systems to be considered highly available, the organization must protect them from disruptions not only due to security threats such as denial-of-service (DOS) attacks but also due to power outages, hardware failures, and system upgrades.

There are two additional goals of security to consider: authentication (or authenticity) and nonrepudiation. **Authentication** is basically making sure that the parties involved are who they say they are and that transactions, data, or communications are genuine. **Nonrepudiation**, which is particularly important in e-commerce, refers to making sure someone cannot renege on his or her obligations—for example, by denying that he or she entered into a transaction with a Web merchant. In e-commerce, digital signatures and encryption are used to ensure that the goal of nonrepudiation is met.

Information Security Threats

Most people think of information security as unauthorized access to individual or organizational data or systems. But it is much broader; it also includes dealing with natural disasters, such as earthquakes and fires, as well as dealing with any threats to computerized systems, such as viruses, hackers, and accidental loss of data or systems. One definition of an information security threat is any event or circumstance with the potential to affect negatively the confidentiality, integrity, or availability of the resources of the organization (including information systems, data, people, and processes). In reality, in today's interconnected world, there are a large number of information **security threats** to individuals and organizations alike.

Examples of threats that affect confidentiality include unauthorized access to systems or data where someone accesses systems and/or data illegally, or theft and fraud. Examples of threats that affect integrity of data or systems also include unauthorized access, for example when someone conducts a password attack to gain access to illegally modify information in a database. Examples of threats that affect availability include **denial-of-service (DOS) threats** that render a system inoperative, limit its capability to operate, or make data unavailable, and also include natural disasters or malicious software with similar consequences. Confidentiality, integrity, and availability are often considered the threat targets, while the methods used to conduct the attack are considered the **threat vectors**. In this section, we discuss threat vectors in the order of confidentiality, integrity, and availability. However, as you will see, each threat vector really can serve multiple purposes.

Threat Vectors for Unauthorized Access

There are many ways information or system confidentiality can be compromised. One of the most prominent is unauthorized access, or illegal access to systems, applications, or data. One of the biggest worries for organizations is theft of data, including customer information, trade secrets, or other important information regarding the organization, its business partners, or its employees.

Hacking. When thinking of unauthorized access, we often imagine a hacker who lives on the margins of society and tries to access an organization's data files to do something good for the world or to prove he can. However, the reality is that organizational insiders often perpetrate unauthorized access; they have the knowledge needed to hack into the systems to obtain information, such as credit card numbers. Some incidences of unauthorized access are difficult to identify because they are passive, such as recording transmissions. By doing so, a hacker can obtain passwords, which can then be used to access information on an organization's systems through a valid account. Another approach is to use a password cracker, which is software used to recover a "lost" password, but it can also be used to conduct a brute-force password attack, which is intended to get unauthorized access to a system or stored password. Some forms of malware, which we discuss later in this chapter, can also be used to perform security breaches. For example, hackers can remotely install a keystroke logger on someone's computer to capture all kinds of information, including credit card numbers and passwords. The keystroke logger simply records every keystroke a user makes on a keyboard in a log file that is accessible by the hacker. Hackers can also install other software for future attacks; this is like having a sleeper agent on one's computer.

The rise of cryptocurrencies has led to a new form of unauthorized access called cryptojacking. The goal of cryptojacking is to hack personal or business computers and mobile devices to install software that will use the device's power and resources to steal cryptocurrencies (also called mining) from unsuspecting victims. There were 41.4 million cryptojacking attacks globally in the first half of 2020 (Sonicwall 2020).

Social engineering and phishing. The threat vectors discussed in the hacking discussion often focus on technology. However, many breaches arise because users

LEARNING ACTIVITY 8.2

Detecting Phishing

There are many ways to detect phishing emails. Look for the following common problems to identify phishing emails:

1. The email is not addressed to you directly.
2. The email is addressed to you using your email account info (without the @provider).
3. The email does not have a personalized salutation (i.e., Dear Bélanger).
4. When you hover the mouse over the hyperlink, the site does not seem to be from the proper company. (Look at the domain name.)
5. When you hover the mouse over the hyperlink, the site seems to be located in another country.
6. The email contains poor grammar and spelling mistakes real companies would not make (we hope).
7. The email makes you feel that your response is urgent or something bad is going to happen.

Read more about phishing at the following sites:

- https://www.consumer.ftc.gov/articles/how-recognize-and-avoid -phishing-scams
- http://www.spamlaws.com/phishing.html

Then test your skills detecting phishing emails here: https://www.sonicwall.com/phishing-iq-test/. You can also identify the problems mentioned above in the sample phishing email in Figure 8.3.

From: customer.center@BankOfAmerica.com
To: belanger@vt.edu
Date: Oct-22–10 2:36 PM
Subject: Online Banking Alert
Your Online Banking Is Blocked! Please Take Immediate Action.

Dear Belanger,

We recently reviewed your account and we suspect that your Bank of America account may have been compromised by an unauthorized person. Protecting the security of your account is our primary concerns. Therefore, as a preventive measure, we have temporarily limited access to important account features.
  To restore your online account access, we need you to confirm your information at the following website within 48 hours of receiving this email. It is very important that you update your information within this time frame if you do not wish your online access to be placed on hold.
  https://www.bankofamerica.com/update/
  We thank you for your rapid attention to this matter.

Figure 8.3 Example of a Phishing Email

have access to someone else's legitimate passwords and accounts. In many cases, individuals are negligent about protecting their passwords. In other cases, individuals can be tricked into giving out this (and other) information. This is referred to as **social engineering**. Examples of social engineering actions include tricking someone into telling you his or her password by claiming to be someone with rights to know it or tricking someone into sending you an important file or opening a link to a dangerous website. Sophisticated social engineering can even involve listening to lunch conversations among unsuspecting professionals with privileged information. Today, most Internet users are familiar with phishing through email messages sent from hackers pretending to be banks or large companies such as eBay.com, PayPal, or Amazon.com. The problem is that for most individuals, it is hard to identify whether such emails are phishing emails when they actually have an account with those companies. Test your abilities in Learning Activity 8.2.

While we discuss hacking and social engineering as attacks on confidentiality, it is important to realize that hacking into a system can also result in threats to integrity and availability. If the hacker actively modifies the system or data being hacked—for example, creating false account information for fraud purposes or changing the content of messages to falsely implicate the company in certain events or to delay negotiations on an important contract—then the integrity of the information and system have been compromised. Similarly, if the hacker breaks into a system and damages the information or applications on the system, then we have an attack on availability. Figure 8.4 shows the differences between passive and

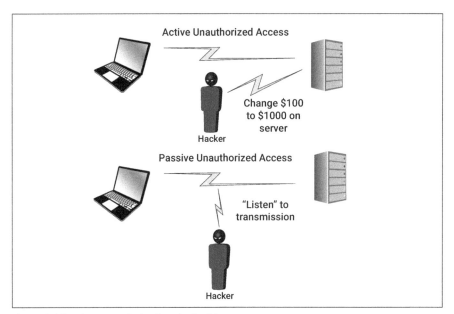

Figure 8.4 Passive versus Active Unauthorized Access

What Do Hackers Break into the Most?

Hackers have been increasingly making the news for stealing information from a number of large companies. For this learning activity, visit https://enterprise.verizon.com/resources/reports/dbir/ and download Verizon's latest Data Breach Investigations Report. As you are reading the report, do the following, and be prepared to discuss in class:

1. List the targets of hackers and how many reported incidents there were.
2. Determine which of these incidents were of the greatest concern to consumers.
3. Come up with various ideas for how you can protect yourself from hackers seeking to gain access to information you have shared with organizations.

active unauthorized access. When listening in to data transmissions, the term *sniffing* is often used.

How is hacking actually performed? It requires a combination of skills and patience. Most of the tools available for hacking are unfortunately easy to find on the Internet, but not everyone has the skills, patience, or intent to perform break-ins. Software, such as **sniffers**, can be used to map information on someone's network and identify weaknesses in computers and networks. When you performed Learning Activity 8.1, the scanning software identified **open ports** on your computer. This means that some specific applications are installed and running on your computer. Once this is known, a hacker can test the targeted system for vulnerabilities (like unpatched software), and if a vulnerability is found, he or she can exploit it. A common type of vulnerability in software is what is also called a *software security hole*. All operating systems and applications have them. The Internet is again a source of information, since vulnerabilities are documented. Software patches (updates) are meant to address these vulnerabilities once they are found. When new software is made publicly available, it usually takes very little time for hackers to identify the security holes. The software developers then have to quickly react by developing fixes, which are made available to users via download on the software developers' website. Unfortunately, many users do not update their operating systems fast enough, or they may even turn off automatic updates.

Finally, hackers often perform break-ins using **spoofing**, which means they pretend to be someone else (or another computer). In an email message, you can easily spoof an address by simply changing the information in the "From" line to reflect the name of the person you want to impersonate. This is not very sophisticated, and many individuals are able to detect such simple spoofing. More advanced spoofing can be done at the IP address level by changing the source (sender's) IP address to look like it comes from a different address. This is done by changing the information in the address field of the IP packet. As discussed in Chapter 7, a **packet** is a

BUSINESS EXAMPLE BOX 8.1

Even Big Technology Companies Can Be Hacked

One would think that large technology companies should have the resources and expertise not to be victims of hackers, right? Well, in September 2017, Equifax, one of the three largest credit bureaus in the United States, reported having been the victim of hackers in March of that year. The cybersecurity breach affected close to 148 million people, mainly in the United States but also in Canada and the United Kingdom, with approximately half of the United States' population affected.

The information that was accessed included names, birth dates, addresses, social security numbers, and even drivers' license information. All of these are considered personally identifiable information. Because of its size and the type of information accessed, many consider this breach to be one of the worst to date in history. An investigation by a House Oversight Committee concluded that the breach was preventable and that Equifax had subpar security practices and policies. This included failing to patch a known vulnerability on a server and failing to renew one of their security certificates for encryption. Hackers sent more than 9,000 queries to Equifax databases and were able to download data 265 separate times. The company informed the public in September but found the breach two months earlier. This is actually considered fairly rapid in terms of disclosing a major breach. Equifax now offers credit file monitoring, credit reports, ID theft insurance, and various scanning services for free to those affected.

Sources: Yurieff, K. 2017. "Equifax Data Breach: What You Need to Know." *CNN Business*, September 10. http://money.cnn.com/2017/09/08/technology/equifax-hack-qa/index.html; Whittaker, Z. 2018. "Equifax Breach Was 'Entirely Preventable' Had It Used Basic Security Measures, Says House Report." *Tech Crunch*, December 10. https://techcrunch.com/2018/12/10/equifax-breach-preventable-house-oversight-report/.

small unit of data that flows through networks, allowing for the transmission of messages.

Web-based attacks. Many recent major data breaches are the results of attackers using web applications to conduct their attacks. A common type of web-based attack is cross-site scripting (XSS), where an attacker is able to upload code to a website that fails to control for malicious input. When users go to the website, the code is executed. This is one of many forms of code injection used by attackers to take advantage of web applications to conduct attacks. One form of Web-based attack that is growing is formjacking. With formjacking, the code that is injected by the cybercriminal is used to control access to the site's form page and collect the user information that is entered there.

Consequences of unauthorized access. Even when a system has been hacked, organizations do not necessarily know that this has happened unless a problem occurs immediately. Some organizations have tools to detect hacking attacks, which we will discuss in the next section of this chapter. Hackers, however, will often start the attack by erasing audit logs that would show what has happened and installing **rootkits**, or software that will allow them to have unfettered access to everything on the system, including adding, deleting, and copying files. Hackers often install **backdoors**, allowing them to access the system again at their will.

Finally, they can install victimization software, such as a **keystroke logger** as discussed before, spyware, attack software for denial-of-service attacks, or other remote tools to control the systems that have been hacked. We discuss most of these in the next subsections.

Why would anyone perform such malicious acts? Some may be disgruntled employees. Others may attack a given target for political reasons—for example, protesting a company's stand on some issue. The groups who perform such actions often use the term **hacktivism**, which involves finding information that, if revealed, will advance human causes. Clearly, many attacks are conducted for financial reasons, either with theft of data or in an attempt to conduct fraud. Finally, espionage can have another purpose, which is attempting to obtain secret information such as new patents an organization is working on. Whatever the purposes, breaches continue to be very damaging to organizations.

Theft and fraud. Beyond theft of data through unauthorized access, companies are also concerned with theft of software or hardware devices that increasingly contain huge amounts of information and theft of data using small hardware devices. Theft of software usually occurs when employees copy legitimate software installed on their company's servers to bring home or to give to someone else. These copies do not carry licenses and are therefore illegal copies. This is a major concern for organizations, and many have started to perform audits of installed software. The best security control for this problem, though, is user education. Many employees do not see a problem with copying software this way, although it is illegal. Some organizations provide only computers that do not allow saving to any external media devices, preventing software from being copied.

As hardware has become more portable and powerful, theft of hardware has become an increasing concern for organizations. Think of how much data can be stored on a laptop, tablet, or even a personal digital device like an iPhone or Android phone. Besides the cost of the hardware itself, organizations have to worry about the information contained on these devices as well as the access that the devices often have to the organization's internal systems. Theft of data is also possible thanks to storage devices with ever-increasing capacities. Today, a small USB drive can contain hundreds of gigabytes of data. Think of a complete list of all of an organization's customers, with addresses and historical purchases; this can fit easily on a USB drive. An individual can install software that automatically copies everything when the USB drive is inserted. There have been reports of such tools being inserted into computers in hotel business centers, in kiosk devices in shopping malls, and others. The hackers then take them back after a while with the information of all persons who used that service center or kiosk. A nonemployee roaming an organization's offices could easily copy tons of files if allowed in with the device and not supervised while there.

Theft and fraud clearly affect the confidentiality of information, in particular when theft of devices is for the data they contain. However, theft and fraud really also affect availability, since the data and or systems (like a laptop) are no longer available to perform work. The biggest threat vectors for availability, however, are those attacks seeking to deny availability of services, systems, or data, which we discuss next.

Threat Vectors for Denial of Service

One of the most prominent types of attacks on availability are denial-of-service attacks. Denial-of-service attacks lead to legitimate users not being able to access a system or data that they should normally have access to. Security threats that can cause denial of service can result from intentional acts, careless behavior, or even natural disasters. While natural disasters such as earthquakes, tornadoes, hurricanes, or fires cannot be prevented and can completely destroy systems or data, a careful security **disaster recovery** plan will include steps to recover from such disasters (backups, duplicate systems, etc.). Careless behavior can include forgetting to perform proper backups of one's computer, not installing security updates to an operating system, or failing to update one's antivirus software. Often, individuals even turn off security features because they find them annoying. Was this the case in Learning Activity 8.1? Did you block some protection features from your computer? No one should underestimate the potential security issues that careless behaviors can generate. What solutions exist to protect against them? Education is the first one, and the second is the automation of security tools and updates so that the user does not have to perform security tasks or is prevented from reducing the effectiveness of the current security of the systems and data. Intentional acts can be performed by organizational insiders or by outsiders. For example, a disgruntled employee can perform vandalism by reformatting a server's main drive or starting a fire in the computer room. Any downtime in computer systems or networks can mean financial losses for companies, including major losses in productivity if employees cannot perform their normal duties.

Denial-of-service attacks. There are many sophisticated tools that can be used to disrupt use of or destroy systems or data. In a flooding attack, the attackers take advantage of a vulnerability in a system to overwhelm the target system with a flood of useless packets or useless network traffic. In a SYN flood attack, the system is tricked into sending numerous synchronization requests that are never answered, flooding the system. In a smurfing attack, the network is flooded with traffic to the target system, which becomes overwhelmed and stops functioning properly or simply crashes. Hackers usually spoof their IP addresses when they launch denial-of-service attacks so that the attacks cannot be traced back to them. In a logic attack, the attacker takes advantage of software vulnerabilities to damage, render ineffective, or crash the target system. This is similar to our discussion of unauthorized access where attackers take advantage of software "holes." If many computers are then used for conducting attacks, we call them **distributed denial-of-service (DDOS) attacks**.

How do DOS and DDOS attacks become possible? Of course, if organizations have failed to patch their systems, attackers can use a variety of sniffing or scanning tools to identify the existence of those vulnerabilities. Other attacks are the result of information that insiders have leaked, often through falling victim to phishing or social engineering attacks. A large number of successful DOS and unauthorized access attacks, however, are possible because the target systems are successfully infected by the attacker with malware.

Malware

Malware refers to malicious software or malicious code. Malware can affect confidentiality, integrity, and availability. There are many forms of malware in existence today. The main ones include traditional viruses and worms, as well as the growing threat of **ransomware**.

Viruses have been around for several decades and continue to be an issue for organizations and individuals. It is likely that at least one person in your class has been a victim of a virus. Viruses are often referred to as **targets of opportunity**, which means they are sent out to find any victim they can. What are viruses? What do they look like? **Viruses** are computer programs designed to perform unwanted functions. Some cause minor harm, such as sending undesirable messages. Others are very destructive, deleting all files on a computer or creating so much traffic on a network that it crashes and cannot be used by its customary users. The lines of code that make up a virus can be embedded into other files, or they can be attached to the initial boot sectors of USB drives (so they automatically infect anyone using that device).

What the virus looks like is called the **virus signature**; it is the particular bit patterns that can be recognized, which is how virus detection software knows your computer has contracted a virus. If viruses are embedded into a legitimate file, they are often called **Trojan horses**. If they can propagate themselves throughout the Internet with no user intervention, they are referred to as **worms**. A **stealth virus** is a more advanced virus that changes its own bit pattern to become undetectable by virus scanners. Some viruses are downloaded when users access certain websites. Browsers today often protect you from downloading what is called **active content** (for example, ActiveX programs). However, when asked, many users simply say yes to downloading any content from a website. Once viruses are installed on a computer, it may be difficult to remove them. (We discuss later how to do this.) Viruses can also modify themselves as they move to other computers, changing their signatures to become less easy to detect. These are called **polymorphic viruses**. A **macro virus** infects documents by inserting commands. You might be familiar with the use of macros in Microsoft Word or Excel, for example. In these cases, the ability to run programs (i.e., macros) also leads to the possibility of having a virus embedded within the documents. There are several other classifications of viruses based on their behavior. Two that have grown in importance in recent years include spyware and ransomware.

Spyware (which we discuss in Chapter 9) is malware that captures everything users do on their computers, unbeknown to them. Ransomware is software that is installed on the victim's computer via viruses, unauthorized access, or phishing (clicking on a link; see Learning Activity 8.2). Once installed, ransomware attacks the computer by encrypting the data on the computer until an appropriate password

> **BUSINESS EXAMPLE BOX 8.2**
>
> ## Costs of Data Breaches
>
> Every year, a number of organizations and institutes conduct surveys to identify the state of cybersecurity. In a recent survey, IBM Security and the Ponemon Institute published their "2020 Cost of a Data Breach Report." They highlighted the growth of cyberattacks in general year-to-year, with ransomware attacks having more than doubled over the previous year. They suggested that the average cost of a data breach globally is $3.86 million, with the United States having the largest average at $8.64 million per data breach. The health care industry had the highest cost per breach in the report, averaging $7.13 million per breach. Of course, not all intrusions are reported because companies do not want to spend the energy on reporting incidents, are afraid of negative publicity or competitors taking advantage of the incidents, or do not believe law enforcement can help or would be interested. In the 2020 report, most respondents believed that the increase in remote work due to the COVID-19 pandemic would result in higher costs per data breach.
>
> The report also identified malicious attacks (intentional attacks like ransomware and other malware, denial of service, web attacks, social engineering, and theft) as the most common and most costly. Companies who are able to identify and contain an attack quickly (i.e., within less than 200 days) can save on average $1 million per breach.
>
> Source: IBM Security. 2020. "2020 Cost of a Data Breach Report." https://www.ibm.com/security/digital-assets/cost-data-breach-report/#/.

is provided. Without the password, the owner of the computer cannot access the data. The hacker who sent the ransomware demands a ransom, generally in the form of **bitcoins**, to provide that password. If the computer owner does not pay the ransom, then he or she cannot access the data on his or her own computer, as it is being held for ransom.

In the first six months of 2020, there were 121.4 million ransomware attacks globally (Sonicwall 2020). It is interesting that hackers demand to be paid in bitcoins. To understand why, it is first necessary to understand what bitcoins are. Bitcoins are cryptocurrency (a currency that utilizes encryption) that relies on the **blockchain** to track the transactions they are involved in. The blockchain provides a secure network where users can conduct transactions anonymously, eliminating the need to have trust between parties. In fact, because the blockchain does not require trust between individuals who use it, it doesn't require individuals to know who each other is. It creates a way to remain anonymous in the execution of transactions. Because bitcoins use this blockchain technology that provides anonymity, criminals feel safe demanding payment this way. Prior to bitcoins, electronically sending payments required banks or financial institutions where the identity of customers was known. Thus, the anonymity provided by bitcoins opened the door for a new way for hackers to make money through computer vulnerabilities. We will discuss bitcoins and blockchain in more detail in Chapter 11.

Other Threats

There are many other information security threats that are not necessarily targeting the CIA (confidentiality, integrity, and availability) of data and systems. We only discuss a few of them here.

Spamming involves sending emails to many individuals at once, often promoting various legitimate or nonlegitimate products. They are problematic for network administrators, as they represent a huge volume of unwanted messages. Of course, many can also include malware. Spimming is a similar issue, but sent via text messaging. There are also **virus hoaxes** on the Internet. These are not viruses but false virus alerts being sent and resent by individuals. Their consequences include either substantial unnecessary network traffic or unwanted actions by users (for example, telling users that certain files must be deleted to avoid the virus). Usually, unsophisticated users who panic at the idea that they might have caught a virus propagate these hoaxes. In case of doubt, the best action is to go to the website of the company that built the targeted application and search for virus hoaxes.

Security Technologies and Solutions

Before managers can evaluate security measures and come up with a proper security plan, they must identify all of the threats to security their organization faces. We presented some of the most common security threats in this chapter. There are, however, other threats to security that can occur, and, most importantly, this is a very dynamic landscape where new threats continue to develop. The good news is that security solutions also continue to evolve.

Preventive, Corrective, and Detective Controls

A large number of security technologies and solutions exist to address the threats we have discussed so far. One way to classify the security solutions and tools (often called *security controls*) is to consider whether they are preventive, detective, or corrective controls. The goal of **preventive controls** is to stop or limit the security threat from happening in the first place. For example, **antivirus software** scans computers and programs and alerts users if potential viruses exist before they can infect files or computers. The goal of **detective controls** is to find or discover where and when security threats occur. For example, **audit logs** can be used to track access to a system to see if multiple false attempts are made from a particular user ID. Finally, the goal of **corrective controls** is to repair damages after a security problem has occurred. An example is the antivirus software previously discussed, which can remove a virus, quarantine the file with the virus, or simply delete the problematic file.

Physical Security versus Logical Security

Security solutions can also be classified as physical or logical. **Physical access controls** are those security solutions that involve protecting physical access to systems, applications, or data, whereas **logical access controls** include security solutions that protect access to systems, applications, or data by ensuring that users or systems are authenticated and allowed to have such access.

Physical access controls can include locks for laptops, locked computer rooms, and secured rooms for backup storage. Once devices are no longer needed, software programs called *drive shredders* should be used to make sure that all discarded disk drives cannot be read again. Physical security also involves ensuring that wiring closets, where all the routers and connections are located, are properly locked and secured. Another important aspect of physical security is selection of personnel. Remember that insider abuse is one of the most significant security threats. Organizations must ensure that only proper personnel have access to key information systems and data. Organizations should run background checks on such personnel and provide appropriate training. Several surveys have identified onsite contractors as a major threat to security for organizations.

Logical access controls use many technologies to require authentication of users and systems trying to access specific applications, networks, data, or computers. These include user profiles, biometrics, firewalls, and encryption.

User Profiles

User profiles are one of the main solutions used to prevent unauthorized access to systems, data, and applications. Users are assigned profiles that consist of a user identification (self-selected or assigned by the organization) and a set of privileges. Once the user confirms his or her identification, he or she can access the system, application, or data with the level of privileges he or she has been provided. For example, faculty can view student data but cannot modify some of those data (such as personal addresses). This is considered the **principle of least privilege**. Give users access to only what they need and no more!

User profiles require that individuals be differentiated from one another using one or several levels of identification: possession, knowledge, or trait.

- *Possession* is when an individual owns a form of identification. For example, your driver's license, your student ID, and your passport are possession forms of identification.
- *Knowledge* is when an individual needs to know something to gain access. Passwords are a good example of knowledge required to access systems. Combining possession and knowledge, such as requiring a personal identification number (PIN) with a banking card, provides more security.
- *Traits* require recognition of physical or behavioral human characteristics, such as a fingerprint or a signature style, to gain access to systems, data, or applications. This is part of biometrics, which will be described in the next section.

User profiles remain one of the most used security solutions for access to systems and data. Yet there are many issues with using user profiles for authentication. First,

LEARNING ACTIVITY 8.4

How Strong Is Your Password?

Passwords are one of the most used security solutions to control access to systems, applications, and data. We have no doubt that you have several passwords you use on a regular basis. But how strong are these passwords? Let's see what a strong password is.

1. Select two passwords you use frequently and test their level of security using one of the following password checkers:
   • https://howsecureismypassword.net/
   • http://www.passwordmeter.com
2. Read the information on how to create a strong password:
   • https://support.google.com/accounts/answer/32040?hl=en
3. Create a strong password. Test it. Do not use it for your accounts!
4. Bring the following to class:
   • The level of security of your two passwords (Be honest, but do not tell the passwords!) Note how long it would take to break them.
   • The strong password you created and what it stands for.

users often use words or numbers that are easy to remember for their passwords. Many systems today will not accept such passwords, which can be easily "broken" with a **dictionary attack**, where all words of several dictionaries in multiple languages are tested as passwords with numbers before and after the words. With this method, it often takes only a few seconds for a password to be discovered. The second problem is that users also write down their passwords on a sticky note taped to their computer or taped to the inside of their desk drawers because they have too many passwords to remember. Alternatively, they use the same password for everything. If someone finds out about it, all systems the user accesses are compromised. Password recovery can be a problem, too, since it is often easy to know the name of a person's pet, family members, school attended, or city of birth, as well as birthday or important numbers. Just think how much of that information can be found on social networking sites like Facebook!

To increase the effectiveness of user profiles, security systems now often require users to change their passwords regularly, such as every one, three, or six months. Many systems will not allow more than a few attempts at entering a password before they block access to the user account. Most systems also require the use of strong passwords. As you discovered in Learning Activity 8.4, strong passwords require a mix of uppercase letters, lowercase letters, alphanumeric characters, and numbers. Tools exist to help manage passwords. For example, **password crackers** are available online for various systems to recover passwords that are forgotten. Of course, they can also be used to crack someone else's password—for example, on a stolen laptop. There are also password generators that can help you generate strong passwords. Even with all these tools, however, most individuals still create poor passwords. SplashData has

TABLE 8.1 Top 20 Worst Passwords in 2020			
Rank	Password	Rank	Password
1	123456	11	1234567
2	123456789	12	abc123
3	password	13	1q2w3e4r5t
4	qwerty	14	q1w2e3r4t5y6
5	12345678	15	iloveyou
6	12345	16	123
7	123123	17	000000
8	111111	18	123321
9	1234	19	1q2w3e4r
10	1234567890	20	qwertyuiop

Source: Wagenseil, P. 2020. "These Are the Latest World's Worst Passwords—Don't Use Any of Them." *Tom's Guide*, July 6. https://www.tomsguide.com/news/worst-passwords-2020.

been analyzing millions of leaked passwords over the years and reporting the most popular each year. Every year, we continue to see that most users still have really weak passwords; the top 20 passwords for 2020 are shown in Table 8.1.

Many security professionals now recommend password managers (previously called wallets) to ensure that users have stronger passwords. Password managers are applications that users can install on their devices (computer or mobile device) to store and organize passwords. The passwords are encrypted, and the user needs to remember only one strong password to have access to the database of his or her other passwords. There is a debate in the field of security whether this is the best approach. Some argue that it finally allows users to make use of strong passwords for all their

TABLE 8.2 Password Manager Examples	
Password Manager	Website
1Password	https://1password.com/
Dashlane	https://www.dashlane.com
Keeper Password Manager & Digital Vault	https://keepersecurity.com/
Password Boss	https://www.passwordboss.com/
Sticky Password	https://www.stickypassword.com/
Free Password Managers (Many also have paid versions.)	
Bit Warden	https://bitwarden.com
LastPass	https://www.lastpass.com
LogMeOnce Password Management Suite	https://www.logmeonce.com

Source: Rubenking, N., and B. Moore. 2020. "The Best Password Managers for 2020." *PC Magazine*, July 1. https://www.pcmag.com/picks/the-best-password-managers.

accounts. Others claim that it renders users vulnerable to having their passwords hacked if the database is hacked (even if encrypted). For example, Windows 10 was bundled with a password manager called Keeper, but in December 2017, Tavis Ormandy reported a bug that would allow someone to break into the password manager (Chirgwin 2017). The security issues were quickly fixed—in fact, within eight days—but this reinforces the debate on whether password managers are the ultimate password security solution. Nevertheless, as password managers grow in popularity, many different solutions are being offered. Table 8.2 shows some password managers recently evaluated by *PC Magazine*. There are also free password managers, with some of the highest ranked ones also shown in Table 8.2. Most free password managers are "toned-down" versions of commercially available ones, but some are also open source.

Two-Factor Authentication

As you can see from the discussion of passwords and password managers, cybersecurity experts are still trying to find the best way to provide security for everyone. One approach that has gained popularity in the past few years is **two-factor authentication** (2FA). With 2FA, you add an extra step to logging into an application or device that goes beyond using your username and password (the first factor of authentication), which is meant to add another layer of security. This extra step (the second factor of authentication) can be an app on a trusted device (such as Google Authenticator, Microsoft Authenticator, or Cisco Duo), a token on a security device users carry with them, a phone number that you trust to receive calls or text messages, or verification codes that are generated for one-time use only.

Many applications you currently use allow for 2FA. For example, Apple uses it for various log-ins with your Apple ID. Google, LinkedIn, Amazon.com, some banks, and most other large organizations from which you access services online use some form of 2FA. For example, the universities of two of this book's authors requires two-factor authentication for access to their systems. It's not even an option; it's mandatory.

Virus Protection

Viruses are one of the most prevalent security threats in organizations. You should not underestimate the amount of money involved in terms of support personnel time involved in fixing virus problems. Fortunately, today antivirus software has become very easy to use and to keep up to date. Recall that a virus has a signature: the bit pattern that represents the virus. Antivirus software looks for these signatures or variations of them in files and systems. If they are found, the software can remove the virus, quarantine the file, or delete the file. The software can even detect viruses in emails and files as they are loaded but before they are opened. Once a new virus is identified, it usually takes fewer than 24 hours for specialists to decode it and write a protection for it. But viruses are easy to write, with several tools for writing viruses available online. Symantec, one of the leading providers of antivirus software, states

LEARNING ACTIVITY 8.5

Biometrics on the PC

For this learning activity, start with the list of biometric solutions provided in Table 8.3 and utilize the Internet to bring the following to class:

1. Identify which of these biometrics could be implemented on your personal computer to provide authentication.
2. Identify additional biometrics that can be used in businesses.
3. Identify additional biometrics that can be used in the context of IoT.

that there are now more than one million computer viruses in existence.

Unfortunately, most antivirus programs are primarily reactive. They only detect existing viruses or ones that look similar to an old virus archived in the software's database. Newer virus protection tools, called **behavioral-based antivirus protection tools**, look for suspicious behaviors in programs instead of just a virus's signature. One of the most important protections, however, remains user education about updating software and not downloading unknown executable files to computers.

Biometrics

Another security method used to authenticate users and control access to systems, data, and applications is **biometrics**, which uses human traits and characteristics to recognize individuals. While they go beyond possession and knowledge, biometrics can be used in conjunction with them to provide strong security—for example, requiring a card with a password and a fingerprint to have access to certain data. **Fingerprint recognition** is a low-cost example of biometric technology. There are several other biometric technologies, which can be classified as **physiological biometrics** (those that use physical traits) or **behavioral biometrics** (those that use behavior). Table 8.3 briefly describes these security solutions.

Law enforcement agencies have used fingerprinting for a long time to identify criminals. Today, many laptops have fingerprinting scanners to restrict access or replace passwords. Iris scanning offers some of the best security because the iris is unique among all individuals. (The iris is the colored portion of the eye.) Immigration services in the United States use fingerprinting for noncitizens and iris scanning for frequent travelers who have special cards identifying them as trusted travelers. The English government in the United Kingdom has a large database of earprints because it is said that no two ears are the same. In recent years, we have seen a significant growth in the use of biometrics with mobile devices. Most recent smartphones allow you to use a fingerprint in lieu of a passcode to turn on your phone. Several devices now even support use of an iris reader as well as **facial recognition**. For example, BioID (mobile.bioid.com) offers authentication on your smartphone via facial recognition; you can log in to applications with "face log-in."

Biometrics are used in one of two ways: authentication or identification. In **authentication**, also called *one-to-one matching*, the goal is to match the individual with his or her stored biometric data. Fingerprint biometrics on laptops are good example of that. The recent mobile biometrics just discussed also fit in that category. In **identification,** also called *one-to-many matching,* the goal is to identify an individual from an entire population of individuals with stored biometric data. An example

TABLE 8.3 Sample Biometric Solutions	
Physiological Biometrics	**Type**
Fingerprint recognition	Analyzes ridges and valleys (minutiae) of human fingertips.
Facial recognition	Analyzes facial features or patterns (faceprints).
Finger geometry recognition	Analyzes 3D geometry of the finger.
Hand geometry recognition	Analyzes geometric features of the hand, such as length of fingers and width of the hand.
Iris/retina recognition	Analyzes features (eyeprints) in the iris or the patterns of veins in the back of the eye (retina).
DNA recognition	Analyzes segments from an individual's DNA.
Odor recognition	Analyzes an individual's odor to determine identity.
Voice recognition (speaker recognition)	Analyzes voice to determine the identity of a speaker (who is speaking); different from speech recognition (what is being said).
Ear recognition	Analyzes the shape of the ear.
Behavioral Biometrics	**Description**
Signature recognition	Analyzes the signature. *Static signature* recognition compares scanned or ink signatures with stored signatures; *dynamic signature* recognition analyzes not only the signature but how it is written using pressure points—how much an individual pushes on the pen when he or she writes his or her name.
Gait recognition	Analyzes the walking style or gait of individuals.
Keystroke recognition	Analyzes rhythm and patterns of keystroke of individuals on keyboards.

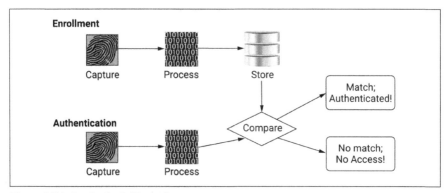

Figure 8.5 Authentication in Biometrics

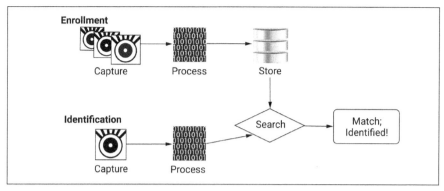

Figure 8.6 Identification in Biometrics

is facial recognition at airports used to identify potential terrorists entering a country from a database of facial prints stored by law enforcement agencies. The processes of authentication and identification are shown in Figures 8.5 and 8.6.

Firewalls

At the beginning of the chapter, you tested your computer security using an online tool. The software was looking for whether your computer was letting out information about itself and whether there were open ports or software applications running. The best protection against these potential problems is a firewall, and if your computer's firewall was running and up-to-date, your computer should have indicated that your computer was "stealth" (not visible to outsiders).

A **firewall** is a computer or a router that controls access in and out of the organization's networks, applications, and computers. The term *firewall* comes from the construction industry, where it represents a wall built with fire-resistant materials. If a fire starts in one room of a building, it won't spread to other rooms if firewalls are used. In computer security, firewalls control the transmissions that are attempting to enter and/or leave the organization's networks or computers, as shown in Figure 8.7.

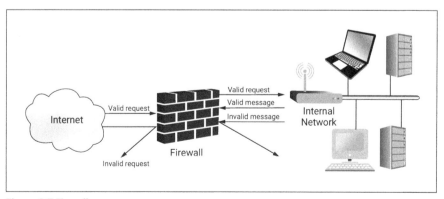

Figure 8.7 Firewall

TABLE 8.4 Firewall Types and Terminology

Firewall type	Description	Features
Packet-level firewall (includes screening router firewalls)	A firewall that works at the network level, looking at information within the data packets (see Appendix H) to see if it follows the set of rules. For example, if the source IP address of the packet is from an "acceptable" computer, the firewall will let the message through, assuming that the destination address is also a valid internal IP address.	Easy to implement and prevents most basic attacks, such as flooding attacks and other denials of service. Sophisticated attacks could bypass these. Packet filtering can be static or dynamic. Static filtering uses predetermined rules to handle transmission requests. Dynamic filtering is able to handle emergent events and update rules as needed.
Stateful inspection firewall	A firewall that examines the current connections between the internal and external systems to make decisions on whether to allow traffic through.	More advanced but requires more processing power.
Application-layer firewall (One form is called a *proxy server*.)	A firewall that works at the application layer to verify access to applications. Users must log into the firewall before they can access applications inside the organization (from outside).	More complex to install and manage. Slows down communication but provides better security than a packet-level firewall.
Firewall deployment	**Description**	**Features**
Screened subnet firewall	A firewall used between departments or divisions inside the organization.	Organizations can use these to isolate departmental data and provide greater defense and in-depth protection.
Border firewall	A firewall used to protect access to the internal network and computers of the organization.	This is the type of firewall usually referred to in most definitions.
Host (personal) firewall	A firewall that is typically in the form of software installed on a computer or server.	Good option for all computers in addition to others to provide defense in depth.

There are several types of firewalls, which vary in how they control access into and out of the organization. Table 8.4 lists various terms used to describe firewalls.

Firewalls are great security tools, but they cannot handle all security threats. For example, they cannot protect an organization from a virus that is attached to a message with an appropriate destination and source address or that is introduced by an individual on a portable storage device. Firewalls also cannot prevent hackers from exploiting an unsecured computer; keep disgruntled employees from copying sensitive information; or stop employees from losing sensitive data loaded on an unse-

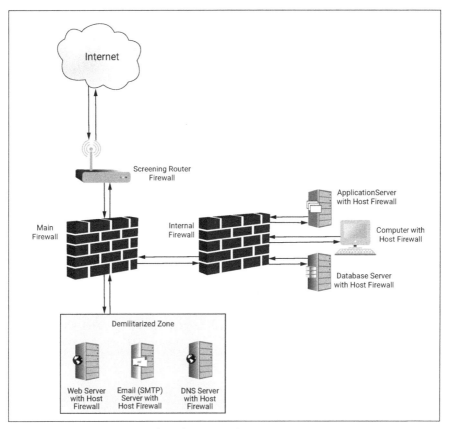

Figure 8.8 A Firewall Architecture for Defense in Depth

cured laptop, thumb drive, or smartphone, for example. Nevertheless, firewalls are part of the main line of defense for organizations. Consistent with the principle of defense in depth, different firewalls should be implemented at different locations in the organization, as shown in Figure 8.8.

Rendering Data Unreadable: Encryption

Encryption, also known as *cryptography,* is the use of mathematical algorithms to convert a message or data into information that is scrambled to make it unreadable. The original message is called **plaintext**. The converted or unreadable text is called **ciphertext**. When the ciphertext is converted back to plaintext, this is referred to as **decryption**. The algorithm used to convert the message is called the **cipher**. The **encryption or decryption key** is information known only by the proper users of the encryption tool.

The ciphers used in Learning Activity 8.6 are very simplistic. Ciphers used in encryption are much more complicated and use complex algorithms (not simply adding

---

STATS BOX 8.1

Security Policy Violations

Even when security is a top priority for organizations, employees can often find ways to not follow or to violate security policies, such as by encrypting email or bypassing security controls such as firewalls, restricted access, or automatic log-off systems. In fact, 20% of employees do not always follow the security policy of their organization. These noncompliant employees often say it is because it is inconvenient to follow policies like strict passwords or it is unrealistic to manage all devices they use at work. Also, they are three times more likely to be millennials and Gen Z employees than over the age of 56.

Source: Anonymous. 2020. "20% of Workers Don't Follow Company Security Policies All the Time." *Security Magazine*, August 5. https://www.securitymagazine.com/articles/92992-of-workers-dont-fol low-company-security-policies-all-the-time.

---

numbers and letters). The strength of an encryption technique is related to the length of the key. This is usually expressed in the number of bits the key has. The larger the size of the key, the more secure the encryption. This is because larger keys are harder to break, since there are more combinations of bits possible. Today, keys in commercial use are 256 bits or longer. It is estimated that with advances in technology, a 64-bit key can be broken in a day or less. Understandably, a 256-bit, or even 1024-bit, key is much more secure.

Types of Cryptography

There are two main types of cryptographic systems used today: asymmetric or symmetric, as shown in Figure 8.9. The systems are based on whether the same key is issued to encrypt and decrypt the data. In **asymmetric encryption**, two keys are used. The public key is used to encrypt messages. It is sent to any person or system with whom one wishes to exchange encrypted messages. Using the public key, anyone can encrypt messages for the intended recipient, who will then use his or her private key to decrypt those messages. The public key and the private key are linked (forming a **key pair**), but only the recipient has the private key. This is also called *public key cryptography,* since one of the keys can be shared with anyone (public).

In **symmetric encryption**, the same key is used to encrypt and decrypt data. In this case, individuals have to be very careful with whom they share their encryption keys because these individuals or companies will also be able to decrypt their messages. It is very efficient, but it requires that both parties know the key and no one else. The problem is therefore how to distribute the key without having it intercepted by others. This is why both types of encryption are often used together, with public key encryption used to distribute a secret key to a party with whom you wish to interact. Once the secret key is received, it can be used for symmetric encryption, since both parties know it. Finally, it is important to realize that encryption can be used for more than transmission of data. It can be used to protect data on servers and

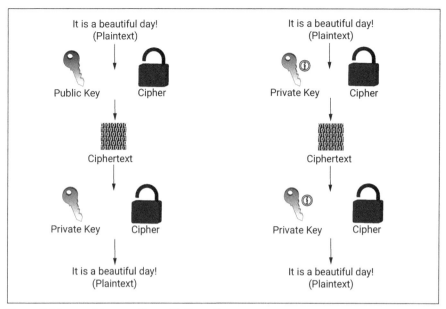

Figure 8.9 Asymmetric versus Symmetric Encryption

computers. For example, you could encrypt your data on your personal computer so that no one could access it without the proper key.

Encryption has many applications for securing data and transmissions. One of them is the **virtual private network (VPN)**. A VPN is a connection that makes use of an open wired network such as the Internet but that provides a secured channel through encryption and other security features. It is often used for employees to securely connect to their organizations from remote locations. VPNs function through the use of encryption. When the employee wishes to connect to the company's internal network, he or she requests a VPN connection. After inputting secret information, the employee's computer establishes an encrypted session with the company's server. (This is called a *handshake*.) Once both computers share the secret key (which is only for that session), all transmissions are encrypted until the VPN connection is dropped.

Web and Credit Card Security

Encryption is also used for protecting information in business transactions online. When you shop online and have to enter your credit card information, you typically look for the lock icon or the https:// in the URL to indicate that the site is secure. If you double-click on the icon, you will see the security certificate that was issued to the website by a certificate-issuing organization such as VeriSign. Figure 8.10 shows a sample certificate located in the Internet options of a browser application. This certificate includes a public key, which is sent to your browser when you connect to

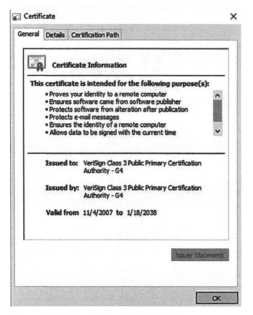

Figure 8.10 Sample Security Certificate

this secure Web page. The browser can then generate a secret key and send it to the server using the public key encryption, since only the server can decrypt this information (via the private key). Once the server receives the secret key, your browser and the server can communicate securely using this secret key (called a *session key*). The transmissions will be encrypted until the current session is dropped.

One major change in electronic payments has been the rise of credit cards with chip technology. Europay, MasterCard, and Visa (EMV) have worked together to develop chip card standards and technology, which involve using chips instead of the old magnetic stripes. The

---

LEARNING ACTIVITY 8.6

Breaking the Encryption

Security technologies are unlikely to be able to prevent all unauthorized access to data, applications, and systems. One way to provide additional protection for important data is through encryption. Before we discuss encryption in depth, let us explore what makes data unreadable.

Part 1
Given the plaintext a b c d e f g h I j k l m n o p q r s t u v w x y z and the resulting ciphertext b c d e f g h I j k l m n o p q r s t u v w x y z a, what is the cipher?

Part 2
Using the cipher, convert the words "computer lab closed."

Part 3
Using the following cipher and your personally selected keys below, convert the text "Fun for Spring Break."
   Key selection:
   1. Pick a number from 1 to 5.
   2. Pick a sign (+ or −).
   3. Pick a symbol (!,@,#,$,%,^,&,*).
   4. Use this cipher to encode: new letter = original letter (sign selected) number, spaces replaced with symbol selected.
   5. Compare your results with classmates.

cards can be used via either contact chips (inserted into a chip reader) or contact-less technology (also often called "tap-and-go" credit cards). The technology allows data to be stored on the chip embedded in the credit card. It is using what is called **Near Field Communication** (NFC). The data are then secured with both hardware and software security measures. For example, a verification method can be programmed into the card itself, allowing accept or decline decisions to be more dynamic.

Wireless Security

Because more and more individuals are connecting their devices to the Internet and connecting to their organizations via wireless networks, we need to briefly discuss the additional issues related to wireless security. In reality, all the security solutions discussed so far can be used in the wireless environment. However, because the wireless networks typically broadcast their communications over the "airwaves," they are more susceptible to security breaches, since they can be intercepted without physical connections to the devices.

BUSINESS EXAMPLE BOX 8.3

Expert Advice on Protecting Your Home Wireless Network

There was a time when the home network was just for that, home computing, although there were some individuals who performed remote work (i.e., telework or telecommuting) as well as entrepreneurs who used their homes as their offices. Today, however, there is huge growth in remote work, due to both the increased availability of broadband connections at home and the need for remote work (e.g., as it occurred with the 2020 COVID-19 pandemic). As a result, there is also an increased need to improve homes' wireless network security. But how do you do that? Of course, the use of encryption via WPA2 is a good start. But what else? Experts suggest 12 main steps. The first six are summarized below.

1. Change the name of your SSID (Service Set Identifier; home network name) because the default name can tell hackers what type of router you have.
2. Set a strong and unique password for the network because the default password for your router could be known to hackers.
3. Use WPA2.
4. Turn off the router when you are not at home.
5. Place your router toward the middle of your home so it does not reach outside as much.
6. Change your administrator password for your router to a strong password. The default passwords are often "admin" or "password."

Additional steps include changing the default IP address on your router, turning off the Dynamic Host Configuration Protocol (DHCP), keeping the router software up-to-date, installing a firewall, and protecting the devices themselves on your network.

Source: Rijnetu, I. 2019. "12 Steps to Maximize Your Home Wireless Network Security." *Heimdal Security*, April 18. https://heimdalsecurity.com/blog/home-wireless-network-security/.

LEARNING ACTIVITY 8.7

IoT Security

For this activity, watch the following video about IoT security and refer to the security threats and the overall security solutions presented in this chapter: https://www.youtube.com/watch?v=IQkRscixagM.
   In small groups:

1. Identify all of the IoT security threats shown in the video.
2. Identify security solutions that can be used for each of these IoT threats.
3. Rank the security solutions in order of importance.

Note that many of the solutions discussed in this chapter can be combined to provide proper IoT security.

The best protection for wireless networks remains encryption. The standards for wireless encryption are numerous, and the most popular ones in America are called WEP and WPA. **WEP, Wired Equivalent Privacy**, is an older encryption algorithm, which today can be easily cracked within minutes (with proper knowledge and software tools). **WPA, Wi-Fi Protected Access**, replaced WEP with a more powerful encryption algorithm. Later, this was replaced by WPA2, which provides an even more powerful encryption algorithm and is widely available in most routers. It uses a 256-bit key for encryption. So when setting up a home wireless network, users should select WPA2 encryption if their devices support it. A further protection for home wireless networks is to disable the broadcasting of the network's ID (SSID). Only with knowledge of the network's name can someone find it and connect to it (with the WPA2, WPA, or WEP encryption key information). Other ideas are shown in Business Example Box 8.3.

Security Policies

Most medium to large organizations today have security policies that describe *what* the general security guidelines are for the organization. **Security policies** tend to be for internal use and include a number of **security procedures**, which are specific statements describing *how* to implement the security policies. For example, a security policy could be "All users must change their passwords every two months." One of its related security procedures could then describe steps to be taken to change one's password. Another procedure could involve an automated system to force users to change their passwords every two months, while an additional one could include actions that should happen if a user attempts to enter an unacceptable (not strong) password. A security policy should have clear goals and objectives, a detailed list of security policies and procedures, and a list of actions for the enforcement of procedures.

Even with the best policies and procedures in place and with the best security technologies available, an organization may not be well protected if its users and employees fail to follow the policies and procedures or fail to use the technologies properly. This is where security education and training become very important. Users

LEARNING ACTIVITY 8.8

Where Is the Security?

Now that we have explored all the security threats and solutions, we need to see how they fit together. For this learning activity, you need to map the threats to security and the goals of information security to the security solutions available. Prepare a table that contains the following information:

1. Column 1: List all the main security threats discussed.
2. Column 2: List all the security solutions that can address each of the security threats you have identified. (There might be many solutions for each threat.)
3. Column 3: For each security solution in column 2, indicate which security goal is met by the security solution.

must be made aware of the risks involved, the tools available, and the consequences of not performing the proper security behaviors that are asked of them. Refer to the book's website (https://www.prospectpressvt.com/textbooks/belanger-information -systems-for-business) for several tips on how to secure your computer, network, and data properly.

Chapter Summary

In this chapter, we discussed the security of information. We first defined information security and described the threats to information security. We then discussed the technologies and solutions related to information security and concluded the chapter with a discussion of information security policies.

Here are the main points discussed in the chapter:

- There are three main goals that security tools and policies are meant to address: confidentiality, integrity, and availability (CIA). Confidentiality involves making sure that only authorized individuals can access information or data. Integrity involves making sure that the data are consistent and complete. Availability involves ensuring that systems and/or data are available when they are needed. Two additional goals are authentication (or authenticity) and nonrepudiation. Authentication means making sure that the parties involved are who they say they are and that transactions, data, or communications are genuine. Nonrepudiation refers to making sure someone cannot renege on his or her obligations—for example, by denying that he or she entered into a transaction with a Web merchant.
- There are several categories of information security threats, which target one or more of the important goals of information security: confidentiality, integrity, and availability. Unauthorized access affects all three goals. Unauthorized access threats refer to someone accessing systems and/or

data illegally. Examples of methods used to gain such access include hacking, password attacks, web-based attacks, social engineering, and phishing. A form of unauthorized access, theft and fraud, can also lead to lack of availability and confidentiality. Denial-of-service (DOS) threats mainly affect availability. They are threats that render a system inoperative, limit its capability to operate, or make data unavailable. They can be caused by intentional acts, such as malware and computer attacks; careless behavior, such as a lack of backups; and natural disasters. IoT also leads to increased security threats, since with interconnectivity come more opportunities for all of the threats discussed in the chapter to occur.

• Security solutions and tools (also called security controls) can be classified as preventive, detective, or corrective controls. The goal of preventive controls is to stop or limit the security threat from happening in the first place. The goal of detective controls is to find or discover where and when security threats occurred. The goal of corrective controls is to repair damages after a security problem has occurred. Security controls can also be classified as physical or logical. Physical access controls are those security solutions that involve protecting physical access to systems, applications, or data, whereas logical access controls include security solutions that protect access to systems, applications, or data by ensuring users or systems are authenticated and allowed to have such access. Physical access controls can include locks for laptops, locked computer rooms, and secured rooms for backup storage. Logical access controls include user profiles, biometrics, firewalls, encryption, virus protection, and wireless security.

• There are many security solutions. The security solution of user profiles requires users to be assigned profiles that consist of a user identification (self-selected or assigned by the organization) and a set of privileges. Two-factor authentication requires an additional step before individuals have access to their devices and applications. Biometrics use human traits and characteristics to recognize individuals to grant them access or to identify them among other individuals. Firewalls are computers or routers that control access in and out of the organization's networks, applications, and computers. Encryption, also known as cryptography, is the use of mathematical algorithms to convert a message or information into a scrambled message or information that makes it unreadable. Antivirus software looks for virus signatures (patterns) in files and systems to prevent viruses from being executed. Several actions can help with wireless security, but the main protection is through encryption, such as WEP, WPA, and WPA2.

• Security policies describe what the general security guidelines are for an organization, including procedures, enforcement mechanisms, and objectives, as well as listing actions for the enforcement of procedures.

- Risk management is the process of identifying, assessing, and prioritizing the security risks an organization may face and deciding whether to accept the risks, mitigate the risks, or share the risks by buying insurance.

Review Questions

1. Briefly explain the three main goals of security and the additional two discussed in the text.
2. What is a denial-of-service attack, and what kind of events can lead to such an attack?
3. Explain how unauthorized access threats can occur.
4. Define preventive security controls, detective security controls, and corrective security controls. Provide two examples of each type.
5. What are the various purposes of biometrics, and how do they work?
6. What is a firewall, and how does it work?
7. What are the two types of encryption discussed in the chapter, and how do they differ?
8. How does antivirus software work?
9. What are the key steps required for wireless security?
10. What are security policies and procedures, and how are they related?

Reflection Questions

1. What is the most important thing you learned in this chapter? Why is it important?
2. What topics are unclear? What about them is unclear?
3. For each of the following security threats, suggest one preventive, one detective, and one corrective security control.
   - A virus is affecting the organization's intranet.
   - A hacker accesses consumer data on the company's database server to commit fraud.
   - A disgruntled employee steals trade secrets from the company's computers.
   - A construction worker breaks the data communication and electrical cables coming into the company's building.
4. Which biometric technologies are better used for identification and which are better used for verification? Explain why.
5. Explain what happens when you request your antivirus software to scan your hard drive.
6. A friend wants to send you encrypted messages. She asks you to send her both your public and private keys. What should your response be? Why? What should you send her? How will you send it? Does it matter?
7. List at least three security threats against which a firewall will not protect a company. Explain why.

8. Is a virtual private network a network or a security tool? Explain.
9. What do you believe are the most serious security concerns for the different types of networking architectures discussed in Chapter 7?
10. If an organization decides to accept a security risk instead of trying to mitigate it, would this still be considered risk management? Why or why not?

Additional Learning Activities

8.A1. Identify all the security threats individuals can face when using social networking sites such as Facebook, LinkedIn, Instagram, and others. Create a table that lists the threats, your estimate of how serious each threat is (low, medium, high), and a suggestion for how such sites should go about addressing these security threats.

8.A2. Perform research on encryption software that individuals can use for emails, their hard drives, or even mobile devices. Create a table that lists the various features of each major vendor and the costs for individual users like you. Then select one of the products and try it. Be prepared to discuss the advantages and disadvantages of the product you selected, both for you as a student and for businesses that might want to use it.

8.A3. Read Appendix I, which discusses how to implement a security plan for an organization. Select a local small business that has at least 10 employees and develop a security plan for this business.

8.A4. Find your university's security policy and identify the main components of this policy. Be prepared to discuss the following:
    a. What threats are referred to?
    b. What security technologies are included in the policy?
    c. Select one example each of a specific policy and a specific procedure.
    d. Who is responsible for ensuring that you, as a student, follow the policy?
    e. What happens if you do not follow the policy?

8.A5. Compare and contrast the security policy you found in Learning Activity 8.A4 with the templates for security policies provided by the SANS organization: http://www.sans.org/security-resources/policies/.

8.A6. Read the article by A. S. Horowitz, "Top 10 Security Mistakes," *Computerworld*, July 9, 2001, http://www.computerworld.com/article/2582953/security0/top-10--security--mistakes.html. Briefly explain each of the 10 mistakes. Then answer the following questions:
    a. Where do we stand today on these security mistakes?
    b. Why have we not improved more?
    c. What is the possible role of technology in handling these "mistakes"?

8.A7. Research the ransomware Wannacry. Prepare a report of its history, its main victims, and what experts suggest companies do to avoid becoming a target of such ransomware.

8.A8. This activity requires you to complete a case analysis regarding security breaches. The case may be accessed through your university website, the Ivey Business School business case website (https://www.iveycases .com), or the Harvard Business School cases website (https://hbsp .harvard.edu/home/), or it may be provided to you by your instructor. Dubé, L. 2016. "Autopsy of a Data Breach: The Target Case." *International Journal of Case Studies in Management* 14(1). Product number: HEC130.

8.A9. As computing technology changes, so do the threats to the security of information. Research the current state of threats, including spear-phishing and advanced persistent threats. Create a table with four columns. The first column should contain the name of the new threat. The second column should contain a definition of this new threat. The third column should identify how this threat builds on existing threats discussed in this chapter. Finally, column four should list what makes this threat different from existing threats discussed in this chapter. Be prepared to discuss these findings in class.

8.A10. In Chapter 7, you learned about Bring Your Own Device (BYOD). There are many benefits and drawbacks to BYOD for companies. One of the drawbacks often mentioned is security. Explore more in depth the security issues that have arisen to date with BYOD for companies and prepare a short report.

8.A11. This activity requires you to complete a case analysis regarding cyberespionage. The case may be accessed through your university library or may be provided by your instructor. Sipior, J. C., D. R. Lombardi, C. A. Rusinko, and S. Dannemiller. 2020. "Cyberespionage Goes Mobile: FastTrans Company Attacked." *Communications of the Association for Information Systems* 46. https://doi.org/10.17705/ 1CAIS.04614.

References

Chirgwin, R. 2017. "Windows 10 Bundles a Briefly Vulnerable Password Manager." *The Register,* December 18. https://www.theregister.co.uk/2017/12/18/ windows_10_bundles_vuln/.

Sonicwall. 2020. "2020 Sonicwall Cyber Threat Report." *Sonicwall.com*. https:// www.sonicwall.com/2020-cyber-threat-report/.

Glossary

Active content: Executable files on websites.

Antivirus software: Programs that look for virus signatures or variations of them in files and systems.

Application-level firewall: A firewall that verifies access to applications by requiring users to log into the firewall before they can access applications inside the organization (from outside).

Asymmetrical security warfare: One party must do everything to protect itself, while the other party (the attacker) only needs to find one security weakness.

Asymmetric encryption: A type of encryption that uses two keys: a public key for encrypting and a private key for decrypting.

Audit logs: Software programs that can scan for unexpected actions to detect potential hackers.

Authentication: A process by which the identity of a transacting party is verified.

Authentication (biometrics): A type of security that matches the individual with his or her stored biometric data.

Availability: System and/or data are available when needed.

Backdoors: Ways for hackers to reaccess the compromised system at will.

Behavioral-based antivirus protection tools: Programs that look for suspicious behaviors in programs instead of just a virus's signature.

Behavioral biometrics: Biometrics that use human behaviors.

Biometrics: Technologies that use human features to recognize individuals and grant them access.

Bitcoins: A cryptocurrency or currency that utilizes encryption based on the blockchain technology.

Blockchain: A network where transactions can be tracked but do not require trust between the transacting parties, thus allowing anonymity.

Border firewall: A firewall used to protect access to the internal network and computers of the organization from outside attacks.

Cipher: An algorithm used to encrypt and decrypt plaintext.

Ciphertext: An encrypted (unreadable) message.

Confidentiality: Making sure that only authorized individuals can access information or data.

Corrective controls: Controls meant to repair damages after a security problem has occurred.

Cybersecurity: Protection of everything that is in electronic form, including information, devices, and networks.

Decryption: Converting ciphertext back to plaintext.

Decryption key: A key used to convert the unreadable text into its original form.

Defense in depth: Multiple layers of security protections in place.

Denial-of-service (DOS) threats: Threats that render a system inoperative or limit its capability to operate or make data unavailable.

Detective controls: Controls meant to find or discover where and when security threats occur.

Dictionary attack: When all words of several dictionaries in multiple languages are tested as passwords with numbers before and after the words.

Disaster recovery: Procedures and tools to recover systems affected by disasters and destruction.

Distributed denial-of-service (DDOS) attacks: When many computers are being used for DOS attacks.

DNA recognition: A type of security that analyzes segments from an individual's DNA.

Dynamic signature recognition: A type of security that analyzes not only the signature but also how it is written using pressure points.

Encryption: The application of a mathematical algorithm to a message or information that scrambles that message or information to make it unreadable.

Encryption key: A key used to convert the text into an unreadable form.

Facial recognition: A type of security that analyzes facial features or patterns (face-prints).

Finger geometry recognition: A type of security that analyzes the 3D geometry of the finger.

Fingerprint recognition: A type of security that analyzes ridges and valleys (minutiae) of human fingertips.

Firewall: A computer or a router that controls access in and out of the organization's networks, applications, and computers.

Hacktivism: A type of hacking in which hackers try to find information that, if revealed, will advance human causes.

Host firewall: A firewall installed on a computer.

Identification (biometrics): Identifying an individual from an entire population of individuals with stored biometric data.

Information security: A set of protections put in place to safeguard information systems and/or data from security threats, such as unauthorized access, use, disclosure, disruption, modification, or destruction.

Integrity: When data are consistent and complete.

Key pair: A linked public key and private key.

Keystroke logger: Software or hardware that records every keystroke a user makes on a keyboard in a log file.

Knowledge (in security access): When an individual needs to know something to gain access.

Logical access controls: Security solutions that protect access to systems, applications, or data by ensuring users or systems are authenticated and allowed to have such access.

Macro virus: Infects documents by inserting commands.

Malware, which means malicious software, is a broad term to refer to all harmful computer programs used by hackers, such as viruses, worms, ransomware and others.

Near Field Communication (NFC): A low-speed connection standard for electronic devices over a distance of 4 cm (1.5 inches) or less.

Nonrepudiation: Making sure a party cannot renege on obligations—for example, by denying that he or she entered into a transaction with a Web merchant.

Open ports: In the context of telecommunications, it means a port (specific application) configured to accept traffic.

Packet-level firewall: A firewall that controls access by looking at the source and destination addresses in data packets. Also called *a screening level firewall.*

Packets: Small units of data that flow through networks, allowing for the transmission of messages.

Password crackers: Software used to recover forgotten passwords.

Personal firewall: A firewall installed on a given personal computer.

Physical access controls: Controls that involve protecting physical access to systems, applications, or data.

Physiological biometrics: Using physical traits to identify individuals.

Plaintext: Original message before it is encrypted.

Polymorphic viruses: A virus that modifies itself each time it infects a computer to avoid detection.

Possession: When an individual owns a form of identification.

Preventive controls: Controls meant to stop or limit the security threat from happening in the first place.

Principle of least privilege: A form of access control in which users are only given permission to access the minimum amount of information required to complete their jobs.

Ransomware: Software that is installed on the victim's computer via viruses, unauthorized access, or phishing that attacks the computer by encrypting the data on the computer until an appropriate password is provided.

Risk management: The process of identifying, assessing, and prioritizing the security risks an organization may face and deciding whether to accept, mitigate, or share the security risks.

Rootkits: Software that allows hackers to have unfettered access to everything on the system, including adding, deleting, and copying files.

Screening subnet firewall: A firewall used inside an organization, between departments or divisions.

Security: Protection against security threats.

Security levels: The layers of protection technologies and policies used to secure stored information.

Security policies: Statements describing what the general security guidelines are for an organization.

Security procedures: Specific statements describing how to implement the security policies.

Security threats: Events or circumstances that can negatively impact the confidentiality, integrity, or availability of information or systems.

Sniffer: Software that monitors transmissions, capturing unauthorized data of interest.

Social engineering: Tricking individuals into giving out security information. Also called *phishing*.

Spamming: Sending emails to many individuals at once, sending unsolicited commercial email to individuals, or targeting one individual computer or network and sending thousands of messages to it.

Spoofing: Pretending to be someone else (or another computer) to enter a system or gain attention.

Spyware: A form of virus that logs everything users do on their computers, unbeknown to them.

Static signature recognition: A security measure that compares scanned or ink signatures with stored signatures.

Stealth virus: A more advanced virus that changes its own bit pattern to become undetectable by virus scanners.

Symmetric encryption: A form of encryption where the same key is used for encrypting and decrypting data.

Theft and fraud threats: Threats related to the loss of systems or data due to theft or fraudulent activities.

Threat vector: A method used to conduct an attack on confidentiality, integrity, or availability.

Traits: Physical or behavioral human characteristics needed to gain access to systems or data.

Trojan horses: Viruses embedded into a legitimate file.

Two-factor authentication: Also known as 2FA, this requires two layers of authentication, with the first typically being a username and password, followed by a second factor such as an app on a trusted device, a security token, calls or text messages on a trusted phone, or one-time verification codes.

Unauthorized access threats: Individuals who access systems and/or data illegally.

User profile: Assigned profiles that consist of a user identification, a password, and a set of privileges.

Virtual private network (VPN): A connection that makes use of an open wired network such as the Internet but that provides a secured channel through encryption and other security features.

Virus: A computer program designed to perform unwanted events.

Virus hoaxes: False virus alerts sent and resent by individuals.

Virus signature: Bit patterns of the virus that can be recognized.

Wi-Fi Protected Access (WPA): A recent and powerful encryption algorithm for wireless security. It has a newer version called WPA2.

Wired Equivalent Privacy (WEP): A protocol that uses encryption established via a pre-shared key known to the router and the wireless device or computer on a wireless network.

Worms: Viruses that propagate themselves throughout networks with no user intervention required.

Chapter 9

Protecting the Confidentiality and Privacy of Information

Learning Objectives

By reading and completing the activities in this chapter, you will be able to:

- Understand the various threats to information privacy
- Identify technologies and solutions used to protect the confidentiality and privacy of information
- Explain how information privacy is a component of the PAPA ethical framework
- Discuss the relationship between information privacy and information security

Chapter Outline

Focusing Story: Perceptions of Anonymity
Introduction and Definitions
Information Privacy Threats
Data Collection
    Learning Activity 9.1: Privacy and Smartphones
    Learning Activity 9.2: Concern for Information Privacy
    Learning Activity 9.3: Privacy Pizza
Consequences of Privacy Violations
Technologies and Solutions for Information Privacy
Anonymous Browsing
    Learning Activity 9.4: Surveillance Societies around the World
    Learning Activity 9.5: Privacy Policy Creation
Government Information Privacy Regulations
Mobile Information Privacy
    Learning Activity 9.6: Why Your Adviser Cannot Talk to Your Parents
    Learning Activity 9.7: PAPA, Privacy Policies, and FERPA
IoT and Privacy
Privacy and Ethics
Ethical Decisions
    Learning Activity 9.8: Making Ethical Decisions
Relationship between Security and Privacy

Perceptions of Anonymity

Internet service providers and email services can and do track a lot of what you say and do online. The big question is whether they use that information or not. Remember those times you are asked to click "I Agree" to access some services online? What are you exactly agreeing to? Even when your personal information is removed before searching or browsing, it is sold to a third party; it may not be as "anonymous" as you think. Moreover, what about the information that is collected about your purchasing habits from your credit card company? What are they allowed to sell to others? All of these situations can be problematic when multiple different pieces of information are joined together. Large data sets of information about you and others, which are called metadata, can be easily "mined" today to find patterns, even if they have been stripped of personal data—which is supposed to make them "anonymous data." Let us see some examples.

In 2020, two students from Harvard University developed a tool to analyze what is claimed as "anonymized" data as a class project. Their tool was able to "de-anonymize" data leaked in several large breaches, including breaches of Experian and MyHeritage.com, by searching for and putting pieces of information together from several breaches. These students were not the first ones to do this (Bode 2020). In 2019, researchers from the United Kingdom developed a statistical model that can identify people from an anonymized dataset. With 15 attributes (e.g., an email, a username, an address), their tool was able to identify 99.98% of the individuals in an "anonymized" dataset of Americans. The researchers suggested that 15 is a small amount of attributes, giving the example of a marketing analytics company that leaked anonymized data accidentally in 2017 with 248 attributes for 123 million households in America (Bushwick 2019).

Location information is now commonly collected, in particular by your smartphones and tablets. One study was conducted to see how much anonymous data from using Google Maps (with settings allowing location information to be sent) would reveal about two users (Pangburn 2017). Within a few days of starting to collect the location information and using publicly available information like mortgage information, LinkedIn, and Google Street View, the expert was able to identify one person exactly (with all kinds of personal information available) and one close enough (where he or she worked and lived). Another study of mobile users in 2013 by two universities found that they could uniquely identify 95% of 1.5 million individuals from 15 months of mobility data (Pangburn 2017).

The same issue exists with anonymized social networking data, as demonstrated by researchers from Stanford and Princeton Universities (Bode 2020). In another study led by a student at M.I.T. of credit card transactions conducted over three months in 10,000 stores by 1.1 million individuals, researchers only had access to the date of the transactions, the amount charged, and the name of the stores (Singer 2015). This means that the data were "anonymized," since there were no names or account numbers. Yet by using four random pieces of information, including publicly available information on sites like Twitter and Instagram, the researchers were able to reidentify 90% of the users of those credit cards.

Sources: Pangburn, D. J. 2017. "Even This Data Guru Is Creeped Out by What Anonymous Location Data Reveals about Us." *Fast Company,* September 26. https://www.fastcompany.com/3068846/how-your -location-data-identifies-you-gilad-lotan-privacy; Bode, K. 2020. "Researchers Find 'Anonymized' Data Is Even Less Anonymous than We Thought." *Vice,* February 3. https://www.vice.com/en_us/article/dygy8k/research- ers-find-anonymized-data-is-even-less-anonymous-than-we-thought; Singer, N. 2015. "With a Few Bits of Data, Researchers Identify 'Anonymous' People." *New York Times,* January 29. https://bits.blogs.nytimes .com/2015/01/29/with-a-few-bits-of-data-researchers-identify-anonymous-people/; Bushwick, S. 2019. "'Anonymous' Data Won't Protect Your Identity." *Scientific American,* July 23. https://www.scientificamerican .com/article/anonymous-data-wont-protect-your-identity/.

Focusing Questions

1. What is the role of publicly available data in the examples provided in this focusing story?
2. What are the possible consequences of being identified from one's browsing or searching habits?
3. What can you do when you are on the Internet that will protect your privacy?

Introduction and Definitions

Privacy of information is the confidentiality of the information collected by organizations about the individuals using their services (Van Slyke and Bélanger 2003). It can also be defined as one's ability to control information about oneself (Bélanger, Hiller, and Smith 2002). From an individual's perspective, it refers to a desire to control or influence data about oneself (Bélanger and Crossler 2011). Privacy of information is a very important concept for individuals and organizations alike. Many students may have faced issues with the privacy of their information when friends posted something about them on Facebook or Twitter without their prior approval. This may have been just for fun or may have caused some tensions between friends. However, as students move to the workplace, they will realize that organizations need to be even more keenly aware of information privacy concepts, since workers, professionals, and managers have access to and deal with great amounts of information on a daily basis. Therefore, everyone has to be concerned not only about their own information privacy but also about the privacy of information of others, including customers, employees, business partners, students, parents, children, and more.

Information Privacy Threats

There are many possible threats to information privacy—for example, dealing with collection and unauthorized secondary use of data, improper access to data, and errors in data (Smith, Milberg, and Burke 1996). Inaccuracies or **errors in data** can have serious impacts on individuals—for example, those applying for jobs or loans. **Improper access to data** deals more with the security of information, as discussed in the previous chapter with the security threat of unauthorized access. The information privacy threats of data collection and unauthorized secondary use, however, are worth discussing in more depth.

Data Collection

The collection of data from individuals is not a new concern, and the concept of privacy has been explored and discussed for centuries. One of the major differences today, however, is that as communications and information have been digitized, it has become easier and faster to collect ever-increasing amounts of information.

LEARNING ACTIVITY 9.1

Privacy and Smartphones

Most students today have smartphones. Depending on the settings on these devices, it is likely that location information is associated with every picture taken. The following video demonstrates how this happens and the possible implications. Please watch the video and answer the questions below: https://www.youtube.com/watch?v=N2vARz vWxwY.

Discussion Questions
1. What could people do with the information that is available with the pictures?
2. Who should educate people about this?
3. What else could be geotagged on your phone?
4. Why would all consumers not disable this feature?
    • Is there a reason not to?
    • What is the trade-off?
5. What IoT devices could also be tracked besides smartphones?

Specifically, with the advent of advanced information and communication technologies, the **data collection threat** means that data can be collected, aggregated, and analyzed at a faster pace and in larger volumes than ever. Of a greater concern sometimes is the fact that data can be collected without anyone's awareness—for example, through the use of cookies. **Cookies** are small text files that store information on your computer about you or your computer when you browse or enter information into certain websites.

Another hidden data collection approach is **clickstream data**, which tracks online browsing behaviors. The tracking is done by software that looks at the IP address of the computer connecting to a site, identifying what page the computer connected from, what page it goes to after the current one, how often it connects to this site, and whether actual purchases are made from this site. Thorough analysis of this data can then give marketers an indication of successful and unsuccessful online ads, referring sites, and other information on the preferences of each user. Customized ads can also be delivered to the consumer's screen. Interestingly, it is likely no one would agree to be followed in his or her buying decision process in a physical retail store. Imagine someone walking with you and observing you as you stroll through a Walmart and look at various products! Yet online tracking of consumers is a regular occurrence.

The proliferation of data sources, tools to manipulate these data, and online sites collecting data in today's interconnected society leads to more data being available and to individuals being increasingly concerned about the privacy of their information. Current technology allows easy loading of information from online forms directly into databases. For companies, this is a major advantage, since the data are loaded immediately (faster) and accurately (no transcribing errors and no problems dealing with unreadable writing) (Van Slyke and Bélanger 2003). However, as more

LEARNING ACTIVITY 9.2

Concern for Information Privacy

A number of reports and studies relate the various concerns that individuals and organizations have about information privacy. For this learning activity, search the Internet for statistics on the level of concern either individuals or organizations have about sharing information electronically. Your instructor may also assign you to search for specific concerns related to IoT devices like fitness trackers, smart home speakers, smart thermostats, contact tracing, and other current uses of IoT. Note any benefits reported as trade-offs that might overcome these concerns for sharing information. Be prepared to discuss your findings in class.

information is being stored online, consumers are becoming increasingly concerned with losing control of how their information is being collected and used.

Information privacy is also an important concern for corporations. The increased ability to store consumer information makes corporations targets for hackers trying to gain access to the vast amount of consumer information that they store, causing an increase in the number of breaches that companies are experiencing. For example, hundreds of millions of consumers who had records with the credit reporting agency Equifax had their information breached in 2017. This included personally identifiable data, like names, addresses, dates of birth, social security numbers, and drivers' license numbers. For some consumers, even their credit card numbers were exposed. Over the years, as discussed in Chapter 8, many other large organizations have been hacked, compromising individuals' personal information.

The rise of the Internet of Things (IoT) also creates new concerns for information privacy, in particular with respect to data collection. Recall that IoT means that all of these devices are collecting information and are highly interconnected. For example, a study by the **Federal Trade Commission (FTC)** in the United States suggested that fewer than 10,000 households could generate 150 million different pieces of information every day (Insider Intelligence 2020).

Secondary Use of Information

Unauthorized secondary use of information refers to the use of data for purposes other than those for which they were originally collected. For example, you enter your information on a website to register for the warranty of your new digital camera, but then the company uses your information to send you marketing ads about its other products. Information privacy refers to having some level of control over these potential secondary uses of one's personal information. *Function creep* is another term used to refer to data being used for other functions beyond those purposes for which they were collected. Often, individuals are given the option to opt in or opt out of secondary use on websites. When you **opt in,** you have to specifically state that you agree that your data can be used for other purposes, often to receive special deals or information from partner companies. When you **opt out,** you have to make sure

LEARNING ACTIVITY 9.3

Privacy Pizza

Your instructor will present a video titled *Privacy Pizza* in class or may direct you to the course website for you to watch the video individually before class. As you watch the video, make notes on the types of information and technologies that are discussed and/or used in the video. The *Privacy Pizza* video can also be found at https://www.aclu .org/ordering-pizza. Be prepared to answer the following questions:

1. Do you think access to the various types of information identified is regulated or not?
2. How likely is it that a pizza shop/company would have the technologies mentioned in the video?
3. What can someone do to avoid this situation from happening to him or her?
4. How do you feel about the central collection of information such as that represented in this video in the context of a pandemic or global need to share information?

you tell the company you do not want your data shared with others or used for other purposes. The trick is that sometimes you have to click a box to opt out and sometimes you have to unclick one. If you forget, by default you are agreeing to let the companies share or use your information.

In summary, when information collection and use are not regulated, individuals have the responsibility of protecting their own (or others') information, particularly in the United States. However, it is often the case that individuals are not aware of the risks or do not know how to protect their own information. Often people make a choice to give this information because it provides convenience or personalization of one's online experience. These preferences do vary around the world.

Consequences of Privacy Violations

Individual Consequences: Identify Theft

Everyone can face serious consequences when people take their personal information for fraudulent activities. This is the concept of **identity theft**. Identity theft cases have been increasing substantially in recent years. The Federal Trade Commission (FTC) reported that in 2019 there were 650,572 victims of identity theft in the United States and that individuals made more reports about identity theft (20.3% of all reports to the FTC) than any other types of complaints they received (FTC 2020). FTC studies also show that as the ability to pay for products electronically using smartphones has emerged, the likelihood of having money stolen where the transaction occurs has also increased. In addition to these high-tech methods of identity theft, it is important to protect yourself from low-tech methods of identity theft as well, including the stealing of wallets, mail, and other physical documents that contain people's personal information. As seen in the previous chapter, trans-

---

S T A T S   B O X   9 . 1

Perceptions of Monitoring

The focusing story showed that we are not as anonymous online as we wish we were. At the same time, it seems many people realize they are being monitored by their government. For example, the Pew Internet and American Life Research Center found that 72% of Americans believe that all or most of what they do online is tracked by companies, and 47% believe the same with respect to governments (Auxier et al. 2019). The survey shows some differences based on age and ethnicity, with more 18- to 29-year-olds (59%) believing the government is monitoring their online or cellphone activities, compared to older adults (44% for 50- to 64-year-olds and 30% for 65-year-olds and older), and 60% of Blacks and 56% of Hispanics compared to 43% for Whites.

Source: Auxier, B., L. Rainie, M. Anderson, A. Perrin, M. Kumar, and E. Turner. 2019. "Americans and Privacy: Concerned, Confused and Feeling Lack of Control over Their Personal Information." *Pew Research Center*, November 15. https://www.pewresearch.org/internet/2019/11/15/americans-and -privacy-concerned-confused-and-feeling-lack-of-control-over-their-personal-information/.

---

mitting information online is relatively secure, but as new mobile technologies are emerging, this is changing. Furthermore, what happens after the information is received by local organizations can sometimes not be as secure. Identity theft can have several negative consequences for individuals, including financial, social, and emotional impacts.

Students are often concerned that their data are being stolen from the Web when they acquire some goods or services online. In reality, as stated in the previous paragraph, the risks are greater that fraudsters or identity thieves access large amounts of information from unsecured databases. Often individuals who illegally access these data then resell them on the black market. It is difficult to have precise numbers on the size of this market, but, as detailed later in this chapter, various researchers and centers have identified the value of your information on the black market. How much would you be willing to pay for the following?

- A valid credit card number with a security code
- Valid bank account details, including the PIN (personal identification number)
- A valid Social Security number
- A completely new (valid) identity

The best way for individuals to protect themselves from identity theft is to follow basic security guidelines and common sense. The Identity Theft Resource Center (http://www.idtheftcenter.org), a not-for-profit organization, has been created to help identity theft victims and offers very useful information about protection against identity theft and what to do if it happens to you.

The following are some examples of the recommended actions:

- Do not use your Social Security number unless it is absolutely needed.
- Shred everything that has any data about you.
- Place outgoing mail at the post office or locked collection box.
- Password protect financial accounts with strong passwords and two-factor authentication, if available.
- Really check the statements you receive.
- Request photo identification when someone asks for your information, and do not give it out over the phone, Internet, or mail unless you initiated the contact.
- Destroy digital data by going beyond a simple delete.
- Limit the information provided on your checks.
- Request your free annual credit report and check it.

Organizational Consequences: Company Reputation

For organizations, information privacy violations also have serious consequences. There are, however, two sides to the issue here. Organizations have the responsibility of protecting their customers' data, but they are sometimes the ones infringing on the privacy of their clients.

If organizations fail to protect the privacy of their customers' information, then their reputations can suffer. There are many examples of companies' reputations being at stake. In 2019, for example, First American Financials reported that 885 million files were exposed in a data breach. The same year, close to 700,000 customer records of Choice Hotels were exposed and 7 million records were breached at Adobe Creative Cloud, to name a few. The Identity Theft Resources Center reports that breaches were up 19% in 2019 over the previous year and that there were 1,473 breaches per month (ITRC 2019).

The costs of privacy breaches can be enormous for the companies involved. First, they may lose current and future customers. Second, they have to repair the breaches. And third, they have to compensate customers for their potential identity theft issues. To deal with the breach they faced in 2017, Equifax incurred close to $1.4 billion in costs by early 2020. They had to provide 143 million breach victims with monitoring services, set aside $1 billion for information security technologies, and set aside $380.5 million to settle lawsuits they faced (Viyajan 2020). Of course, this does not include losses from companies withholding their business from them.

Finally, but not least, privacy breaches in organizations can directly impact the organization if the information that is leaked concerns the organizations' patents, trade secrets, or other intellectual property.

Technologies and Solutions for Information Privacy

Now that we understand the broad concepts of information privacy threats, it is important to further discuss the particular roles that technology can play in both infringing on information privacy and as tools to protect one's information privacy.

STATS BOX 9.2

Fraud Type from Identity Theft

Every year, the Federal Trade Commission (FTC) collects information on identity theft complaints and on fraud committed with someone else's identity in the United States. The figure below shows the top seven identity theft types reported to the FTC in 2019. As can be seen, credit card fraud is the number one identity theft issue.

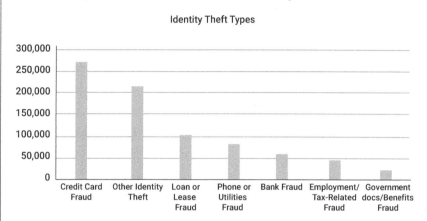

Identity Theft Types

Source: Based on data published by the Federal Trade Commission. 2019. *Consumer Sentinel Network*

Types of Fraud Committed through Identity Theft

Data Book 2019. https://www.ftc.gov/system/files/documents/reports/consumer-sentinel-network -data-book-2019/consumer_sentinel_network_data_book_2019.pdf.

Some of the tools that we describe in this section, however, are more like approaches to information privacy rather than tools. For example, a privacy policy on a website is more about the content and whether it is applied properly than about technology to create a policy. Similarly, privacy seals and government regulations toward privacy are ways to protect you online, but they are more like tools and techniques than they are technologies.

Cookies and Cookie Managers

Websites often use cookies, small text files located on your computer, to temporarily (or sometimes for as long as you have your computer) store information about you, your accounts, and your computer. Information is stored in a cookie every time you enter information and are asked, "Do you want this to be remembered?" Cookies provide convenience to individuals. Because the information is stored on the computer, you do not have to retype all the account information or addresses or even remember your password to access a website that has stored information in the cookie file. Information not typed in can also be stored in cookies. For

---

STATS BOX 9.3

The Black Market for Stolen Identities

Identity theft is an increasingly prevalent issue because of the value of people's information. However, this information cannot be sold through traditional business channels. Therefore, a black market has emerged for people's information, with different values for different pieces of information. It is often called the "dark web." For example, when Yahoo was hacked in 2013, three entities are believed to have paid $300,000 each for a copy of the database that was stolen. Two of them are believed to be spammers (Goel and Perlroth 2016). While getting these kinds of data from the dark web is not easy, Sandorf (2020) used data from the FBI and a company called Top10VPN to develop an estimated value of what your stolen information is worth, which is presented in the table below.

Stolen Information	Value
Credit card	$33.88
Bank details	$259.56
Amazon account	$30.36
Passport	$18.45
Craigslist account	$4.66
Uber account	$11.22
Facebook account	$9.12
Match (dating) account	$7.86
Gmail account	$5.87

Sources: Goel, V., and N. Perlroth 2016. "Hacked Yahoo Data Is for Sale on Dark Web." *New York Times,* December 15. https://www.nytimes.com/2016/12/15/technology/hacked-yahoo-data-for-sale-dark-web.html; Sandorf, B. 2020. "Identity Theft Report: The Cost of Your Identity on the Dark Web." *Reviews .org.* June 16. https://www.reviews.org/internet-service/how-much-does-it-cost-to-become-you/.

---

example, the website can record information such as your IP (Internet Protocol) address, the domain name of your computer, the length of time you stay on a specific Web page, and so on. When the user accesses one of the addresses in the address range of the website that stored the cookie, the browser transmits all the information retained in the cookie to the server of the website the user is currently browsing. Any item used to personalize your experience on the Web uses a set of cookies.

In terms of technologies to deal with cookies, there are many options. The simplest option is the settings you use on your Internet browser. The privacy settings allow you to state whether you allow cookies and what types of cookies should be automatically accepted. If you set your computer to be superprivate, most cookies will not work, and then many websites will not allow you to use their pages. If you set your privacy settings too low, then most information about you and your com-

TABLE 9.1 Sample Cookie Management Tools	
Name	**Link**
Cookiebot	https://www.cookiebot.com/en/
CookieMetrix	https://www.cookiemetrix.com/
Cookie Crawler	https://www.consentmanager.net/cookies/
Cookie-Editor	https://cookie-editor.cgagnier.ca/

puter will be stored and can be used by websites. The right balance depends on your privacy preferences.

Different operating systems and browsers have different ways to deal with cookies. For example, in the 2020 version of Microsoft Edge, you would go to "Settings" and select "Privacy, search, and services." There, you can use "Clear browsing data" to remove cookies as well as your download and browsing history. You can also decide how much tracking of your browser history you are comfortable with. In the 2020 version of Google Chrome, you would go to "Settings" and click on "Privacy and security." There, you would see a setting for "Cookies and other site data."

There are also **cookie managers** that can be downloaded for free or for a fee. These give you more details about which cookies are stored and what is stored in them and allow you to delete them. Other tools allow you to scan websites to see what cookies they collect, which companies can use to see if they are compliant with privacy laws, such as the General Data Protection Regulation in Europe or the California Consumer Privacy Act in the United States. Table 9.1 shows you a sample downloadable cookie manager (Cookie-Editor) and several cookie scanners (such as CookieMetrix). Many more tools exist on the Web. In addition, many new antivirus, antimalware, and other security programs automatically scan for and delete unnecessary cookies.

Anonymous Browsing

Given individuals' mounting concerns over information privacy, one approach that has been suggested is to use anonymous browsing. This can be done on computers, laptops, tablets, and even smartphones. The idea is to not allow websites to track what you are doing online, which is often used by advertisers to create targeted ads for you.

There are a few things you can do to leave less of a trail when you are browsing the Web. First, as previously discussed, you can make sure that cookies are either blocked or deleted. You can also adjust the settings so your browser does not send location information to websites you are surfing. Chrome, Safari, and Firefox have specific settings for this. For Microsoft browsers, you can use your computer's privacy settings. Third, you can use an anonymous search engine (like DuckDuckGo) or turn off the settings that "personalize" your searches. By getting more "personal" search information, you are also giving your search engine your preferences. Just think of how much Google knows about you already! You can also make sure your

Surveillance Societies around the World

We discussed privacy as having consequences for individuals and organizations. Privacy can also have global aspects as well. However, the concern for privacy and how it influences people's decisions is different in various countries. Visit the EMC Privacy Index (http://infographicsagency.com/project/emc-privacy-index/) and explore the different infographics available. As you do so, be prepared to answer the following questions:

1. In what contexts are people most willing to trade their privacy? Least willing?
2. In which countries are people most willing to trade their privacy for services? Least willing?
3. What is a paradox, and how does it affect people in different information-sharing contexts?

social networks, like Facebook, are not allowed to provide you with personalized ads, but that does not mean they will not track you. One of the authors of this book always uses anonymous browsing, which is available for most browsers (for example, there is InPrivate browsing on Internet Explorer and Incognito browsing in Chrome). There are also private browsers like the Tor browser (https://www.torproject.org/), which uses a complex open network so that websites cannot track and analyze your Web traffic. For smartphones and tablets, there are also different ways to use private browsing. For example, the Safari app on iOS devices can be set to use private browsing, but you do this within Safari instead of in the privacy settings. When looking at your opened Safari windows, you click the multi-window button at the bottom right, then you can see a "Private" option on the bottom left; clicking this allows you to set Safari in private browsing mode.

Unfortunately, even when trying to cover your tracks with anonymous browsing, you may still be tracked. A journalist and some experts in Europe acquired "anonymous" browsing history from three million German individuals and started analyzing them (Hern 2017; Price 2017). Close to three billion URLs revealed the identity of most of those individuals and specific information on their browsing habits, including some that most people would really not like to reveal.

Privacy Statement or Policy

Most serious companies who do business online have privacy policies. A **privacy policy** is a statement that describes what the organization's practices are with respect to the privacy of its customers. In general, most privacy policies include some or all of the following information:

• What data are collected
• How the data are collected

LEARNING ACTIVITY 9.5

Privacy Policy Creation

This activity will allow you to explore the contents of privacy policies. Based on the following scenario, create an appropriate policy.

Scenario

Two of your best friends from class have asked you to join them in a new venture. They are starting to sell customized, high-quality, university-branded apparel to fellow students and mostly to wealthy alumni. They want you to be the chief technology officer, and you have enthusiastically joined with them. As you oversee the development of the website for the company, which you are outsourcing, you have been asked to provide a privacy policy regarding your handling of customer data. This policy will be on the website at the bottom of the pages in small print.

Activity

1. Go to one of these sites (or another tool that you find online) to help generate a privacy policy:
   a. PrivacyPolicies.com: http://privacypolicies.com/
   b. Free Privacy Policy: http://www.freeprivacypolicy.com
   c. Shopify Policy Generator: https://www.shopify.com/tools/policy-generator
2. Bring your resulting policy to class (if the activity is performed before class).
3. Be prepared to discuss how you created your policy and which decisions you had to make.

- What the organization is doing with the data
- Why it is collecting the data
- What other companies it may or may not share the data with
- What actions the company is taking to protect the data
- How you can access the data the company has about you and how you can fix errors
- Whether you can opt out

The information contained in the privacy policies of companies usually follows the **fair information practices (FIP) principles** set forth by the FTC and a number of other organizations. These principles are derived from a number of studies, guidelines, and reports coming from a variety of countries. The FIP principles provide guidance for how to deal with personal information and include (1) notice/awareness, (2) choice/consent, (3) access/participation, (4) integrity/security, and (5) enforcement/redress. Table 9.2 shows how the FIP principles can be mapped to what privacy policies include.

It is important to note that privacy policies, like the privacy seals we will discuss below, are considered self-regulation mechanisms for information privacy. This means that most companies are not legally required to have a policy. **Self-regulation**

| TABLE 9.2 Mapping Fair Information Practices to Privacy Policies ||
Fair Information Practices	Privacy Policies
Notice/awareness	What data we are collecting How the data are collected What we are doing with the data Why we are collecting the data Which other companies we may or may not share the data with
Choice/consent	How you can (or cannot) opt out of us collecting these data about you
Access/participation	How you can access the data we have about you
Integrity/security	What actions we are taking to protect the data
Enforcement/redress	How you can fix errors in our data about you

is often an attempt by industry leaders to avoid government regulations. In this case, businesses do not want the government to impose privacy rules they would have to follow. Instead, they suggest that privacy policies provide sufficient protection for individuals. Do you think this is true? Well, the first question you should ask is, who reads privacy policies? Research clearly shows that most individuals never read privacy policies on websites. Now that you know what policies contain, do you feel protected? What would you do if a company failed to follow its own policies? The FTC is in charge of ensuring that privacy policies are respected. However, there is no "privacy policy police." It is up to consumers to make complaints to the FTC in the United States. With enough complaints about a particular site, the FTC will investigate. So if you find that a company fails to follow its own policy, you must turn to the FTC with your complaint in the United States or the privacy and data protection regulators of European countries, the European Commission or the EU Commission Data Protection Working Group. Companies dealing with customers in Europe have to be aware that the privacy laws are more stringent there than in the United States.

Privacy Seals

Privacy seals offer companies another attempt at self-regulation regarding privacy of consumers and a way to reassure consumers about transacting with them online. The way it works is that some company or organization, such as the Better Business Bureau, develops a seal program with a **security seal**—a logo that companies can post on their website if they follow certain rules, such as BBBOnline. There are several **seal programs** that businesses can participate in to show their commitment to trustworthiness (i.e., this website can be trusted). These seals are also called **reputation seals** and are issued by such sites as TrustArc (previously called TRUSTe) (http://www .trustarc.com) and WebTrust. Security seal programs assure consumers that the website follows a set of security procedures protecting their data; examples include those offered by VeriSign (http://www.verisign.com) and McAfee Secure (https://www .mcafeesecure.com/). Privacy seals ensure that the website follows certain privacy

BUSINESS EXAMPLE BOX 9.1

Privacy and Facebook

Which students (and their parents) are not on Facebook? Not many, right? Everyone seems to be on Facebook. Why? It is easy to communicate with so many people, so fast, from anywhere, and at any time. It is a great place to share pictures and keep in touch. It is addictive. Facebook is currently the most commonly used social networking site by adults.

Facebook likes to collect information on their users because it is the company's business model. Using profiles, Facebook is able to offer targeted marketing for companies. Facebook clearly says that it does not provide the information directly to its clients but rather uses its access to your information to provide targeted advertising for its clients. Think about the information available: what you like; where you shop; when you date; when you break up; when someone in your family gets a job, gets fired, or gets married; and so on. So is information privacy an issue on Facebook? Some say, "Yes, but I don't care." Others say, "No, not at all." And few, very few, say, "Yes, and I quit." Where do you stand?

Let us look at some of the complaints against Facebook. In 2006, Facebook released a news feed application that displayed changes to a friend's page directly to your computer without you having to access his or her page. In 2007, Facebook released an application called Beacon that showed information on your page (available to all your friends) when you performed actions on partner sites, such as what movie you rented on Netflix or what you bought on eBay. Following class actions and a settlement, Facebook discontinued that application but launched similar programs.

In spring 2010, Facebook changed settings to allow private information from users to be displayed to all users regardless of the initial choice of keeping this private, including job title, music preferences, school attended, and home city. The information is also shared with selected business partners of Facebook through the "instant personalization" application. After complaints, Facebook offered ways to control this setting (opt out), but many users never changed it. It is apparently not an easy task to change the settings. As a result, European Union regulators filed complaints against Facebook, alleging that users did not have effective control over their information. The Federal Trade Commission also started to investigate and later asked the company to simplify privacy controls for users (which the company did) and to provide more opt-in rather than opt-out options (which the company did not do).

Lawsuits abound against Facebook regarding privacy and its tracking and use of data it collects. For example, in 2014, Facebook began battling a class-action lawsuit in Vienna due to its privacy practices. In 2015 and 2016, lawsuits were filed in Austria and Illinois, to name a few. In 2017, Facebook was linked to Russian propaganda during the 2016 United States presidential elections and was blamed for selling user data that was used to promote Brexit in the UK (Binder 2019). In the wake of these events, some individuals have argued that with the amounts of detailed personal data Facebook has about each user, there are huge challenges for the privacy of these users (Binder 2019; Parakilas 2017). In 2019 alone, The Federal Communications Commission (FCC) in the United States fined Facebook $5 billion for privacy violations (Binder 2019). Adding to concerns over how Facebook sells user data, it has acquired several popular companies over the years, including Instagram and WhatsApp.

Sources: Best, J. 2014. "Facebook Forced to Respond to Privacy Complaints of 25,000 Europeans." *ZDNet,* August 21. http://www.zdnet.com/article/facebook-forced-to-respond-to-privacy-complaints-of-25000-europeans/; Cowan, J. 2010. "Why We'll Never Escape Facebook." *Canadian Business* 83(10): 28–33; Gelles, D., and M. Palmer. 2010.

"Facebook in Privacy Action." *Financial Times*, July 8:16; "Getting Personal. But Too Personal? Facebook's Mark Zuckerberg on What Information People Should Share—and Who Gets to Decide." 2010. *Wall Street Journal*, July 7; Gibbs, S. 2015. "Class Action Privacy Lawsuit Filed against Facebook in Austria." *The Guardian*, April 9. https://www .theguardian.com/technology/2015/apr/09/class-action-privacy-lawsuit-filed-against-facebook-in-austria; Guynn, J. 2016. "Facebook to Face Privacy Lawsuit over Photo Tagging." *USA Today*, May 5. https://www.usatoday. com/story/tech/news/2016/05/05/facebook-photo-tagging-lawsuit-faceprints-privacy-illinois/83999984/; Lenhart, A., K. Purcell, A. Smith, and K. Zickuhr. 2010. "Social Media & Mobile Internet Use Among Teens and Young Adults." *P. I. a. A. L. Project;* Lyons, D. 2010. "Facebook's False Contrition: A Business Built on Your Data." *Newsweek* 155(23); Lyons, D. 2010. "The High Price of Facebook: You Pay for It with Your Privacy." *Newsweek* 155(22); Menn, J. 2010. "Technology and Society: Virtually Insecure." *Financial Times*, July 28: 7; Parakilas, S. 2017. "We Can't Trust Facebook to Regulate Itself." *New York Times*, November 19. https://www.nytimes.com/2017/11/19/opinion/facebook -regulation-incentive.html; Binder, M. 2019. "Here's Why You Should Finally #DeleteFacebook in 2020." *Mashable*, December 26. https://mashable.com/article/delete-facebook-2020/.

guidelines, such as ESRB Privacy Online (http://www.esrb.org/privacy/). Even general trust seals often deal directly with privacy. For example, TrustArc states to consumers, "Certified companies that display our privacy seal demonstrate compliance with our Certification Standards and their commitment to privacy protection, instilling confidence and trust in users of their products and services" (http://www.trustarc .com). Seals are supposed to instill consumer confidence in the website, but studies have shown that even experienced Web users are less familiar with privacy and security seals than with security technologies (Bélanger et al. 2002). There are many other seal programs, some specific to other areas of the world, such as TrustUK for the United Kingdom and eTICK for Australia.

Government Information Privacy Regulations

Even though information privacy is largely our responsibility as individuals, there are specific situations where governments have created regulations to protect information privacy. In fact, there are actually many such regulations throughout the world. For example, the General **Data Protection Regulation** (GDPR) regulates how personal data are processed and protected in the European Union. In the United States, there are many privacy laws as well. Some that students might be more familiar with include the **Gramm-Leach-Bliley Financial Services Modernization Act of 1999 (GLBA)**, which regulates use of private information by financial institutions; the **Family Educational Rights and Privacy Act (FERPA)** discussed in Learning Activity 9.6; the **Children's Online Privacy Protection Act of 1998 (COPPA)**, which prevents websites from collecting **personally identifiable information** from children without parental consent; **HIPAA, the Health Insurance Portability and Accountability Act**, which protects your medical information (and requires you to sign a form at the doctor saying you are aware of their privacy practices), and the **California Consumer Privacy Act of 2018 (CCPA)**. Table 9.3 describes a few of these privacy protection regulations. Note that there are many more privacy regulations around the world that are not included in the table. Business Example Box 9.2 provides more details on the GDPR.

Table 9.3 Sample Privacy Regulations	
Law	Source
Children's Internet Protection Act of 2001 (CIPA)	https://www.fcc.gov/consumers/guides/childrens-internet-protection-act
Children's Online Privacy Protection Act of 1998 (COPPA)	https://www.ftc.gov/enforcement/rules/rulemaking-regulatory-reform-proceedings/childrens-online-privacy-protection-rule
Electronic Communications Privacy Act of 1986 (ECPA)	https://epic.org/privacy/ecpa/
Family Educational Rights and Privacy Act (FERPA)	http://www2.ed.gov/policy/gen/guid/fpco/ferpa/index.html
General Data Protection Regulation	https://gdpr-info.eu/
Gramm-Leach-Bliley Financial Services Modernization Act of 1999 (GLBA)	https://www.ftc.gov/tips-advice/business-center/privacy-and-security/gramm-leach-bliley-act
Health Insurance Portability and Accountability Act (HIPAA)	http://www.hhs.gov/hipaa/index.html
California Consumer Privacy Act of 2018 (CCPA)	https://oag.ca.gov/privacy/ccpa

BUSINESS EXAMPLE BOX 9.2

GDPR in Practice

The General Data Protection Regulation (GDPR) is a fairly recent law enacted in the European Union (EU) to protect the data of EU citizens anywhere in the world (https://gdpr-info.eu). Companies that do not comply, including companies not in the EU but that use, collect, or store EU consumer data, can be fined up to €10 million or 2% of their global revenues. It is very important for all managers to realize that if any of their customers are in the EU, GDPR will apply to them.

The law originally was supposed to be in full effect by May 2018; however, it is not simple to implement. It has 11 chapters and 99 articles. In fact, by July 2019, Greece, Portugal, and Slovenia had not yet been able to get their laws in line with GDPR. By August 2020, 347 fines were issued, with the largest fine going to Google, which received a €50 million fine from France. The smallest fine was €90 for a hospital in Hungary. In July 2019, the United Kingdom issued an intent to fine British Airways for €204 million and Marriott International for €110 million, but as of August 2020 those fines were still not final. Facebook and WhatsApp were also fined but for smaller amounts. The biggest fines to date besides Google were two Italian companies, the Austrian Post, and a German company.

Sources: Uzialko, A. 2020. "How GDPR is Impacting Business and What to Expect in 2020." *Business News Daily*, February 4. https://newsoncompliance.com/how-gdpr-is-impacting-business-and-what-to-expect-in-2020/; *Data Privacy Manager*. 2020. August 18. https://dataprivacymanager.net/5-biggest-gdpr-fines-so-far-2020/.

Mobile Information Privacy

The threats to information privacy and the related solutions discussed in this chapter apply to most digital environments. However, many researchers have claimed that information privacy is even more of an issue in mobile settings, such as in the use of your smartphone or tablet. Why? Because in the mobile environment, additional information is often being collected without users being aware of this. In addition, the mobile environment has fewer regulations, most users have enormous amounts of personal information on their devices, and users often download apps without considering what the apps can do (Bélanger and Crossler 2019). If you think about it, most smartphones and smartphone apps collect location information on a constant basis, knowing exactly where the phone is at all times of the day, unless you change the settings to limit this sharing of information. For example, camera apps on your smartphone will tag where and when each picture was taken, creating a trail of where you have been.

Experts suggest that protecting your privacy on your smartphone involves several steps (Bushnell 2019). Examples of various steps are shown next. First, use only a

---

LEARNING ACTIVITY 9.6

Why Your Adviser Cannot Talk to Your Parents

Legislation exists to protect the information privacy of individuals in a number of specific cases, such as individuals' health and financial information, as well as information about children and students. The purpose of this activity is to allow you to explore how your information privacy is protected and to better understand the difference between legitimate and inappropriate uses of information.

1. Go to the Family Educational Rights and Privacy Act (FERPA) website for students: https://www2.ed.gov/policy/gen/guid/fpco/ferpa/students.html. FERPA protects the privacy of student education records. This link is available on the book's website.
2. Read the information on the main page. Pay particular attention to the types of information covered by FERPA and individuals to whom protected information can be released.
3. Answer the following questions and be prepared to discuss them in class:
   a. What types of information are protected under FERPA?
   b. Under what conditions may school officials provide protected information to parents? In your opinion, how do these conditions relate to the concept of "owning" your personal data?
   c. Why do you think FERPA was created? What problem did it solve?
   d. Compare the protections afforded by FERPA to the privacy policies you examined in Learning Activity 9.5. What elements do they have in common? How are they different? Which has stronger protections?

recognized app download provider (Apple App Store, Google Play, etc.) by using a setting that does not allow downloads from unknown sources. Second, use settings that limit what apps have access to. Third, install security apps on your smartphone, particularly for Android phones (e.g., Bitdefender, F-Secure, etc.). Fourth, secure your lock screen with a PIN or password. Most experts believe a swipe is too easy to break. Fifth, use features like Find My Phone and Remote Wipe to do exactly that: find your phone if you lose it and remote wipe all information if your phone is not returned to you. Finally, do not use public networks, at least for anything important.

IoT and Privacy

How does one protect information privacy in the world of IoT given the huge amounts of data collected by Internet-connected devices? Because a lot of those data are very personal (i.e., think of your fitness tracker or your intelligent home speaker) and can now be connected to other data about you, IoT amplifies the issues with information privacy we discussed in the focusing story. It also adds new challenges, like how to obtain informed consent and how to provide transparency given that the IoT devices can be everywhere and unnoticed. On a broad scale, the Internet Society (http://internetsociety.org) suggests the following:

- Enhance user control regarding the data collected by IoT devices and services and how they are managed.
- Improve transparency and notification by providing clear, accurate, relevant, and detailed information to users.
- Privacy laws and policies should keep up with technology and include the use of IoT sensors and continuous monitoring.
- Involve a broad variety of stakeholders in IoT privacy discussions to address the diversity of IoT risks and benefits (Internet Society 2020).

TABLE 9.4 PAPA Ethical Framework: For Students

PAPA Component	Description	Questions to Ask Yourself
Privacy	How the confidentiality of the data collected about you is maintained	• What information about you must you reveal to others? • What information should others be able to know about you—with or without your permission? • How is your information protected?
Accuracy	Whether the data about you is what it is supposed to be and does not include errors	• Who is responsible for the accuracy of your information? • Who is accountable for errors about your information? • How do you remedy errors about your information?
Property	Who owns the data about you	• Who owns your information? • Who has the legal rights to your information? • How is the distribution of your information regulated?
Accessibility	Who has access to the information systems and the data that they hold about you	• Who can have access to your information? • Which companies can have access to your information? • What safeguards are in place when someone accesses your information?

Source: Mason (1986).

Privacy and Ethics

As discussed at the beginning of the chapter, information privacy and confidentiality are not new issues, but the increased use of technology has created new concerns about the use of data, and not only information privacy concerns. In the mid-1980s, Mason (1986) suggested that the increased use of information technologies, which was labeled the Information Age, would lead to four major concerns about the use of information: privacy, accuracy, property, and accessibility (PAPA). Table 9.4 shows the **PAPA framework**. The framework can be used by managers to consider areas of concern about information they handle in their organizations.

Ethical Decisions

"A man without ethics is a wild beast unleashed on the world." This quote from the French writer Albert Camus nicely sums up the importance of ethics. While philosophers may argue about the definition of ethics, at its core is the idea of doing the right thing—standards of behavior that tell us how we should act. Since decision-making is choosing among alternatives, ethical decision-making is considering stan-

LEARNING ACTIVITY 9.8

Making Ethical Decisions

The following scenarios describe some decisions that have ethical issues. For each question, state what decision you would make and why you would make that decision. Be prepared to discuss your answers in class.

1. You need some special software for a school project. The software is available in the school's computer labs, but going to the lab is inconvenient. You can purchase the software at the bookstore, but your friend has the software and is willing to loan you the installation DVDs. Would you borrow the DVDs?

2. One of your coworkers is using his work computer for his part-time business. You become aware of this. Should you report your coworker?

3. You are the chief information officer for your company. As part of your duties, you are in charge of safeguarding data about your customers. The vice president for marketing is interested in trading information about your customers with another company. (In exchange, your company will receive information about the other company's customers.) The marketing VP thinks this trade will increase your company's revenue. Should you provide the customer information?

dards of conduct when making choices. Ethical decision-making requires work, as doing the right thing often does. The work is worth the effort. If you always strive to make ethical choices, you will be a success. Many authors have proposed various procedures for making ethical decisions. One of them, the PLUS framework, was originally proposed by the Ethics Resource Center, which is now the Ethics & Compliance Initiative (https://www.ethics.org).

The PLUS framework is a set of questions to help in making ethical decisions. The questions ask whether an alternative

- [P] is consistent with organizational policies, procedures, and guidelines
- [L] is acceptable under applicable laws and regulations
- [U] conforms to universal values, such as empathy, integrity, and justice
- [S] satisfies your personal definition of what is good, right, and fair

Keeping the PLUS questions in mind, follow these steps (adapted from the original framework):

1. Define the problem.
   a. Why do you need to make a decision? What outcome do you want from the decision?
   b. Does the current situation violate any PLUS criteria?
2. Identify alternatives.
   a. Try to identify three or more alternatives.

3. Evaluate the alternatives.
   a. What are the positive and negative consequences that flow from each alternative? Consider "positive" and "negative" from the perspective of those impacted by the decision.
   b. How probable is it that those outcomes will occur?
   c. Consider whether each alternative will create or resolve any violations of the PLUS criteria.
4. Make the decision.
5. Implement the decision.
   a. Minimize negative consequences to the greatest extent possible.
   b. Communicate the reasoning behind the decision to those affected by the decision.
6. Evaluate the decision.
   a. Did the decision improve the situation that led to the need for the decision?
   b. Were any new problems created by the solution?
   c. Are any PLUS considerations resolved or introduced by the new situation?

If you keep the following in mind throughout the decision-making process, you will be much more likely to make ethical decisions:

- Who are the stakeholders for this decision or its outcomes? Try to think broadly about this. Often, stakeholders may not be readily apparent.
- Could someone be damaged by this decision? If so, consider whether there are alternatives that can reduce this harm. Sometimes your decisions will harm people; tough decisions often do. However, by considering the impact on others and trying to develop ways to minimize the harm, you will often be able to soften the blow.
- What alternatives are available? How do the outcomes from those alternatives impact each stakeholder group? For each alternative, consider which option
  - Produces the greatest good and the least harm
  - Best respects stakeholders' rights
  - Leads you to act as the type of person you want to be
- How can you implement the decision with the greatest attention to stakeholders' concerns? Decisions often have uncomfortable outcomes. This applies to ethical decisions as well. The way you carry out the decision can have a significant impact on the level of harm to any individual stakeholder.

While information systems cannot make you an ethical decision maker, they can help reduce some of the extra work that sometimes comes from ethical decision-making. For example, suppose you are tasked with cutting expenses for your company. A spreadsheet can help you analyze various alternatives for doing so. Information systems also contain the data and information you may need for developing and analyzing alternative courses of action.

Ethical decision-making is worth the effort. Leadership requires followers, and if you always strive to make ethical decisions, the people you lead will trust you to have their interests in mind, even when making difficult decisions. When those you lead trust you, you can be a great leader.

Relationship between Security and Privacy

In the previous chapter, we discussed information security. After reading this chapter on information privacy, you should realize that these concepts are very much related in practice. Security is the protection of information against threats such as unauthorized access to data, falsification of data, or denial of service. A company can provide every security protection possible against these threats to your information without necessarily having the intent of protecting the confidentiality of your information. Thus information privacy is different from information security, even if these concepts are often used interchangeably.

Chapter Summary

In this chapter, we discussed information confidentiality and privacy. We first described the threats to information privacy and the consequences of privacy violations for individuals and organizations. We then discussed the technologies and solutions related to information privacy. We concluded the chapter with a discussion of the PAPA framework and of the relationship between security and privacy.

Here are the main points discussed in the chapter:

- There are four main categories of threats to information privacy: data collection, unauthorized secondary use of data, improper access to data, and errors in data.
- We identified several technologies used to infringe on and/or protect information privacy, such as cookies, cookie managers, privacy statements and policies, trust seals, and government regulations.
- Information privacy is one of the four components of the PAPA ethical framework, which includes privacy, accuracy, property, and accessibility. The framework can be used to identify concerns about the use of information.
- Information privacy and information security are related concepts, since it is mandatory for the information to be secured before it can be private. The reverse is not necessarily true, since information that is protected from a security standpoint can still be shared with others, infringing on the privacy of the information.
- Ethical decision-making considers standards of behavior when making decisions.

Review Questions

1. Explain briefly the four main threats to information privacy.
2. What is the difference between opt-in and opt-out options in privacy policies?
3. What are some threats to information privacy in the context of IoT devices?
4. What are some of the possible consequences of privacy violations?
5. What is the purpose of privacy policies or statements? Who regulates them?
6. What are the fair information practice principles? What is their purpose?
7. What are privacy seals, and what is their purpose?
8. Briefly describe some of the key regulations protecting information privacy in the United States and Europe.
9. What is the PAPA framework? Explain its components. What is its purpose?
10. Describe the PLUS approach to ethical decision-making.

Reflection Questions

1. What is the most important thing you learned in this chapter? Why is it important?
2. What topics are unclear? What about them is unclear?
3. Are there any overlaps between the information privacy threats discussed in the chapter and the components of the PAPA framework?
4. How are privacy policies, cookies, and privacy seals related?
5. Imagine that a well-known online retailer reveals that a hacker successfully accessed the personal information of 500 of its consumers. Is this a threat to privacy or security for the consumers? Explain.
6. Give examples of five types of data that could be stored in a cookie set up by your university when you browse the main student pages.
7. What is the difference between information privacy and information security? Can one exist without the other?
8. Why are data collection and secondary use of information threats to information privacy? Why are these threats enhanced in the context of IoT?
9. Can businesses be victims of identity theft? Why or why not?
10. How can information and communication technologies help make ethical decisions?

Additional Learning Activities

9.A1. This activity requires you to complete a case analysis regarding information privacy. The case may be accessed through your university

website, the Ivey Business School business case website (https://www
.iveycases.com), or the Harvard Business School cases website (https://
hbsp.harvard.edu/home/), or it may be provided to you by your
instructor. McGee, H., N.-H. Hsieh, and S. McAra. 2016. "Apples:
Privacy vs. Safety (A)." *Harvard Business School Case*, February 10.
Product number: 9-316-069.

9.A2. This activity allows you to explore two privacy seal programs in greater
depth. First, visit five websites and note which privacy seals are
common to the websites. Identify trends. Do better sites use particular
seals? Then select the two most popular privacy seals and find their
websites. Identify the requirements and costs for organizations to
participate in each of the two seal programs and create a comparison
table that includes whether a fee is charged (high/low), policies for
affixing the seal, disclosures required, consumer redress options, and
any other important information.

9.A3. This activity requires you to listen to an interesting series of broadcasts
on information privacy presented on the National Public Radio's *All
Things Considered* program titled "The End of Privacy." Listen to each
segment and then write a short essay on one of these questions:
   a. Is privacy still possible? Why?
   b. Is real-time tracking a concern for privacy or a security advantage?
   Why or why not?
   c. Is social networking worth the privacy risks? Why or why not?
   d. Can online information about you come back to haunt you? How?
   • October 26, 2009—Online Data Present a Privacy Minefield (7
   minutes, 49 seconds): http://www.npr.org/templates/story/story
   .php?storyId=114163862
   • October 27, 2009—Is Your Facebook Profile as Private as You
   Think? (5 minutes, 47 seconds): http://www.npr.org/templates/
   story/story.php?storyId=114187478&ps=rs
   • October 28, 2009—Digital Bread Crumbs: Following Your Cell
   Phone Trail (7 minutes, 50 seconds): http://www.npr.org/
   templates/story/story.php?storyId=114241860
   • October 29, 2009—Digital Data Make for a Really Permanent
   Record (7 minutes, 46 seconds): http://www.npr.org/templates/
   story/story.php?storyId=114276194

9.A4. Select two websites you regularly transact with or send information to.
Obtain their privacy policies and map them to the fair information
practices and privacy policy contents as done in Table 9.2. Do all
privacy policies include the proper content? Explain.

9.A5. Perform a comparison of two of the most recent information privacy
laws: GDPR and CCPA. Create a table of similarities and differences.
Which of the two provides the best protection to consumers, in your
opinion?

9.A6. Much of the discussion in this text focused on laws and individual behaviors related to privacy that were mainly in the United States and Europe. Research laws and privacy behaviors for individuals in another part of the world (e.g., Asia, Africa, or South America). Compare the privacy laws and behaviors of individuals to those discussed in this chapter. How are they the same? How are they different? Draw a table that shows how two of those laws address the fair information practices principles.

9.A7. With the permission of one of your good friends, search the Web to find everything you can about him or her. Then share your findings with your friend. How much information were you able to collect? When adding all the pieces of information together, what have you learned about your friend that you did not know before? How did your friend react to the information you found? Was your friend aware that so much was available about him or her on the Web?

9.A8. Compare and contrast the information privacy threats discussed in the chapter with the components of the PAPA framework. Identify any overlaps and differences.

9.A9. Think of a time when you had to make a decision that was unclear from an ethical perspective. Apply the PLUS model to the decision. Prepare a two-page reflection paper that comments on the following:
• Why the decision was unclear ethically
• The effectiveness of the PLUS model

References

Bélanger, F., and R. E. Crossler. 2011. "Privacy in the Digital Age: A Review of Information Privacy Research in Information Systems." *MIS Quarterly* 35(4): 1017–42.

Bélanger, F., J. S. Hiller, and W. J. Smith. 2002. "Trustworthiness in E-Commerce: The Role of Privacy, Security, and Site Attributes." *Journal of Strategic Information Systems* 11(3/4): 245–70.

Bushnell, M. 2019. "10 Ways to Secure Your Smartphone Against Hackers." *Business News Daily*, June 7. https://www.businessnewsdaily.com/11197-protect-your-smartphone-from-hackers.html.

Crossler, R. E., and F. Bélanger. 2019. "Why Would I Use Location-Protective Settings on My Smartphone? Motivating Protective Behaviors and the Existence of the Privacy Knowledge–Belief Gap." *Information Systems Research* 30(3), 995–1006. https://doi.org/10.1287/isre.2019.0846.

Hern, A. 2017. "'Anonymous' Browsing Data Can Be Easily Exposed, Researchers Reveal." *The Guardian*, August 1. https://www.theguardian.com/technology/2017/aug/01/data-browsing-habits-brokers.

ITRC. 2019. "2019 End-of-Year Data Breach Report." *Identity Theft Resource Center*. https://notified.idtheftcenter.org/s/resource#annualReportSection.

Insider Intelligence. 2020. "The Security and Privacy Issues That Come with the Internet of Things." *Business Insider*, January 6. https://www.businessinsider.com/iot-security-privacy.

Internet Society. 2019. "Policy Brief: IoT Privacy for Policymakers." September 19. https://www.internetsociety.org/policybriefs/iot-privacy-for-policymakers/.

Mason, R. O. 1986. "Four Ethical Issues of the Information Age." *MIS Quarterly* 10(1): 5–12.

Price, E. 2017. "Your 'Anonymous' Browsing Data Is Not Very Anonymous." *Lifehacker.com*, August 3. https://lifehacker.com/your-anonymous-browsing-data-is-not-very-anonymous-1797490806.

Smith, H. J., S. J. Milberg, and S. J. Burke, 1996. "Information Privacy: Measuring Individuals' Concerns about Organizational Practices." *MIS Quarterly* 20(2): 167.

Van Slyke, C., and F. Bélanger, 2003. *Electronic Business Technologies.* New York: John Wiley & Sons.

Vijayan, J. 2020. "2017 Data Breach Will Cost Equifax at Least $1.38 Billion." *Dark Reading*, January 15. https://www.darkreading.com/attacks-breaches/2017-data-breach-will-cost-equifax-at-least-$138-billion-/d/d-id/1336815.

Glossary

Accessibility: Refers to who has access to the information systems and the data that they hold.

Accuracy: When data are what they are supposed to be and do not include errors.

California Consumer Privacy Act of 2018 (CCPA): A law that provides consumers in California with added control of how their personal information is collected and used by businesses.

Children's Internet Protection Act of 2001 (CIPA): A law that regulates access to offensive content over the Internet on school and library computers.

Children's Online Privacy Protection Act of 1998 (COPPA): A law that prevents websites from collecting personally identifiable information from children without parental consent.

Clickstream data: Tracking of online browsing behaviors.

Cookie manager: A software application that allows you to view which cookies are stored on your computer and what is in them and gives you the ability to delete them.

Cookies: Small text files located on your computer to store information about you, your accounts, and your computer.

Data collection threat: A privacy threat resulting from the fact that data can be collected, aggregated, and analyzed at a faster pace and in larger volumes than ever, without individuals' awareness.

Electronic Communications Privacy Act of 1986 (ECPA): A law that regulates access, use, disclosure, interception, and privacy protections of electronic communications.

Errors in data: A privacy threat where there are inaccuracies in data.

Fair information practices (FIP) principles: Guidelines for how to deal with personal information, which include notice/awareness, choice/consent, access/participation, integrity/security, and enforcement/redress.

Family Educational Rights and Privacy Act (FERPA): A law that protects the privacy of student education records.

Federal Trade Commission (FTC): A government agency responsible for (among other things) ensuring that privacy policies are respected.

General Data Protection Regulation: A law that regulates how personal data are processed and protected in the European Union.

Gramm-Leach-Bliley Financial Services Modernization Act of 1999 (GLBA): A law that provides regulations to protect consumers' personal financial information held by financial institutions.

Health Insurance Portability and Accountability Act (HIPAA): A law that provides regulations to protect personal health information held by covered entities and gives patients an array of rights with respect to that information.

Identity theft: Using another person's personal information for fraudulent activities.

Improper access to data: A privacy threat where unauthorized individuals have access to one's private information.

Opt in: A privacy option where individuals state they agree that their data can be shared with others or used for other purposes, often to receive special deals or information from partner companies.

Opt out: A privacy option where individuals must state that they do not want their data to be shared with others or used for other purposes.

PAPA framework: A framework that identifies four major categories of concerns about the use of information: privacy, accuracy, property, and accessibility.

Personally identifiable information (PII): Also called sensitive personal information, is defined by U.S. law as any information that can be used to identify, contact, or locate an individual on its own or combined with other information.

Privacy: One's ability to control information about oneself.

Privacy policy: A statement that describes what the organization's practices are with respect to the privacy of its customers.

Privacy seal: A seal that a business can post on its website to show its commitment to privacy.

Property: Refers to who has ownership of the data.

Reputation seal: A seal that a business can post on its website to show its commitment to trustworthiness.

Seal program: A program offered by an organization that posts a set of rules that companies must follow to be a part of the seal program.

Security seal: A seal that a business can post on its website to show its commitment to security.

Self-regulation: An attempt by industry leaders to avoid government regulations by suggesting (rather than requiring) that companies have privacy policies—for example, with privacy seals and privacy policies.

Unauthorized secondary use of information: A privacy threat resulting from the use of data for purposes other than those for which they were originally collected.

Developing Information Systems

Learning Objectives

By reading and completing the activities in this chapter, you will be able to:

- Explain the benefits of a software development methodology
- List and describe the phases in the traditional systems development life cycle
- Compare and contrast the traditional systems development life cycle with alternative methodologies
- Explain factors that influence the decision to build custom software or purchase commercial, off-the-shelf software
- Define open source software
- Describe major outsourcing models
- List the benefits and risks of outsourcing

Chapter Outline

Focusing Story: The $6 Billion Software Bug
Time, Cost, and Quality
Software Development Methodologies
    Learning Activity 10.1: Determining Requirements
Traditional Systems Development Life Cycle
    Learning Activity 10.2: Advantages and Disadvantages of the SDLC
Alternative Methodologies
    Learning Activity 10.3: Developing by Modeling
    Learning Activity 10.4: Comparing the Methods
Build or Buy Decision
    Learning Activity 10.5: App Developers Needed!
Using Open Source Software in Business
    Learning Activity 10.6: How Open Source Software Impacts Build versus Buy
Outsourcing Information Systems
    Learning Activity 10.7: What to Outsource

In this chapter, you will gain a basic understanding of the software development process. In addition, the chapter will help you learn about alternatives to building software from scratch. Outsourcing information systems is also discussed.

The $6 Billion Software Bug

In August 2003, the largest power blackout in North American history hit the northeastern United States and southeastern Canada. More than 50 million people lost power, some for up to two days. At least 11 people died as a result of the blackout. Total cost: $6 billion. Investigators concluded that the initial problem started when a power line in Ohio sagged and hit an overgrown tree. Normally, this is an isolated incident. When such an incident occurs, alarm systems alert power systems operators of the problem so they can take appropriate steps to make sure that the problem remains isolated. Unfortunately, in this case, a software bug caused the alarm system to malfunction. This started a cascade of errors that led to more than 50 million people losing electrical power. The lack of power also led to water supply problems and the shutdown of major transportation systems, including railroads and airlines. Gas stations could not dispense fuel. Cell phone service was interrupted (although wired phones continued working).

While lack of proper maintenance was the root cause of the blackout, without the software bug, the impact would have been small and limited to relatively few people. With the software error, tens of millions of people were affected, and there were billions in economic costs.

There are many other examples of the serious consequences of faulty software. A few recent examples include:

- In February 2020, hundreds of flights were disrupted at Heathrow airport in London, UK, when a technical issue made the check-in systems and departure boards defective.
- In July 2019, Facebook users were unable to upload photos to their news feed, see stories on their Instagram account, or use WhatsApp to send messages because of a software glitch that was accidentally generated when the company performed routine maintenance on software.
- In April 2018, a UK bank called TSB was planning to close Internet and mobile banking services for a weekend to perform a system upgrade. However, when the new system was switched on, millions of customers could not access their accounts for up to two weeks.
- In 2017, Dodge Ram recalled 1.25 million Ram trucks because of a software glitch that caused the airbags and seatbelts to fail during a rollover accident.
- In 2017, a major software failure at British Airways caused a significant number of local flights to be canceled and several delays to international flights.

Sources: Computerworld UK Staff. 2020. "Top Software Failures in Recent History." *Computerworld*, February 17. https://www.computerworld.com/article/3412197/top-software-failures-in-recent-history.html.

Focusing Questions

1. Have you ever experienced software that did not operate correctly? What consequences did you experience?
2. What can be done to limit software errors?
3. Why is it important to catch software errors early in the development process?

Time, Cost, and Quality

Every software development project faces a tension among time, cost, and quality. This reality, which applies to all sorts of projects, is known by several names, including the *project triangle*, the *engineering triangle*, and the *design triangle*. The basic idea is that there are trade-offs among the three dimensions. The three constraints are as follows:

- *Time:* How long will the project take to complete?
- *Cost:* What resources are required to complete the project?
- *Quality:* How well does the completed project meet user requirements?

Some versions of the triangle substitute *scope* (what is done as part of the project) for quality. Scope and quality are closely related, especially when considering an overall project. Quality is related to user requirements, and increasing the scope of a software development project adds new requirements that need to be met. So increasing the scope of a project also makes it harder (more costly and/or longer in duration) to meet quality goals.

There is an old saying: "You want it fast, good, and cheap? Pick any two." Increase the speed with which a project is delivered and you will have to compromise on either quality or costs. Decreasing the cost requires a longer completion time or accepting lower quality. It is important to understand these trade-offs when making software development decisions. Typically, one of the three constraints is fixed (for example, you may have a strict budget or delivery date). As a result, the other two constraints must be managed against each other. For example, if you have a fixed budget, moving to an earlier delivery date requires changes in quality or scope. You can learn more about this in relation to project management in Appendices J and K on funding and managing projects.

Software Development Methodologies

Developing software is a complex, time-consuming, and costly endeavor. Because of this, it is often useful to have a framework for planning, structuring, and controlling software development projects. Such frameworks are commonly called **software development methodologies**. These methodologies provide discipline to the software development process by defining processes, roles, and deliverables related to software development. Typically, the methodologies break down software development into phases, each of which has its own set of processes, roles, and deliverables. In some methodologies, the deliverables of one phase provide the inputs for the next phase.

Most large organizations have a standard software development methodology that is used across the organization, although many organizations may allow different methodologies for particular projects. For example, a Web-based system development

LEARNING ACTIVITY 10.1

Determining Requirements

Determining system requirements is a major step in the software development process. One method for uncovering requirements is to interview the intended users of the software. Suppose that a club you belong to wants to create a system for keeping track of members. Get in pairs. One of you will act as the interviewer who is trying to discover requirements. The other will act as the user. Document at least 10 requirements for the system.

project may follow a different methodology than a transaction processing system (TPS) development project.

The main reason to use a formal software development methodology is that it provides discipline by clearly defining how the project should proceed, what work should be done, how that work should be done, and what outputs should be produced. Having a standard methodology in an organization also allows individuals to more easily join a project in progress. The standardized roles, processes, documentation, and deliverables make it easier for a new team member to become familiar with what has been done on the project. It also makes it easier for the new member to find relevant information.

Traditional Systems Development Life Cycle

Traditionally, many systems development projects have followed a semisequential, phased approach, which is typically called the **systems development life cycle** (SDLC). The SDLC provides a disciplined approach to systems development. Projects follow a well-defined set of phases that have related objectives, processes, and deliverables. The traditional SDLC is sometimes called the *waterfall method*, since

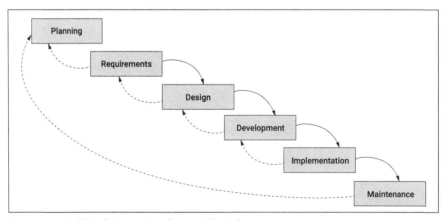

Figure 10.1 Traditional Systems Development Life Cycle

the results of one phase flow as inputs into the next phase. It is important to realize that there are many different versions of the "traditional" SDLC. The differences boil down to how the phases are divided; some versions have more phases than others. The versions with fewer phases collapse multiple phases from the more complex versions into a single phase. In this chapter, we present a relatively typical version with six phases, as shown in Figure 10.1. Note the dashed feedback lines on the left-hand side of the diagram. Sometimes you need to go back and refine an earlier phase based on new information that emerges.

Planning Phase

The goal of the planning stage is to define the overall goal and scope of the system and determine the feasibility of the project. Essentially, you want to answer the question "Why build the system?" In addition, you need to establish whether the project is feasible. There are several different aspects to feasibility:

- *Technical feasibility* concerns whether the proposed system can be completed from a technical standpoint. It is important to consider whether the system fits with the existing technical infrastructure.
- *Economic feasibility* relates to whether the proposed system is affordable and worth the investment.
- *Organizational feasibility* concerns whether the project fits with the organization's existing strategies and practices. If the system does not fit with current ways of operating, the organization needs to determine what needs to be changed and whether these changes are feasible. Organizational feasibility should also consider whether the organization has access to the human resources necessary to complete the project.
- *Legal feasibility* concerns whether the proposed project violates any laws or regulations.
- *Ethical feasibility* considers whether the system fits within the ethical guidelines and practices of the organization and its industry.

The major deliverables of the planning phase are the feasibility analysis, project initiation document, and the project management plan. The project initiation document specifies the high-level goals of the system and describes the business case for the system. (A business case describes the system and why it is needed.) The feasibility analysis results are often included in the business case. The project management plan documents the scope of the project, identifies major tasks and resources, and describes any interrelationships with other projects.

Requirements Phase

The goal of the requirements phase is to uncover and document the functions that the system should provide and desired levels of performance. In other words, in the requirements phase, the task is to figure out and document what the system should do. There are two types of requirements: functional and nonfunctional. **Functional requirements** describe how the system should interact with users and other systems. An example of a functional requirement is "The system shall allow customers to look up the price of a product." **Nonfunctional requirements** are related to constraints on the system—for example, how well it should perform. An example of a nonfunctional requirement is "The system shall be available at least 23 hours per day."

Requirements elicitation involves gathering requirements from various stakeholders and is a primary task in the requirements phase. Gathering the requirements for a system is often a difficult, time-consuming, and error-prone process. To make things worse, requirements may change during the process. Despite the difficulties, it is very important to fully understand the system's requirements; they form the foundation for the rest of the development project.

The output of the requirements phase is a specification of the system's functional and nonfunctional requirements. Typically, this takes the form of a set of requirements statements (sometimes called *system shall* statements) or a set of use cases, in either text or diagram form. A *system shall* statement simply defines a requirement, such as "The system shall allow teachers to enter student grades." A **use case** describes a sequence of actions that results in an outcome for some actor. An actor can be a user or another external system. Figure 10.2 shows a use case diagram. (Use case diagrams are part of the Unified Modeling Language, which was developed as a standard way to specify models related to software systems.)

It is important to understand that requirements should specify *what* the system must do but not *how* the system should meet the requirements. This allows the system designers flexibility in the design of the system, which is covered in the next phase.

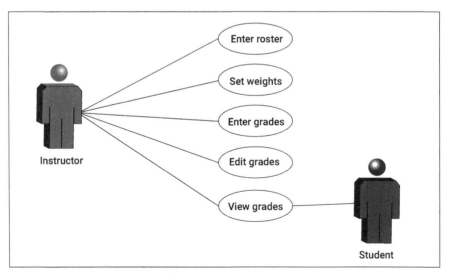

Figure 10.2 Use Case Diagram Example

Design Phase

The main goal of the design phase is to describe in detail how the system will meet the requirements specified earlier. The requirements form the foundation of the system design. The focus here is on *how* the system will satisfy the requirements, although there are elements of the design phase that still deal with what the system is intended to do. These elements are known as the *logical system design*, which includes abstract models of what the system is intended to do. An example of such a model is the database diagram we saw in Chapter 5 (Figure 5.4). In addition to the database model, the user interface and business rules enforced by the system need to be designed. The design phase also specifies the plans for implementing and testing the system. The system design document is a major deliverable of the design phase. This document describes, in detail, the system requirements, operating environment, database design, user interface, and processing logic. Other deliverables include an implementation plan and a training plan.

Development Phase

During the development phase, the outputs of the design phase are converted into the actual information system. This is the phase in which the system is actually constructed. The software is written, databases are created, and any necessary hardware is installed. It is often necessary to integrate the new system with existing information systems; if needed, the integration occurs during the development phase. System testing is a critical part of the development phase. Testing occurs both during the development of the software and when the entire system is complete and integrated. Software testing is a complex, time-consuming, but critical aspect of systems development. It is necessary to thoroughly test the system prior to actual implementation. Errors are very expensive to correct when they are not discovered until after the system is implemented. The errors may also lead to business disruptions. The development phase's deliverables include the application software, data and files for testing, and documentation on how the software components interact with one another and external systems.

Implementation Phase

The goal of the implementation phase is to make the system operational in a production environment. One important implementation phase activity is to notify users of the rollout of the new system. This notification should include the implementation schedule, differences in work processes and responsibilities brought about by the new system, and the process for obtaining technical support for the new system. User training is also an important activity in this phase. Often, it is also necessary to convert data from the existing system into the new system. For example, a new product cataloguing system might need to have data from the old system migrated into the new system. Finally, it is a good practice to have a postimplementation review that documents the project to date and makes recommendations for improvements for future projects. The main deliverable of the implementation phase is the system itself, installed and running correctly. The postimplementation review report is another deliverable from this phase.

There are four common approaches to system migration:

- *Direct changeover (also called immediate changeover):* At a specific point in time, the old system is no longer used and is replaced by the new system. While this has the advantage of being rapid, it is very risky. If the new system has problems, the functions available in the old system are also unavailable.
- *Parallel operation:* Both the old and new systems operate for some period of time. The parallel approach is less risky than direct changeover but also takes longer and is more expensive. The parallel approach is necessary for systems that cannot be taken out of service while changing from the old to the new system.
- *Phased implementation:* The new system is implemented in stages. This approach is an attractive compromise between direct and parallel approaches. However, it is best suited for systems that have clear modules that can be implemented in a phased manner. Systems with significant interdependencies between modules are not suitable for phased implementation.
- *Pilot operation:* The new system is implemented in a business unit (such as a store or department) or location (such as a state or country). The initial pilot is followed by other pilot changeovers or a full changeover to the new system. Each pilot changeover must use one of the other approaches.

Maintenance Phase

The final phase in the SDLC is the maintenance phase. This is usually the longest, most costly phase of the life cycle. It is not unusual for more than half of the total system cost to occur in the maintenance phase. The goal of this phase is ensuring that the system operates properly to meet current needs. This involves fixing any problems that are uncovered and may also involve modifying or adding to the programs to meet new needs. In addition, the system may be modified to improve performance or take advantage of new technologies. For example, an online ordering system may be modified to allow customers to order from their smartphones.

Although we presented the SDLC as a very linear process, where one phase follows the next in an orderly fashion, reality is rarely so neat. During the systems development process, it is often necessary to revisit an earlier phase when new needs arise or problems are discovered.

As you will discover in Learning Activity 10.2, there are advantages and disadvantages to the traditional SDLC. These drawbacks led to the development of alternative methodologies. In the next section, we describe several of these alternative approaches.

LEARNING ACTIVITY 10.2

Advantages and Disadvantages of the SDLC

The traditional SDLC has been around for a long time (which is why it is called "traditional") and has been used successfully for many systems development projects. However, it is not without its drawbacks. With a partner, research the SDLC approach and develop a list of three advantages and three disadvantages to the traditional SDLC. Be prepared to share your list.

Alternative Methodologies

Prototyping

Prototyping is an approach to deal with aspects of a full methodology; it is not a complete, stand-alone methodology. In a prototyping approach to systems development, you begin with an initial investigation of the main requirements of the system. It is not necessary to have a full understanding of all requirements. After the initial investigation, a small-scale mock-up of the system (called the **prototype**) is built and reviewed by stakeholders. The stakeholders point out flaws, which reveals new or misunderstood requirements. The prototype is refined and reviewed. This cycle repeats until the stakeholders are satisfied that all requirements are met. Then the full system is implemented. Some prototypes are "throwaway" prototypes, which means that they are not intended to be part of the final system; their purpose is to uncover requirements and problems. Sometimes, however, prototypes become part of the final system. This is known as *evolutionary prototyping*.

Increased stakeholder involvement is a major benefit of the prototyping approach. The stakeholders know the particular business area addressed by a system much better than the systems developers and having them involved throughout the development process can lead to systems that better meet their needs. In addition, stakeholder involvement may increase satisfaction with the final system. Prototyping is also potentially more effective for uncovering system requirements. This is especially important when requirements are unclear, ambiguous, or difficult to define. In some cases, a prototyping approach may be less expensive than a traditional SDLC approach.

There are also drawbacks to the prototyping approach. Prototyping can lead to poorly designed systems that do not perform well when scaled up for actual use. It is also difficult to estimate how many iterations are required, which makes it difficult to estimate costs and schedules. Usually, prototyping results in less-well-documented systems, which may cause problems later (for example, when the system needs to be revised or integrated with new systems). Finally, nonfunctional requirements (such as performance requirements) are difficult to establish using prototypes.

Prototyping is best for systems that require extensive user interaction and for which the analysts have some understanding of the business issues to be addressed,

---

Information Technology Failures

Canceled or failed information technology (IT) projects and incorrect software cost the global economy billions of dollars each year. In addition to the software failures presented in the focusing story, here are a few more examples of major project failures:

- In March 2015, a German basketball team had to start a game 17 minutes late because the computer connected to the scoreboard started an automatic Windows update. As a result, the delayed game meant that the winning team actually lost and was thus relegated to a lower division.
- In 2015, Target closed its operations in Canada after an attempt to implement SAP throughout its stores failed and the resulting supply chain issues left shelves empty. When the 133 Canadian stores were closed, 17,600 employees lost their jobs.
- In August 2016, Australians were unable to complete their census on their designated day because the online system failed.
- In August 2018, 400 homeowners found that their mortgages from Wells Fargo were foreclosed accidentally because of a calculation error in the accounting software the company was using.
- In March 2018, a self-driving car from Uber Technology made a wrong decision about "an object" and ended up killing a pedestrian in Phoenix, Arizona.

While these failures and those presented in the focusing story mostly show bigger businesses, software and project failures affect all types of organizations worldwide, including many governmental agencies. The Consortium for Software Quality estimated that in 2018, the cost of poor quality software in the United States was approximately $2.26 trillion, composed of losses from software failures (37.46%), issues with legacy systems (21.42%), finding and fixing defects (16.87%), and troubled or canceled projects (6.01%). The remaining 18.22% is assigned to what they call technical debt, which refers to what is the estimated cost of fixing problems still in the software code. Another study suggests that in 2016 companies had to spend on average $97 million to resolve failures for every $1 billion they invested in IT projects (Florentine 2017). Why do projects fail? In some cases, the initial estimates of the costs, resources, and delivery date are unrealistic. Other causes include misunderstanding requirements, "feature creep" (expanding requirements), and inadequate testing.

Sources: Florentine, S. 2017. "IT Project Success Rates Finally Improving." *CIO Magazine*, February 27. https://www.cio.com/article/3174516/it-project-success-rates-finally-improving.html; Krasner, H. 2018. *The Cost of Poor Quality Software in the US: A 2018 Report.* Consortium for IT Software Quality, September 26. https://www.it-cisq.org/the-cost-of-poor-quality-software-in-the-us-a-2018-report/The-Cost-of-Poor-Quality-Software-in-the-US-2018-Report.pdf; Hamrouni, W. 2017. "5 of the Biggest Information Technology Failures and Scares." *Exo*, August 1. https://www.exoplatform.com/blog/2017/08/01/5-of-the-biggest-information-technology-failures-and-scares/; Head, B., and D. Walker 2016. "The Enormous Cost of IT Project Failure." *In the Black*, November 1. https://www.intheblack.com/articles/2016/11/01/enormous-cost-it-project-failure.

LEARNING ACTIVITY 10.3

Developing by Modeling

Systems development can be very labor intensive for large scale organizational applications. However, today's new easy-to-use development modelers allow even those with basic development knowledge to create their own applications. In this learning activity, you will use the Mendix (www.mendix.com) modeler to create an app assigned by your instructor.

1. Watch the tutorial (https://docs.mendix.com/howto/tutorials/) for the type of app your instructor has assigned.
2. Register with your EDU account (to obtain free access) at https://signup.mendix.com/link/signup/.
3. Develop your assigned app.

especially when functional requirements are not well understood. Prototyping is less well suited for systems that have well-defined requirements, when nonfunctional requirements are especially important, or when future scalability is important.

Rapid Application Development (RAD)

As the name implies, rapid application development (RAD) is intended to develop systems more quickly than traditional methods. It is an iterative method that uses rapid development of prototypes rather than engaging in extensive up-front planning. RAD typically breaks software development projects into smaller chunks to reduce overall project risk. This also facilitates making changes as the project's requirements emerge. The focus is usually on filling organizational needs rather than creating technically perfect systems.

Active user involvement is a key aspect of RAD. The idea is to work with users to determine high-level requirements and then build a prototype based on those requirements. Developers review the prototype with the users, who provide feedback. The prototype is refined based on the feedback. This process continues until the full system is developed. Interestingly, some of the same methods used in the traditional SDLC approach can also be applied in RAD. The RAD process is illustrated in Figure 10.3.

Development tools are integral parts of a RAD system. Computer-aided software engineering (CASE) tools, for example, automate many software design and development tasks. Tools for quickly developing graphical user interfaces are often included in a RAD environment. These tools allow developers to build user interfaces by dragging and dropping interface elements onto a screen instead of having to write the software code.

In addition to delivering systems more quickly than the traditional SDLC, RAD often results in lower costs. Because user involvement is such an important aspect of RAD, systems developed using RAD often have greater buy-in from users. RAD also facilitates evolving requirements well. However, the speed of RAD may result in

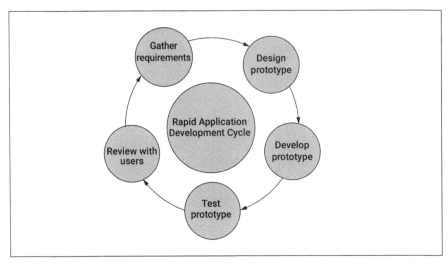

Figure 10.3 Rapid Application Development (RAD) Cycle

lower quality software. Because users often want more and more features, "feature creep" may be more likely with RAD. Since there is less focus on nonfunctional requirements, systems may not perform well when scaled to more users or greater volumes of processing.

RAD is best suited for smaller-scale projects with well-focused organizational objectives and committed users. RAD does not work as well for very large projects or systems where performance and quality are critical. For example, you might not want to use RAD for a safety-critical system such as an air traffic control system.

DevOps

DevOps is an approach to software development that focuses on rapid development of applications and services. The idea is to rapidly build, test, and release software products while continuously monitoring the released software. Figure 10.4 shows the main elements of DevOps. It is called DevOps (Development-Operations) because the operations of the software are no longer separated from the development of the software as in traditional approaches.

The success of DevOps in recent years is based on several principles, such as delivering frequent but small updates to the software (continuous delivery), breaking large projects into smaller projects (this is also called microservices—building a set of small services to build an application), and, importantly, the concept of continuous integration. Continuous integration is when developers place their newly developed code or changes to code into a central repository where tests are run to more quickly identify possible software errors. It is also very important to note the role of logging and monitoring of the released software. As part of DevOps, companies constantly collect data about the software's performance and analyze and log those data so they can better understand how any changes to the software are affecting its users. The benefits of DevOps are numerous. First, because employees working within operations and within the development enterprise are jointly involved in the development, operations, and monitoring of the software, companies experience better collaboration and communication. Of course, the goals of DevOps being speed of development and rapid delivery are also benefits. In addition, many argue that DevOps leads to more reliable and secure software.

In a 2019 report on DevOps by Dora and Google Cloud (https://services.google .com/fh/files/misc/state-of-devops-2019.pdf), it is suggested that there are four key metrics to measure the performance of a company's DevOps usage: deployment frequency, lead time for changes, time to restore service, and change failure rate. The report suggests that the best "elite" DevOps performers are able to deploy on-demand services with multiple "deploys" per day, that it takes less than a day to go from code that is ready for deployment to actual deployment, that it takes less than one hour to restore service after an incident such as an outage, and that less than 15% of the changes that are released result in lower service.

This is just a sampling of some of the more common systems development methodologies. There are many others. Several of these fall into the category of agile development methodologies. *Agile development* occurs in self-managed, cross-functional teams whose members include information systems professionals and user representatives. Rather than being process oriented, agile methods are people oriented. Agile methods emphasize multiple iterations and continuous feedback from user representatives. Many specific methodologies fall into the family of agile methods. These include extreme programming (XP), which uses frequent releases with short development cycles, and Scrum, which features work "sprints" of one to several weeks, at the end of which the development team meets with stakeholders to assess progress and plan what needs to be done next. Another approach is called the Spiral Model, which is like an iterative SDLC that focuses on identifying and reducing risks when developing critical systems. As software and

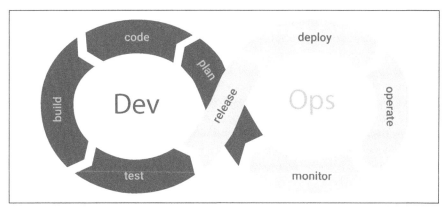

Figure 10.4 DevOps

its uses continue to evolve, we expect that software development methodologies will do the same.

As you can see, there are many different methods for developing systems; choosing the correct methodology depends on many factors. However, developing a system from scratch is not the only option available when an organization needs a new system. It may be preferable to buy rather than build the system. This choice is the topic of the next section.

Build or Buy Decision

When the need arises for a new software-based system, managers face a long-standing decision: whether to create the software or purchase existing commercial, off-the-shelf (COTS) software. This decision is complex and costly if you get it wrong. On the surface, it may seem like it is always preferable to purchase an existing system rather than going through the time, cost, and effort of creating a new system. However, the decision is not that easy. Often, the COTS software will need significant customization to meet business needs. In these cases, it may be more effective to build a new system. In between the extremes of custom and COTS software lie hybrid approaches. These hybrid approaches involve modifying COTS software to better fit organizational needs or combining COTS components to create the desired solution.

Custom-built software is software that is built from scratch using a few commercially obtained components. The main benefit of the custom approach is that the resulting system will likely be a better fit with business requirements. As a result, the custom system may be used to create a competitive advantage. In addition, the organization will have greater control over adding or modifying functions after the initial system is implemented. However, custom solutions usually take much longer and are more costly to develop than COTS or hybrid solutions. In addition, custom solutions are more prone to errors, which can further increase costs and time to delivery.

COTS software is a complete (or nearly complete) solution obtained from a third party. Examples range from office productivity systems to ERP systems. Often, COTS software is less expensive than custom or hybrid approaches. COTS software may also be higher quality, especially if it is a widely used system. Usually (although not always), COTS-based solutions can be provided in a much shorter time frame than custom or hybrid solutions. There are downsides to COTS software, however. First, COTS software often requires changing existing business practices to fit the COTS software. Second, companies are largely at the mercy of the vendor or consultants for support, including enhancements and fixes. In addition, a financially weak vendor may be unable to provide proper support or maintenance. In extreme cases, the vendor may go out of business.

Hybrid solutions may use components from different vendors to create the complete solution or may involve customizing COTS software by modifying programming code or adding small programs called *scripts*. In ideal cases, the hybrid solutions offer the best of custom and COTS approaches. The hybrid solutions usually better fit business requirements and existing processes but can be delivered less expensively and more quickly than custom solutions. However, hybrid solutions also suffer from some of the same drawbacks as custom-built systems and COTS solutions. In addition, it may be difficult to find the expertise necessary to modify or combine the COTS components. Figure 10.5 provides an overview of some of the characteristics of the three approaches.

When making the build versus buy decision, a number of questions must be addressed:

- *Is there COTS software that will meet the business's needs?* If the answer to this question is no, your choices are narrowed to custom development or a hybrid approach. If no COTS software can meet your organization's

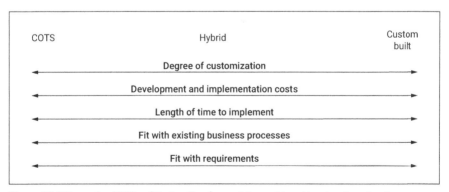

Figure 10.5 Comparing COTS and Custom Development Source: Adapted from http://www.baselinemag .com/c/a/Application-Development/Buy-vs-Build-Software-Applications-The-Eternal-Dilemma.

needs, you have to create the software or customize COTS software. In these cases, you may have the option of outsourcing the development or customization. (We discuss outsourcing later in this chapter.)

- *Will the software give your company a competitive advantage?* If COTS software is an option, the next question is to consider whether the system will give your company a competitive advantage. If so, the scale is tipped toward building custom software. If not, COTS software may be a better choice. To understand why, think about the example of a hotel chain. To such a company, a payroll system would not be likely to provide any competitive advantage. Payroll is a routine function, so it is a good candidate for COTS software. In contrast, having a custom system for analyzing the behaviors and preferences of frequent guests may give the company an advantage over its competitors and is therefore a better candidate for a custom-built system.

- *Does the COTS software fit with business processes?* COTS software will rarely match exactly with existing business processes. As a result, you will usually have to modify at least some business processes to use COTS software. Often, this is not a problem, but sometimes the changes necessary are so extensive that it may be better to build a custom system. The key issue is whether it is practical to alter existing processes to fit the way the COTS software operates.

- *Does the COTS software fit with existing IT architecture and staff?* To be successful, COTS software must fit with your organization's existing IT. Here is a simplistic example. Suppose that your organization used only Windows-based PCs. Macintosh-only COTS software would not be a good fit with your existing IT. It is also important that your IT staff be able to support any COTS software you use. If the existing staff cannot provide support, you will need to either contract with an outside company for support, train your existing staff, or hire new staff. All of these add to the cost and complexity of implementing the COTS software.

- *Do you have the capability to create the required system?* There are times when an organization simply does not have the capability to create a system that meets requirements. In these cases, COTS software is often an attractive option. Outsourcing the development of the system is another option.

- *When is the system needed?* A COTS solution can almost always be implemented faster than a custom-built system. As a result, you should consider the "go-live" goal for the new system when making the build versus buy decision.

- *What is the total cost of ownership for each approach?* Finally, you should compare the total cost of ownership (TCO) for both approaches. The cost of purchasing or building the system is not the only cost you should consider. There are also support and maintenance costs, costs to integrate the system into the existing architecture, and costs related to enhancing the system in the future. One important point to keep in mind is that COTS systems usually are changed much more often than custom-built systems. When the vendor releases a new version of the software, you may incur additional costs related to deploying the new version. Of course, you could stay with the old version, but vendors often drop support for older versions, which may lead to problems.

Using Open Source Software in Business

Open source software is software that allows users to access the underlying source code for an application. Having the source code for a program allows users to modify the program's functions. The **source code** is the human-readable form of a software program. Typically, this source code is translated into computer instructions through a process called *compiling*. The compiled code is not in a form that is easily understood by a human. The licensing for open source software also allows users to modify the program, subject to some conditions. The Open Source Initiative provides a detailed definition for "open source" at http://www.opensource.org/docs/osd.

Open source software is usually available free of charge, although it is common for companies to distribute (for a fee) open source software along with support contracts or other enhancements. Open source software has been in existence for many years but has been largely unknown to the general population until recently. The rise of popular open source applications such as Mozilla Firefox (a Web browser), OpenOffice (a productivity suite), and operating and systems software such as the Linux operating system and Apache Web server has increased the awareness of open source software. In the world of smartphones, Google's open source Android operating system has been credited with a rapid rise in the number of smartphones running Android. Business Example Box 10.1 provides some statistics in this regard.

Businesses are adopting open source software for a variety of uses. For example, as of August 2020, the Apache Web server was used on 36.5% of websites globally, followed by the Nginx web server software at 32.5% of websites (W3Techs 2020). Both Apache and Nginx are open source software. The highest ranked non–open source web server software is Microsoft Internet Information Services, used for 7.9% of websites.

LEARNING ACTIVITY 10.6

How Open Source Software Impacts Build versus Buy

Open source software is software that provides access to the program's code, which means that it can be freely modified. Often organizations must decide whether to build custom software or purchase a commercial software package. In pairs, discuss how open source software might impact the build versus buy decision. Prepare a bullet list of your findings.

BUSINESS EXAMPLE BOX 10.1

Open Android versus Closed iOS

Which smartphone do you own? There are many on the market, but the two clear leaders are Google's Android-based phones and Apple's iOS-based phones. Besides the aesthetics of the phones, there is a significant difference in their operating systems. Android is open source, while iOS is not. According to Statista, a company that analyzes market and consumer data, iOS had a 13% global market share of smartphones in 2019, while Android held 87%. This can be explained by more companies using the Android operating system for their phones, including Samsung, Motorola, and LG. iPhones were representing roughly half of the smartphone sales in the United States as of May 2020, with 105 million subscribers. In 2019, there were 2.5 million Android apps and 1.8 million iPhone apps.

Sources: https://www.statista.com/statistics/266572/market-share-held-by-smartphone-platforms-in-the-united-states/; https://www.digitaltrends.com/mobile/android-vs-ios/.

Why would businesses use open source software? There are a variety of reasons. The most obvious is the potential cost savings. Most open source software can be acquired without charge. It is important to note that this does not mean that the software can be implemented without cost. There are ongoing costs associated with installation, support, and continuous training; these costs should not be ignored. Another reason businesses use open source software is to avoid being dependent on proprietary software vendors for ongoing development and support. For example, users of Microsoft Office are completely reliant on Microsoft for current developments (such as fixing errors or enhancing the software). With open source software, the larger community can carry out these functions. This means that continuous development and support are possible even if the creator of the software goes out of business. There is also some evidence that mature open source software is often more reliable than proprietary software. In addition, open source software is more likely to use open standards for things such as file formats. This reduces the reliance on any particular software vendor. Additional flexibility comes from the fact that users have the ability to modify the software as needed. This is not possible with proprietary software. (It should be noted that most users lack the technical skills necessary to modify software, which reduces the value of this benefit.)

One interesting aspect of open source software is that a community of volunteer programmers often does ongoing development. The structure of an open source development community is like an onion. Core developers are at the center, and passive users are at the outermost layer. The core developers do most of the programming for both the original creation and ongoing development. Developers at outer layers report errors and request features and may also contribute fixes for the errors. Active users also report problems but do not contribute any programming code, and they often help less experienced users in community forums. Thus the community not only supports current users but also provides a platform for ongoing enhancements.

Outsourcing Information Systems

Information technology **outsourcing** occurs when an organization goes to an outside company to provide IT services that were previously provided internally. (Outsourcing is not unique to IT. For example, some companies outsource the manufacturing of products that they sell.)

Outsourcing Models

There are many different outsourcing models. These differ in the degree of outsourcing and in what is outsourced. Some organizations contract with a service provider to provide all IT-related services. This is known as **full or complete outsourcing**. Full outsourcing allows an organization to focus on its core activities and may also reduce IT costs. Service providers may be able to provide services less expensively because of economies of scale.

In contrast, in **selective or partial outsourcing**, only certain aspects of IT are outsourced to a service provider. For example, AT&T at one point outsourced the maintenance of its ERP system to Accenture. There are different approaches to deciding what to outsource. Under **process-based outsourcing**, an organization outsources a particular function or business process. Functions that are often outsourced include help desk support, payroll and benefits processing, and telecommunication services. Outsourcing specific functions is not new. For example, ADP (adp .com) has been providing payroll services to organizations for more than 70 years.

Personnel outsourcing is another form of outsourcing that has a long history. This form of outsourcing allows organizations to meet staffing needs without the long-term costs associated with hiring permanent workers. Bringing in temporary,

---

LEARNING ACTIVITY 10.7

What to Outsource

Outsourcing occurs when a company turns to an outside vendor to provide some service. While the definition of outsourcing is easy to understand, deciding what to outsource is much harder. For each of the following business scenarios, decide whether the system or operation should be outsourced. Explain your answers.

Organization	System/Function
1. Engineering company	Payroll and benefits management system
2. Regional beer brewery	Customer relationship management
3. University	Learning management system (e.g., Blackboard, Moodle)
4. Financial portfolio management company	Investment analysis software
5. Public radio station	Donor management system

contract-based workers may be effective when an organization needs specific expertise on a short-term basis or when the current workload is more than the internal IT staff can handle. In these cases, outsourcing can speed time to deployment for a system and avoid costs related to training the current staff.

Another approach is **project-based outsourcing**. As the name implies, this involves contracting with a vendor for a specific project. Since projects have specific start and end points, when the project is finished, the outsourcing engagement ends. Organizations outsource projects when they lack the internal resources necessary to complete the project. Earlier we mentioned outsourcing as an alternative for developing a custom system. This is an example of project-based outsourcing. When the system is developed and deployed, the outsourcing engagement ends. Interestingly, upgrades to the system may lead to future outsourcing engagements.

Application outsourcing involves a service provider handling activities related to a specific software application for a fee. There are different forms of application outsourcing. With application hosting, the client organization is responsible for operating the IT infrastructure associated with the application, but the client must purchase the software and pay any annual license or maintenance payments to the software publisher. Amazon's Web Services is an example of application hosting. Organizations can deploy applications on Amazon's systems. Amazon handles keeping the servers up and running but is not responsible for maintaining the application. This arrangement is often very cost effective, especially when there are demand spikes that may require additional server capacity. Software as a service (SaaS) provides on-demand Web-based access to an application on a per-use or per-user basis. Salesforce.com is a well-known provider of SaaS. Salesforce.com provides organizations with customer relationship management software for a monthly per-user fee. Different features lead to different fees. For example, Salesforce pricing for Customer Relationship Management (discussed in Chapter 12) in 2020 varied from $25 to $300 per user per month depending on the desired level of service. The SaaS model provides a large degree of flexibility and can be cost effective. The client organization does not have to worry about maintaining the software or the servers. SaaS also provides very good scalability. However, there are security and privacy concerns with SaaS.

Benefits and Risks of Outsourcing

Outsourcing, when done well, can result in a variety of benefits. A major benefit of outsourcing is that it can give organizations access to specific expertise not available in-house. There can also be cost savings with outsourcing. These are normally associated with the vendor having lower cost structures, through either economies of scale or lower wage rates. Outsourcing also shifts employee-related fixed costs to variable costs. In addition, outsourcing may allow for more rapid delivery of projects through additional expertise and personnel and can be an effective means of managing capacity. It is generally easier to contract for additional capacity than to add it through other means. Selective outsourcing allows management to focus on core

BUSINESS EXAMPLE BOX 10.2

Reversing the Outsourcing Decision

In 2005, Houston–based utility Reliant Energy outsourced some of its IT jobs to Accenture. As a result, Reliant cut approximately one-third of its 340 IT jobs. In 2009, Reliant reversed this decision and began hiring IT workers so it could bring the outsourced work back in house. Neither Reliant nor Accenture would comment on the reason for the reversal.

There are numerous other examples of "backsourcing" (bringing a previously outsourced function back in house). In 2012, the Royal Bank of Scotland canceled a contract with an IT provider after the updates they were doing to their systems left millions of customers without access to their accounts to view them or even perform transactions like withdrawing funds. Another example of backsourcing comes from General Motors in 2012, which backsourced its IT from HP by hiring 3,000 HP workers who had been providing outsourced services to GM. This provided GM the ability to hire the expertise back into their company to provide the skills they lost when they chose to outsource their IT department.

Why do companies backsource? Most of the time it is because of time or cost overrun for products that often do not meet expectations. For example, in 2007, IBM entered into a contract with the Queensland health department for $6 million but realized by mid-2008 that it would cost $27 million. The reality was even worse, since it ended up taking many more years and costing $1.2 billion, which is 16,000% over cost, for a faulty product to be developed.

Sources: Flinders, K. 2013. "Insourcing + Outsourcing - Outsourcing = Backsourcing–The Academic's View." *Computer Weekly*, April 30. https://www.computerweekly.com/blog/Investigating-Outsourcing/Insourcing-outsourcing-outsourcing-backsourcing-the-academics-view; Flinders, K. 2012. "General Motors Recruits 3,000 HP IT Workers." *Computer Weekly*, October 19. http://www.computerweekly.com/news/2240167895/General-Motors-recruits-3000-HP-IT-workers; Nead, N. 2020. "6 Examples of Outsourcing Failure." *Hacker Noon*, April 22. https://hackernoon.com/6-examples-of-outsourcing-failure-yv223y4u; Thibodeau P. 2010. "Reliant Energy Takes Back Outsourced IT Project." *Computerworld*, January 29. https://www.computerworld.com/article/2523204/reliant-energy-takes-back-outsourced-it-projects.html; Sealock, A., and C. Stacy. 2013. "Why Some U.S. Companies Are Giving Up on Outsourcing." *Forbes*, January 16. https://www.forbes.com/sites/ciocentral/2013/01/16/why-some-u-s-companies-are-giving-up-on-outsourcing/#4975b0bb65af.

business activities. Finally, when outsourcing sites are spread across time zones, offshore outsourcing can lead to 24-hour productivity.

There are also risks associated with outsourcing. In some cases, outsourcing can lead to a loss of competencies within the organization. This is especially problematic when this loss is related to the core activities of the organization. There are also risks related to the outsourcing relationship. There is a risk of becoming dependent on the service provider, which can lower bargaining power in later negotiations. In addition, there is a danger that the vendor may behave opportunistically and take advantage of the organization's lack of expertise or inadequate monitoring. Risks from opportunism are especially high when the outsourcing contract is not well structured. Outsourcing may also hurt the morale of the remaining workforce. Finally, there are risks related to confidentiality. Outsourcing vendors must ensure that sensitive information is properly protected. This is both a technical and human

TABLE 10.1 Benefits and Risks of Outsourcing	
Benefits of Outsourcing	Risks of Outsourcing
• Access to expertise not available in-house	• Loss of internal competencies
• Potential cost savings	• Dependence on service provider (lowers bargaining power)
• Shifting fixed costs to variable costs	• Opportunistic behavior by service provider
• Effective way to manage capacity	• Lower morale of remaining workers
• Focus on core activities	• Compromises to confidentiality
• Potential for 24-hour productivity	

problem. Even when the proper technical security measures are in place, the employees of the service provider can compromise confidentiality. The benefits and risks of outsourcing are summarized in Table 10.1.

Making the Outsourcing Decision

The decision of whether to outsource is very complex and depends greatly on the particular set of circumstances. However, there are some consistent factors that should be considered (Grover and Teng 1993):

- *How mature is the IT system in question?* Systems that have been around a long time are better candidates for outsourcing. There are several reasons for this. First, if the system is mature, organizations are less likely to be able to gain a competitive advantage from the system. Second, it is easier to find competent vendors for mature systems. This also means that organizations are more likely to have a choice of vendors, which is helpful during negotiations, both initially and in the future.
- *How significant is the system to the organization's competitive advantage?* One common reason for outsourcing is to allow the organization to concentrate on core activities. Outsourcing is more suitable for systems that do not provide a competitive advantage. There is also a greater risk of loss of control when you outsource strategic systems; problems with strategic systems usually have serious consequences.
- *How does the organization's IT capability compare with that of its competitors?* In cases where an organization has relatively weak IT capabilities, it may be better to outsource. This is particularly true if the organization is weak in IT related to the system in question.
- *Are there cost advantages to outsourcing?* Of course, the question of costs is also important. Even if there are no savings through outsourcing, it may still be a good idea if outsourcing allows the organization to focus on critical areas or if the organization lacks the necessary capabilities.

Geographic Considerations

Where to outsource is an important aspect of outsourcing. (Note that choosing a vendor is also important, but here we focus on location.) **Offshore outsourcing** involves using a vendor that provides services from a location outside the client organization's region. A U.S. computer manufacturer outsourcing technical support to India is an example of offshore outsourcing. This is often done to reduce costs. Developing countries have much lower wages than developed countries, so offshore outsourcing can be very cost effective. However, there may be a backlash to offshore outsourcing. This can result from customers being upset about jobs sent overseas. Resistance can also come from poor language skills or cultural differences. Language and cultural differences, along with distance, can make offshore outsourcing arrangements difficult to manage.

Nearshore outsourcing is similar to offshore outsourcing. The difference is that nearshore outsourcing occurs close to the client's home location. If a U.S. company outsources to a Canadian firm, this would be a nearshore arrangement. The reduced travel times may make management easier. Cultural and language differences may be reduced, but this is not always the case. A U.S. company outsourcing to Mexico may still face considerable language and cultural barriers. A nearshore arrangement often results when a company wants to gain some of the cost savings of offshore outsourcing while reducing some of the management, language, and cultural difficulties.

Onshore outsourcing is outsourcing to a firm located in the same country. Onshore outsourcing may offer significant cost savings while eliminating the cultural and language barriers of offshore outsourcing. In addition, onshore outsourcing is less likely to bring about a customer backlash, since the jobs are staying in the country. Cost savings result from differences in wage rates for different areas of a country. Wages in rural areas are often considerably lower than in urban areas. As a result, rural onshore outsourcing saves significant money over urban areas and is only slightly more expensive than offshore outsourcing.

Chapter Summary

In this chapter, we discussed various approaches to developing information systems. We began by describing the time, cost, and quality trade-off. Then we talked about different software development methodologies, including the traditional systems development life cycle and its alternatives. We also discussed the build or buy decision and advantages and disadvantages of using open source software. The chapter concluded with a discussion of outsourcing. Here are the main points discussed in the chapter:

- Software development methodologies provide discipline to the development process by defining processes, roles, and deliverables.

- The traditional systems development life cycle (SDLC) is a semisequential, phased approach. There are different versions of the traditional SDLC.
- The SDLC described in this chapter consists of six phases: planning, requirements, design, development, implementation, and maintenance.
- Alternatives to the traditional SDLC include prototyping, rapid application development, and DevOps, as well as other agile methodologies.
- The decision of whether to build custom software or purchase commercial, off-the-shelf software is complex.
- Open source software allows users to access the underlying source code for the application, which allows users to modify the program.
- Open source software is usually free, although some companies charge a fee for ongoing support.
- Outsourcing occurs when an outside organization provides IT services that were previously provided internally.
- Outsourcing models include full (complete), process-based, personnel, project-based, and application outsourcing.
- There are both benefits and risks associated with outsourcing.
- Factors affecting the outsourcing decision include the maturity of the system, the system's significance to the organization's competitive advantage, the organization's IT capability, and cost.
- Offshore outsourcing occurs when an organization receives services from a company outside the organization's geographic region.
- Nearshore outsourcing occurs when an organization receives services from a company within the organization's geographic region but outside its home country.
- Onshore outsourcing occurs when an organization receives services from a company in the same country.

Review Questions

1. What is the major benefit of a software development methodology?
2. What aspects of software development does a methodology specify?
3. Name and briefly describe the phases of the systems development life cycle described in the chapter.
4. Name and briefly describe the aspects of feasibility described in the chapter.
5. Name and describe three alternatives to the traditional SDLC.
6. Name the advantages and disadvantages of each of the software development methodologies described in the chapter.
7. List the advantages and disadvantages of custom; commercial, off-the-shelf (COTS); and hybrid approaches to software.
8. List the main characteristics of custom-built and COTS software.

9. What are the main factors to consider when deciding whether to build or purchase software?
10. Define open source software.
11. What advantages does open source software offer to businesses?
12. List and briefly describe the outsourcing models presented in the chapter.
13. List the major benefits and risks of outsourcing.
14. What factors should organizations consider when deciding whether to outsource?
15. How do offshore, nearshore, and onshore outsourcing differ?

Reflection Questions

1. What is the most important thing you learned in this chapter? Why is it important?
2. What topics are unclear? What about them is unclear?
3. What relationships do you see between what you learned in this chapter and what you have learned in earlier chapters?
4. How is the traditional systems development life cycle related to the major functions of an information system that you learned about in Chapter 2?
5. What is the difference between functional and nonfunctional requirements? Why are they both important? Which do you think is harder to determine? Why?
6. Your organization needs to build a system to help users carry out a business process that is very unique to your organization.
   a. Should your organization build this system from scratch or use a COTS system? Why? What additional information would be helpful in making this decision?
   b. Do you think this should be outsourced? Why? What additional information would be helpful in making this decision?
7. Commercial, off-the-shelf (COTS) software is often of higher quality than custom-built software. This is especially true if the COTS software is widely used.
   a. Why might COTS software be of higher quality?
   b. Why would this be more likely if the COTS software is widely used?
8. What ethical issues are related to the decision to outsource a function? Are these issues different for offshore, nearshore, and onshore outsourcing?
9. There have been several instances of software development companies making formerly proprietary software open source. (See http://en.wiki pedia.org/wiki/List_of_formerly_proprietary_software for a partial list of proprietary software that has been shifted to open source.) Why would a software development company do this? What would the company potentially gain? What would it potentially lose?

10. Select two software development methods and discuss their possible impacts on information security (Chapter 8) and privacy (Chapter 9).

Additional Learning Activities

10.A1. Form groups of two or three students:
- One of you (the explainer) will answer the following question: What are the possible consequences of incorrect or poorly stated requirements?
- When the first person finishes, the other person (the questioner) will seek clarification and provide her or his own ideas.
- Take notes on the exchange you had and be prepared to share your discussion with the class.

10.A2. You need to recommend a software development methodology for an upcoming project. From the methodologies presented in this chapter, choose a methodology for each of the following projects. Briefly provide the rationale for each choice.
- a. A system for monitoring electrical surges in an electrical power grid. The requirements for the system are well understood. Users will be required to use the system.
- b. A contact management system. The system's potential users are unsure of exactly what they want the system to do. Use of the system will be voluntary.
- c. A relatively small order-tracking system. The objectives of the system are well known, and potential users very much want such a system.

10.A3. For each of the following, recommend whether the organization should build or buy the system in question. Explain your recommendation.
- a. A medium-size company that sells roofing materials to contractors needs a system that will allow customers to check product inventory through a website.
- b. A large (Fortune 500) retailer with a large, well-skilled IT department needs a system that will track the purchasing habits of its frequent customers.
- c. A large (Fortune 500) defense contractor needs a system that will track payments to vendors.

10.A4. For each of the systems described in Learning Activity 10.A3, what additional information would be useful in making your recommendation? Explain why the information would help in the build versus buy decision.

10.A5. Locate open source alternatives for each of the following:
    a. Microsoft Access database management system
    b. Microsoft Outlook (email)
    c. Adobe Photoshop
    d. Salesforce.com customer relationship management system
    e. iTunes media player

10.A6. Find examples of software development projects for IoT. What are some of the additional issues developers face when developing software for sensors, personal devices, and interconnected devices like those used in IoT?

10.A7. Find two examples of outsourcing (other than the examples in the book). Provide a brief (one-paragraph) description of the outsourcing arrangements. State why the companies outsourced the work. Indicate which outsourcing model best describes the example.

10.A8. This activity requires you to complete a case analysis of an information systems development project that was outsourced. Your instructor can provide you with the case, or you may find it using the reference below. Nuwangi, S. M., and D. Sedera. 2020. "A Teaching Case on Information Systems Development Outsourcing: Lessons from a Failure." *Communications of the Association for Information Systems* 46. https://aisel.aisnet.org/cais/vol46/iss1/29/.

References

Boehm, B. 1988. "A Spiral Model of Software Development and Enhancement." *IEEE Computer* 21(5): 61–72.

Grover, V., and J. T. C. Teng. 1993. "The Decision to Outsource Information Systems Function." *Journal of Systems Management* 44(11): 34–38.

W3Techs. 2020. "Usage Statistics of Web Servers." August 26. Retrieved from https://w3techs.com/technologies/overview/web_server.

Glossary

Application outsourcing: An outsourcing arrangement that involves a service provider handling activities related to a specific software application.

DevOps: An approach to software development that focuses on rapid development of applications and services by integrating the operations of the software with the development of the software.

Full or complete outsourcing: Contracting with a service provider to provide all IT-related services.

Functional requirements: Describes how a system should interact with users and other systems.

Nearshore outsourcing: Using a vendor that provides services from a location close to the client's location.

Nonfunctional requirements: Constraints on a system.

Offshore outsourcing: Using a vendor that provides services from a location outside the client organization's region.

Onshore outsourcing: Outsourcing to a firm located in the same country.

Open source software: Software that allows users to access the underlying source code for an application.

Outsourcing: An arrangement in which an organization contracts with a service provider to provide IT-related services.

Personnel outsourcing: An outsourcing arrangement in which a service provider places workers into an organization on a temporary contract basis.

Process-based outsourcing: An arrangement in which an organization outsources a particular function or business process.

Project-based outsourcing: An outsourcing arrangement that involves contracting with a service provider to complete a specific project.

Prototype: A small-scale mock-up of a system or a portion of a system.

Requirements elicitation: Gathering system requirements from various stakeholders.

Selective (partial) outsourcing: An outsourcing arrangement in which only certain aspects of IT are outsourced to a service provider.

Software development methodology: A framework for planning, structuring, and controlling software development projects.

Source code: Text-based computer programming language statements that can be read by humans.

Systems development life cycle (SDLC): A semisequential, phased approach to systems development.

Use case: Describes a series of actions that results in an outcome for an actor.

Information-Based Business Processes

Learning Objectives

By reading and completing the activities in this chapter, you will be able to:

- Explain what a business process is
- Perform business process modeling
- Discuss the impacts of technology on business processes
- Explain how business process improvement can be achieved with business process reengineering (BPR)

Chapter Outline

What Is a Process?
Focusing Story: Improving Processes Is for Everyone!
    Learning Activity 11.1: How Many Steps in This Process?
Process Modeling
    Learning Activity 11.2: Model This!
Technology and Processes
Process Improvement
    Learning Activity 11.3: Informal Business Processes
    Learning Activity 11.4: Redesign This!

What Is a Process?

A **process** is a series of steps or tasks required to achieve a specific goal. In organizations, everything that happens is based on a series of processes or key activities. They are often called *business processes*, but they could more precisely be labeled *organizational processes*, since processes exist not only in businesses but also in government agencies, charities, schools, and other noncommercial organizations. Figure 11.1 shows an example of a process: the process of acquiring materials needed for the production of an enterprise's main goods. It should be noted that while the process appears to be linear, it is not necessarily the case, since several activities can take place at the same time (concurrently).

Improving Processes Is for Everyone!

This book is about information, and many of our examples seem to showcase technology companies. However, when we talk about business processes, or the way things are done in organizations, everyone can benefit from improvements—even companies that are in basic manufacturing industries like the cement industry.

Cemex is a 100-year-old multinational cement corporation based in Monterrey, Mexico. For years, it functioned pretty much like every other cement company, with limited communication among its various plants, constant changes in order information (customers cannot pour concrete in bad weather), and difficulty confirming precise deliveries to customers because of weather, traffic, and labor problems. Needless to say, customer satisfaction is not always high in this industry.

When Lorenzo Zambrano, the grandson of Cemex's founder, was appointed CEO, he looked for ways to improve processes. To handle unpredictable demands and changing conditions, the company realized it needed better real-time communications. It therefore implemented a satellite communication system (which it called Cemexnet) to connect all its cement plants. It also linked all offices via satellite and the Internet and created a central office for coordination. This allowed plants to be constantly aware of fluctuating supply and demand. The second main aspect of the overhaul was to develop a new tracking, scheduling, and routing system (which the company called Dynamic Synchronization of Operations). This logistics application uses GPS transmitters in each delivery truck, allowing dispatchers to allocate trucks to the plants closest to where they are at a given time. With detailed information on traffic conditions, inventory, and customers' locations, trucks can be rerouted as needed. Finally, suppliers, distributors, and customers of Cemex can use an online application to check order status.

The results have been impressive for Cemex: decreased delivery time from 3 hours to 20 minutes (98% of the time), increased customer satisfaction, substantially increased net sales, decreased number of delivery trucks (35% decrease in the first year), and substantial cost savings overall (e.g., fuel, maintenance, and payroll).

Source: Kaplan, S. 2001. "Business Process Improvement at Concrete Co. Cemex." *CIO Magazine,* August 15. http://www.cio.com/article/2441369/process-improvement/business-process-improvement-at-concete-co--cemex.html.

Focusing Questions

Identify specific ways the new technologies at Cemex changed the following:

1. How customers do business with Cemex
2. How delivery truck drivers do their work
3. How office staff work at Cemex
4. How plant managers handle business at Cemex

Processes are not limited to one functional area of the organization. For example, in Figure 11.1, it is likely that a production department will request materials, while an accounting department will pay vendors. This is an important point: processes extend across many organizational boundaries and tend to involve several individuals. Another key characteristic of processes is the **all-or-nothing concept**. This implies that a process needs to be completed or not done at all. Looking again at Figure 11.1, requesting materials requires purchase orders, receiving materials, and so on. The process is not complete if there is no payment request or no payment made.

Every business process can usually be decomposed into several tasks, also called **subprocesses.** This is because processes usually require several decisions and actions to be made along the way. In general, there can be several levels of subprocesses, all the way down to specific activities or tasks. For example, in the process shown in Figure 11.1, the subprocess of verifying the invoice involves viewing the invoice, viewing the ordering information, and comparing what was ordered with the charges. It can involve making sure that the proper discounts are given, which can require several steps (searching for discounts, comparing quantities, etc.). (See Figure 11.2 for some possible subprocesses.)

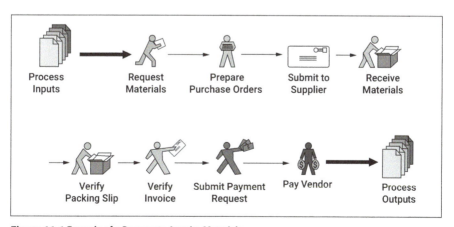

Figure 11.1 Example of a Process to Acquire Materials

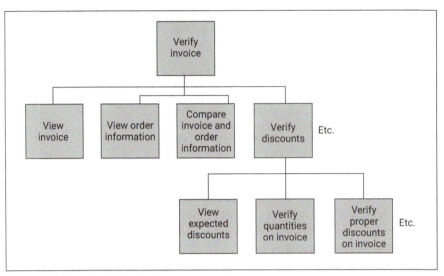

Figure 11.2 Examples of Subprocesses

Mapping the processes and subprocesses is called *process modeling*, which we describe in more detail below. There are many examples of processes you are familiar with even if you do not realize they are considered processes. Think of course registration. What are the various steps involved? You need to identify the courses required for your program and the requirements for each course. You must then log into the registration system, enter your information, and make the requests. Next, you need to react to those requests that were denied. You can appeal them, try to force-add the course, or decide to take a different course. There are also processes for requesting transcripts, updating your personal information, or for instructors to enter grades, just to name a few.

Process Modeling

Process modeling is the mapping of processes and subprocesses used in an organization or division. Examine the process of acquiring materials needed for the production of an enterprise's main goods in Figure 11.1. Only the steps involved are identified. In process modeling, we need to expand the steps to identify when they

---

LEARNING ACTIVITY 11.2

Model This!

Following up on Learning Activity 11.1, your group of managers must now model the steps you identified in the inventory management process for your company. Make your process model as complete as possible.

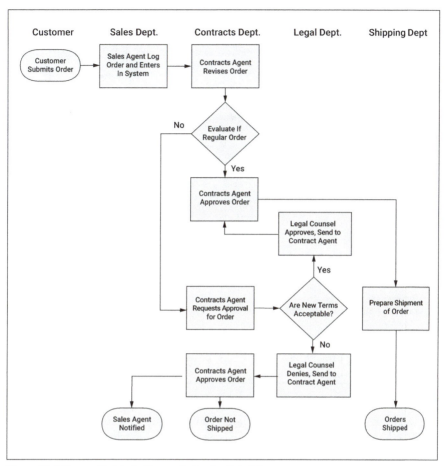

Figure 11.3 Process Model of a Process to Acquire Materials

are occurring and which decisions are involved. We must also recognize if there are any relationships between processes. The bottom line is that understanding which processes occur at different levels in the organization and how they interrelate is necessary to really understand how an organization works. Figure 11.3 shows a process model for customers acquiring goods from a firm.

A process model should clearly describe what happens as a process is performed. As analysts are modeling the processes, they should not analyze the process for improvements (yet); the goal is to describe the process as is. Process modeling can also be explanatory, in that the process modeling effort can identify the rationale for why processes are performed this way. Individual processes can be modeled, although organizations often model several processes simultaneously. Recall that processes are made of several subprocesses, which can also be modeled. As a result,

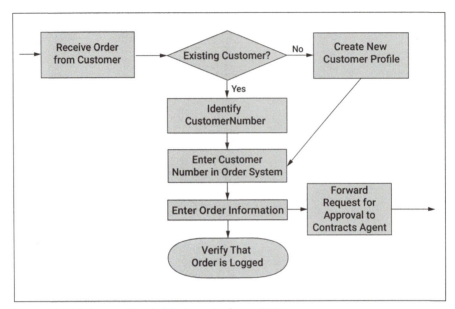

Figure 11.4 A Subprocess Model of Customer Goods Acquisition

process modeling usually results in "levels" of process models. Figure 11.4 shows a refinement of the order entry subprocess from the process shown in Figure 11.3.

Business Process Modeling Tools

There are many **process modeling tools** and software available to help analysts describe business processes and create business process models or diagrams. Mili et al. (2010) identify four types of such tools: traditional process modeling languages, object-oriented modeling languages, dynamic process modeling languages, and process integration languages. Table 11.1 describes these various categories and shows some examples of these languages.

Analysts can become well versed in some of the languages presented in Table 11.1, but to facilitate business process modeling, several companies have developed software specifically targeted at creating business process models. Table 11.2 shows a few of the popular process modeling companies and software.

Technology and Processes

Business processes have existed since organizations started to organize work activities into tasks. Today, however, information technologies play an ever-increasing role not only in how business processes are performed, since technology is embedded in most work environments, but also in how business processes are modified or improved. We discuss business process improvement in the next section. Before we

TABLE 11.1 Examples of Business Process Modeling Tools

Category of Modeling Tool	Description	Sample Languages
Traditional process modeling languages	Software engineering and information systems modeling languages mostly used in business process engineering	• Event-Driven Process Chains (EPC) • Role Activity Diagrams (RAD) • Resource-Event-Agent (REA) • Architectural Modeling Box for Enterprise Redesign (AMBER)
Object-oriented modeling languages	Process modeling languages supporting the object-oriented approach to software development	• Object-Oriented Role Analysis Methodology (OORAM) • Enterprise Distributed Object Computing (EDOC) • Unified Modeling Language (UML) 1 and 2
Dynamic process modeling languages	Process modeling language standards proposed by various—mostly industrial—organizations	• Workflow Process Description Language (WPDL) • Business Process Modeling Language (BPML) • Web Services Business Process Execution Language (WS-BPEL) • Business Process Modeling Notation (BPMN) • Business Process Definition Metamodel (BPDM)
Process integration languages	Process modeling languages for the purpose of integrating the processes of two or more business partners in electronic business	• RosettaNet • Electronic Business Extensible Markup Language (ebXML) • Web Services Choreography Description Language (WS-CDL)

Source: Mili et al. (2010).

TABLE 11.2 Sample Business Process Modeling Companies and Software Products

Company	Product	Address
CSG International	Ascendon Communications	https://www.csgi.com/portfolio/ascendon/
Interfacing Technology	Enterprise Process Center	http://www.interfacing.com
KBSI, Knowledge Based Systems Inc.	Business Process Solutions (several)	http://www.kbsi.com/solutions-and -services/improve-business-processes
Microsoft Corporation	Visio Modeling Software	https://www.microsoft.com/en-us/ microsoft-365/visio/flowchart-software
OpenText	Business Process Management	https://www.opentext.com/products-and -solutions/products/digital-process -automation/business-process -management-bpm
Oracle Corporation	Oracle Business Process Management	https://www.oracle.com/middleware/ technologies/bpm.html
Grandite	Silverrun Modeling Tools	https://www.grandite.com/
Software AG	ARIS Business Process Analysis	http://www.softwareag.com

discuss how things can be done better, however, it is important to realize that technology not only enables new business processes but can also be a barrier or constraint to business process improvement.

Think about how much the tasks individuals perform in organizations are dependent on how the technology they use functions. It might be more efficient to input data a certain way, but the system will not let you do it. It could be useful to create a new report, but the system does not allow the data to be collated appropriately. This is even more apparent when we talk about legacy systems. **Legacy systems** are often older, large applications that may be mission critical. They are hard to replace and can be somewhat inflexible. In addition, they are often still used to support key business processes. Even if an organization creates a new, graphically friendlier interface to access information, what can be done with the information can be highly dependent on the type of data warehouse or application performing the back-office processing.

Process Improvement

Once business processes are properly modeled, they can be analyzed for improvement. Improving business processes is an important goal for managers, and this is labeled either **business process reengineering (BPR)** or **business process improvement (BPI)**, depending on the extent of change necessary. BPR involves the redesign of business processes to improve how work is done across many functions within the organization. BPI, on the other hand, involves the redesign of a business process that is simple enough that it can be done within one function of the business or without significantly involving other business functionality. Regardless of whether an organization is improving its processes utilizing BPR or BPI, the process requires that business process modeling be done well first. It is important to realize that both BPR and BPI often involve changing the way tasks are performed instead of just automating the tasks. Improving processes requires rethinking what is done and how it is done (and even whether it should be done!). In one of their books on improving business processes, Michael Hammer and James Champy (2003) suggest that this process involves the fundamental rethinking (really asking why an organization performs tasks in a certain way) and radical redesign of business processes (meaning that there is reinvention or re-creation, not just simple minor changes). This is done to obtain dramatic improvements in performance (meaning that the goal is to achieve significant improvements for the organization, whether in terms of cost, efficiency, satisfaction, or other important measures). There are usually four key areas for business process improvement: effectiveness, efficiency, internal control, and compliance. These goals are summarized in Table 11.3.

Summarizing our discussion so far, it is clear that improving business processes involves several phases, and just like our strategic planning effort did in Chapter 4, this effort starts with a clear vision of the goal of the redesign. It follows with an understanding of what the current processes are, how they work, who performs them, and who uses their outcomes. At the same time, redesign involves rethinking

TABLE 11.3 Typical Goals for Business Process Reengineering	
Effectiveness	The extent to which the outputs of the process are obtained as expected
Efficiency	The average time it takes for the process to be completed
Internal control	The extent to which the information and data used in the performance of the process cannot be changed by error or illegally
Compliance	The extent to which the process follows the regulatory or statutory obligations of the organization

business processes, so analysts should not end up limiting themselves by focusing too much on how things are done as opposed to how they should be done. Finally, a very important part of the redesign process is ensuring that there are measurements to make certain that improvements are indeed being made. As you can expect, the BPR effort is an iterative process in which what is done or found in later stages may result in the need to revisit earlier stages. Figure 11.5 shows a sample BPR process.

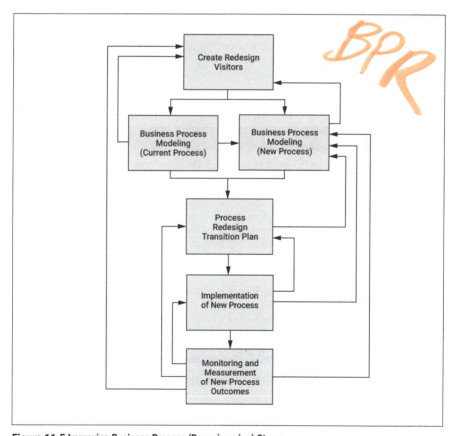

Figure 11.5 Improving Business Process (Reengineering) Stages

BUSINESS EXAMPLE BOX 11.1

Banking Processes Improved via Blockchain

When we mention blockchain improving banking processes, the first thought that likely comes to mind is that of Bitcoins. While Bitcoins are one way that blockchains are used, it is not the only way. To understand the opportunity that blockchain provides to improve business processes, it is necessary to understand what blockchains provide. A blockchain can be used to allow people to openly communicate even when there is little to no trust between the required parties. This is possible using encryption and distributed storage of the blockchain information with many parties. Because of this combination of encryption and distribution, it is not possible for anyone to change the information that is being stored. However, it is possible for all parties involved to see what was communicated by one party to the other. This provides assurances that the information stored in the blockchain is reliable.

In a banking setting, transactions often occur between multiple entities, including other banks, private individuals, and governments. Because of regulations, it is also necessary for banks to ensure that proper guidelines are followed. By adding rules to the processing of transactions, banks can make sure that these guidelines are properly followed, ensuring compliance. Prior to utilizing blockchain as part of banks' transaction processing systems, it was possible for guidelines not to be followed as well as for individuals to commit fraud by manipulating the data within the system. By adding blockchain to the existing transaction processing system, banks have improved their business process.

Source: ProcessMaker. "How Blockchain is Reinventing BPM for Banking." March 26. https://www.processmaker.com/blog/how-blockchain-is-reinventing-bpm-for-banking/.

LEARNING ACTIVITY 11.3

Informal Business Processes

In an organization, there are two types of business processes: official business processes and informal business processes. An official business process is one that has been documented and defined, as discussed in this chapter. An informal business process is a process that the business relies on but has not been properly documented. Based on these definitions, one could claim that an official business process is not the same as an informal process. For this learning activity, think about these definitions as you answer the following questions. Have your responses ready for discussion in class.

1. Is it true that these two processes are not always the same?
2. Why is that?
3. Which one should you model? Why?

LEARNING ACTIVITY 11.4

Redesign This!

This activity builds on what you have accomplished in Learning Activity 11.3. Now that your team of managers has become more familiar with the inventory management system and has gained knowledge on improving business processes, your team should perform a redesign of the business process model you previously completed.

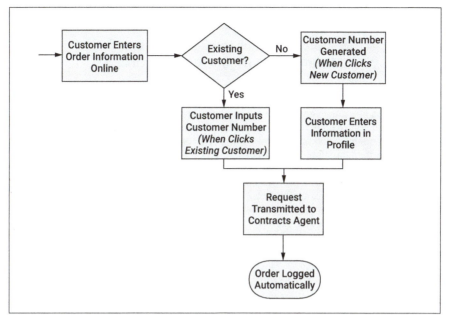

Figure 11.6 Online Ordering Redesign

Remember that, in the end, managers improving business processes need to ask themselves key questions such as the following:

- What are our key business processes?
- Do we have to execute each process?
- How can we use information technologies to perform these processes in better ways?
- What are the likely impacts of the business process redesign?

The result of a business process improvement effort should be a document that defines how the new business processes should or could be performed, indicating clear rules and guidelines that the processes must or should follow. For organizations, the results of a redesign of business processes are often streamlined ways of doing business. Figure 11.6 shows an example of the redesign of the business process from Figure 11.4 after the organization offers online ordering.

Levels of Change

In Chapter 4, we introduced the idea that information technology can be implemented in organizations for strategic purposes—in other words, to create competitive advantages. When new information technology (IT) is implemented, it can have differing effects on organizations, depending on the role it plays in changing or supporting business processes. One way to categorize these impacts is whether they

> ## BUSINESS EXAMPLE BOX 11.2
>
> ### Quicken Loans Rocket Mortgage:
> ### Moving the Mortgage Process Online
>
> At some point, many students will eventually purchase a home. To do so, they will likely take out a mortgage, as this is a very large purchase. The traditional way of completing this process has been very paperwork intensive: applicants need to provide mortgage lenders with two years or more of tax documentation, two months of bank statements, credit statements, paper applications, and more. This process is very time consuming.
>
> To solve this situation, Quicken Loans created a product called Rocket Mortgage. They embraced the concept of financial technology (fintech) and moved this process online. While mortgage lenders still need the same information as they did before, Quicken realized that much of this could be provided seamlessly online. They claim that in only eight minutes, a person can apply for a loan using either the Rocket Mortgage app or their website. Getting the new loan may still take the same amount of time as a traditional mortgage, given other processes that require in-person services like appraisals. However, the amount of time to refinance a mortgage is greatly decreased, and applying for the mortgage is simpler. As technologies are increasingly looked at to improve processes, further improvements in the financial industry are also likely to occur.
>
> Source: Greene, M. 2016. "Will Rocket Mortgage Flameout?" *Forbes*, February 14. https://www.forbes.com/sites/markgreene/2016/02/14/will-rocket-mortgage-flameout/.

represent changes that automate, informate, or transform organizations (Dehning, Richardson, and Zmud 2003; Schein 1992).

- *Automate*, also called *first-order change*, involves using technology to perform a process more efficiently, often by replacing human labor with technology. It is usually fairly easy to justify because managers can quantify the benefits (e.g., reduced wages, reduced errors).
- *Informate*, also called *second-order change*, involves using technology to better inform people, which, as a result, can change how they perform business processes. Whenever people have to change how they perform business processes, it is more difficult to implement systems. Recall from our earlier discussions that people often resist technology implementations when they have to change their work habits.
- *Transform*, also called *third-order change*, requires a fundamental redefinition of the business processes, or how work is done. Obviously, since improving business processes involves the redesign of business processes, it is a third-order change. This is the most difficult level of change to deal with when implementing information systems.

Chapter Summary

In this chapter, we explained in greater depth what business processes are, how they can be modeled, and how organizations seek to improve business processes through business process reengineering (BPR) and business process improvement (BPI). Here are the main points discussed in the chapter:

- A business process is a series of steps or tasks required to achieve a specific business goal of an organization.
- Business process modeling is the mapping of processes and subprocesses used in an organization or division. We demonstrated business process modeling with graphic models of a materials acquisition process and its subprocesses. Business process modeling helps us understand the various processes that occur at different levels in the organization and how they interrelate.
- Technology impacts business processes because it is embedded in how work is done in most work environments. Technology could be a constraint to business processes, especially when legacy systems are involved.
- Improving business processes can be achieved with business process reengineering (BPR) or business process improvement (BPI), which involve the redesign of business processes to improve how work is done. BPR and BPI often involve changing the way tasks are performed instead of just automating the tasks. There are usually four key areas for business process improvement: effectiveness, efficiency, internal control, and compliance.

Review Questions

1. Explain what a business process is and give an example from your school environment.
2. Identify some of the important characteristics of processes.
3. What is business process modeling, and what is the purpose of business process modeling?
4. Identify some of the important characteristics of a process model.
5. What kinds of tools are available for business process modeling?
6. How does technology impact business processes?
7. What is business process reengineering (BPR), and what does it involve?
8. What are the four key areas for business process improvement?
9. Briefly explain the main steps involved in performing business process redesign.
10. What are some of the results expected from a business redesign effort?

Reflection Questions

1. What is the most important thing you learned in this chapter? Why is it important?
2. What topics are unclear? What about them is unclear?
3. Identify all the processes a student encounters during the first week of class as a freshman.
4. How does the business process-related material in this chapter relate to the security topics in Chapter 8?
5. How does the business process-related material in this chapter relate to the privacy topics in Chapter 9?
6. Can all business processes be redesigned? Why or why not?
7. What are the similarities and differences between process modeling as discussed in this chapter and data modeling as discussed in Chapter 5?
8. Should informal business processes (representing how things are really done, even if they do not follow recognized or documented procedures) be modeled or just formalized business processes (those documented by the organization)?
9. What is an example of technology being a barrier to redesign in a school setting?
10. Why do you think business process reengineering is an important concern for CIOs?

Additional Learning Activities

11.A1. This activity allows you to get some additional practice at business process redesign. First, you need to download the template of one of the business processes (your instructor will tell you which one) from Global 360: https://www.smartsheet.com/workflow-templates. In small groups, examine the template and then, with business process redesign in mind, modify the business process you downloaded to improve it.

11.A2. This activity requires you to complete a case analysis regarding business processes. The case may be accessed through your university website, the Ivey Business School business case website (https://www.iveycases.com), or the Harvard Business School cases website (https://hbsp.harvard.edu/home/), or it may be provided to you by your instructor. Sousa, K. 2017. "BrightView Plumbing and Heating: A New Business Model." *Ivey School of Business*, August 17. Product number: 9B17D012.

11.A3. Examine Figure 11.2, identify one of the subprocesses of the order entry process, and model it, as was done in Figure 11.4 for the order entry part in the sales department.

11.A4. Select one of the key processes needed at your school and model it. You can select one from the following list, or you can identify another important process and model it:
- Class registration
- Final grade assignment
- Student enrollment
- Tuition payment
- Requesting transcripts

11.A5. Use the process modeling tools in the Visio software to create your models in Learning Activities 11.A3 or 11.A4.

11.A6. Identify a business process within the university environment. Design what you believe is how this business process works. Then think about how blockchain could change how this business process occurs. Design what you believe the process would look like if you included block-chain as a key component of the process.

11.A7. Think about the university environment. Identify a process that crosses multiple functions of the university (e.g., alumni relations and financial aid) and a process that is fairly stand-alone in nature (e.g., human resources and department course scheduling). Describe each of these processes, being sure to identify the various functionality that is involved. Compare how extensive these two processes are and which would require BPR and which BPI.

References

Dehning, B., V. J. Richardson, and R. W. Zmud. 2003. "The Value Relevance of Announcements of Transformational Information Technology Investments." *MIS Quarterly* 27(4): 637–56.

Hammer, M., and J. Champy. 2003. *Reengineering the Corporation: A Manifesto for Business Revolution.* New York: HarperCollins.

Mili, H., G. Tremblay, G. Bou Jaoude, É. Lefebvre, L. Elabed, and G. El Boussaidi. 2010. "Business Process Modeling Languages: Sorting through the Alphabet Soup." *ACM Computing Surveys* 43(1): 7–62.

Schein, E. H. 1992. "The Role of the CEO in the Management of Change: The Case of Information Technology." In *Transforming Organizations,* edited by T. A. Kochan and M. Useem, 325–45. Oxford: Oxford University Press.

Glossary

All-or-nothing concept: The idea that a process needs to be completed or not done at all.

Business process improvement (BPI): The redesign of business processes to improve how work is done within one function of the organization.

Business process reengineering (BPR): The redesign of business processes to improve how work is done across many functions within the organization.

Compliance: A BPR goal that involves the extent to which the process follows the regulatory or statutory obligations of the organization.

Dynamic process modeling languages: Process modeling language standards proposed by various (mostly industrial) organizations.

Effectiveness: A BPR goal that involves the extent to which the expected outputs of the process are obtained as expected.

Efficiency: A BPR goal that looks at the average time it takes for the process to be completed.

Internal control: A BPR goal that involves the extent to which the information and data used in the performance of the process cannot be changed by error or illegally.

Legacy systems: Older, large applications that may be mission critical.

Object-oriented languages: Process modeling languages supporting the object-oriented approach to software development.

Process: A series of steps or tasks required to achieve a specific goal.

Process integration languages: Process modeling languages for the purpose of integrating the processes of two or more business partners in electronic business.

Process modeling: Mapping of processes and subprocesses used in an organization or division.

Process modeling tool: Tools and software available to help analysts describe business processes and create business process models or diagrams.

Subprocesses: Decomposition of business processes into several tasks.

Traditional process modeling languages: Software engineering and information systems modeling languages mostly used in business process engineering.

Chapter 12

Enterprise Information Systems

Learning Objectives

By reading and completing the activities in this chapter, you will be able to:

- Identify the main characteristics and types of enterprise systems
- Discuss the purposes of enterprise resource planning (ERP) systems
- Explain the purposes and functioning of supply chain management (SCM) systems
- Identify the components of customer relationship management (CRM) systems
- Apply the customer life cycle concepts to CRM

Chapter Outline

Focusing Story: Supply Chain Innovations at Walmart
Enterprise Systems
Learning Activity 12.1: Finding the Components of an Enterprise System
Enterprise Resource Planning (ERP)
Learning Activity 12.2: The Online Beer Game
Supply Chain Management Systems
Learning Activity 12.3: Blockchain and Supply Chain Management
Customer Relationship Management Systems
Learning Activity 12.4: Customer Relationship Management
Learning Activity 12.5: Self-Servicing
Customer-Managed Interactions (CMI)
Warehouse Automation and Robotics

This chapter is about enterprise systems, which are systems that are used by a large number of individuals throughout an organization. We will discuss various ways enterprise systems are developed and explore several such systems. One of these enterprise applications is supply chain management (SCM), which is the topic of interest in the focusing story.

RFID

FOCUSING STORY

Supply Chain Innovations at Walmart

You need a new pair of inexpensive jeans or maybe just a new set of travel speakers for your iPhone. Where would you go to buy them? Many students would turn to a local Walmart store, since the company is well known throughout the world as a low-cost retailer of goods. What makes it possible for the company to offer prices that often seem to be well below what competitors can afford to offer? Of course, there are many answers to this, but one of the most important factors is how Walmart manages the way it gets the products sold in its stores. This is the supply chain. Walmart calls this the *productivity loop*: lower cost products lead to lower prices for consumers, which lead to more sales, which increase profits, and then it repeats the loop as quickly as possible. However, there are many other benefits for Walmart and its partners in the supply chain, as explained below.

Walmart is well known for its innovative strategies regarding its supply chain and its tight management of its distribution channels. By 2010, Walmart had established four very large global merchandising centers that are responsible for ordering goods and clothing products (also called *sourcing*) for the company's stores in 15 countries. It also uses direct purchasing of fresh fruits and vegetables from producers on a global basis, bypassing traditional suppliers.

Walmart's supply chain strategies cover many aspects, but in a speech given at the Council of Supply Chain Management Professionals (CSCMP) annual conference in 2009, Gary Maxwell, then senior vice president of international supply at Walmart, explained that effective inventory management may be the most critical element to master for an efficient supply chain.

To improve the effectiveness of its supply chain, Walmart mandated in 2003 that its largest suppliers use **RFID (radio-frequency identification) tags**. This required suppliers to insert electronic tags into pallets of goods to be shipped to Walmart. Using readers, both Walmart and the supplier can then track the pallets as they move through the various distribution centers all the way to the store. Walmart has been a leader in the development of the RFID technology. However, while it has successfully reduced out-of-stock items (8% worldwide) and improved its overall inventory data (it can resupply three times faster), the adoption of the technology was slower than expected. For many companies, the cost was quite high (from $100,000 to $300,000 for small companies to $20 million for larger ones). In 2003, first-generation chips sold for $1.25 each. In 2009, second-generation chips cost 7¢ to 10¢. By late 2008, 600 of Walmart's 60,000 suppliers, plus 750 Sam's Club (owned by Walmart) suppliers, deployed RFID to some degree. On Walmart's end, RFID is deployed at about 1,000 of the roughly 4,000 Walmart and Sam's Club stores in the United States.

Today, suppliers and Walmart reap many other benefits from RFID, such as knowing in real time how long it takes for items in the promotional areas of stores to sell. They can even evaluate where in the store items sell better. This is a win-win situation for Walmart and its suppliers. Some suppliers have even built the use of RFID tags into their whole information technology (IT) infrastructure, from acquisition to sales to replenishment.

Further innovations to Walmart's supply chain have allowed them to compete better with Amazon.com in the e-commerce marketplace. These innovations include requiring suppliers to deliver orders to Walmart within a one-day window as opposed to the prior four-day window. This allows Walmart to carry less inventory and react more quickly to the marketplace. However, to successfully work with their suppliers, Walmart had to increase their ability to forecast demand and integrate this insight into their suppliers' supply chain.

Walmart continuously strives to innovate in its supply chain. Its most recent effort is convincing its suppliers that reducing gas emissions and becoming more "green" in their

My productivity app

delivery of products to Walmart can result in profitability and sustainability. Walmart is asking some suppliers to report how much energy they use in manufacturing the products Walmart buys from them and requires them to meet stricter quality and environmental standards.

Much of Walmart's original focus on the supply chain was on what could be referred to as the back end of the process. However, Walmart more recently has given employees on the sales floor access to supply chain data through an app on employees' smartphones called "My Productivity." My Productivity allows employees the ability to see inventories and where they are at in the supply chain in real time. This improves the ability to keep shelves stocked and allows employees to answer customers' questions more accurately. This app saves Walmart thousands of hours of employees' time, as they no longer have to go in the back storeroom to look for merchandise that is not there.

Sources: Compiled from Berman, J. 2009. "Walmart's Maxwell Cites Keys to Developing Best-in-Market Global Supply Chains." *Logistics Management* 48(10): 18; Birchall, J. 2010. "Walmart Takes Aim at Supply Chain Cost." *Financial Times,* January 4: 1; Cassidy, W. B. 2010. "Walmart Squeezes Costs from Supply Chain." *Journal of Commerce,* January 5; Lopez, E. 2017. "Behind the Scenes of Walmart's New On-Time, In-Full Policy." Supply Chain Dive. https://www.supplychaindive.com/news/Walmart-OTIF-inventory-flow-ecommerce-supply-chain/507301/; McCrea, B. 2010. "This Is Why 'Green' Equals Good Business." *Supply Chain Management Review* 14(2): 56; Webster, J. S. 2008. "Walmart's RFID Revolution a Tough Sell." *Network World* 25(36): 34–36; Ibbotson, M. 2016. "How Real-Time Data Is Putting Success at Our Fingertips." Walmart. https://corporate.walmart.com/newsroom/innovation/20160602/how-real-time-data-is-putting-success-at-our-fingertips.

Focusing Questions

1. What do you believe are the key success factors for the supply chain implementation at Walmart? What would those factors be for other companies trying to imitate Walmart's supply chain efforts?
2. What is your assessment of Walmart's move into RFID technology as a leader? Could there be an issue being at the leading edge of technology?
3. How can technology be used to help Walmart in its next endeavor—its green supply chain?
4. How can worker efficiency improve when employees get easy access to real-time information about where products are within the supply chain?

Enterprise Systems

Personal versus Enterprise Systems

There are many information system applications used by individuals and companies to perform tasks. For example, if you think of applications you use that are for individuals only, you might find that Quicken is used to manage your finances or that Microsoft Word is used to write your term papers. Those are considered **personal information systems**. Conversely, **enterprise systems** are information systems that serve the needs of the organization or parts of the organization. For example, the organization may use SAP's enterprise resource planning (ERP) applications to manage its finances, including paying bills, receiving invoices, and paying employees. In Learning Activity 12.1, you will identify many other components of an ERP system. As you recall, many modules are available to perform functions needed across several departments in the organization.

LEARNING ACTIVITY 12.1

Finding the Components of an Enterprise System

The supply chain management system at Walmart represents only one type of enterprise system companies can use for running their businesses efficiently. One of the most popular enterprise systems in organizations is enterprise resource planning (ERP). We will discuss enterprise systems and ERP later, but for now, we want you to identify the components of an ERP system offered by the one of the larger companies, such as SAP, Oracle PeopleSoft, or Microsoft Dynamics.

Please prepare a short report for class that identifies the company that sells the ERP system, a brief background on the company, the components of the ERP system, and a brief description of each component. The report should be fewer than two pages long.

Strategic = executives

Three perspectives on developing enterprise systems have emerged: the hierarchical perspective, the functional perspective, and the process perspective (Piccoli 2008).

In the **hierarchical perspective**, software developers build information systems to mirror the organization so that applications are built to meet the needs of individuals at a given level in the organization. The idea is that executives need strategic types of information they can use for decision-making, whereas middle managers and front-line managers need tactical and operational information, respectively. Table 12.1 shows the type of information needed at each level and the characteristics of that information, which is also reproduced in Figure 12.1.

As a result, systems are built for operations, tactical, or strategic management, with sometimes limited interaction and consistency across these systems. Redundancy is also a problem, with different systems sometimes performing fairly similar functions.

In the **functional perspective**, information system applications are developed to meet the needs of individuals in a functional area, such as human resources or marketing. With this perspective, information system developers build a "financial" system, a "marketing" system, a "production" system, or a "human resources" system. The systems tend to serve the functional area well, as they are customized to the needs of the department. Again, when thinking about the various functional areas in an organization (often called *departments*), you can see that information system applications

TABLE 12.1 Types of Information Required by Hierarchical Level			
Activity	**Time Horizon**	**Hierarchical Level**	**Characteristics**
Strategic	Long term	General management Functional management	• Externally focused • Highly unstructured
Tactical	Mid term	Middle management	• Semistructured • Recurrent
Operational	Short term	Operation management Employees	• Highly structured • Transaction focused
Source: Adapted from Piccoli (2008).			

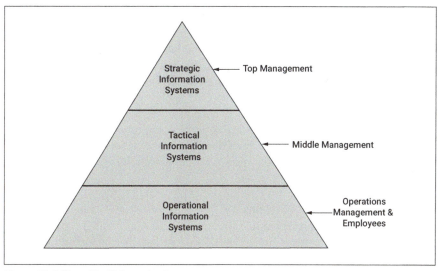

Figure 12.1 Hierarchical Information Systems

built this way could result in redundancy and inaccuracy of data. (Think of having to update customer information in five different departments.) This perspective also limits communication across functional areas.

Finally, in the **process perspective**, information systems applications are developed to support the task being done (or process) regardless of the organizational level or functional area of the individual using the system. A *process* is a series of steps required to perform a task.

Integration and Enterprise Systems

The discussion of the three perspectives for the development of enterprise systems highlights some of the issues that arise when systems are not developed with the organization as a whole in mind. In particular, we discussed the issues of redundancy (identical data stored in multiple locations) and lack of communication across departments. To solve these problems, the process perspective focuses on integration. Integration is the unifying, or joining together, of resources and assets. This is clearly one of the primary advantages of enterprise systems, which are designed to integrate either across business processes or across information systems.

Business integration is the unification of business processes previously performed as separate activities to create cohesive and streamlined business processes. In other words, business integration refers to ensuring that processes are performed the best way they can be, without having barriers along the functional areas (departments) or hierarchical levels. This is accomplished by making sure the processes are performed at the best location and by allowing access to everyone who needs the information. This is also called **data integration**. Examples of enterprise systems that provide business integration include customer relationship management (CRM) and enterprise

↗ ERP, CMS, SCM

knowledge management. **Systems integration** is the unification of information systems and databases that were previously operating as separate systems. Systems integration focuses on the technological components of information systems. Examples of enterprise systems that are used for systems integration include enterprise resource planning (ERP), content management systems (CMSs), supply chain management (SCM), unified communications, and enterprise application integration. As you can see, most enterprise systems are designed to facilitate the flow of information among all business functions inside the boundaries of the organization and to manage connections to outside stakeholders. A brief description of some of these enterprise systems is provided in Table 12.2. The three most popular (ERP, SCM, and CRM) are further described in the chapter.

Characteristics of Enterprise Systems

The enterprise systems presented in Table 12.2 have some common characteristics. First, many of these systems are modular. This means that an organization can implement the "pieces" it needs one at a time. By implementing modules as needed, the organization can limit its efforts and costs and limit disruptions for employees. However, because the modules are designed to be integrated with one another, they function more efficiently together. This concept, called **application integration**, works because an event that occurs in one module automatically triggers events in one or more of the other modules. In addition, because data are stored in one central database, by offering data integration, companies avoid the data redundancy and inaccuracy issues found in hierarchical or functional information systems.

Enterprise Resource Planning (ERP)

As previously discussed, an **enterprise resource planning (ERP)** system is a set of information systems tools that are used to manage an organization's resources and enable information to flow within and between processes (and departments). Before ERP, organizations developed different applications to manage the information processing and management needs of the different functional areas of the organization. With ERP, various modules offer integrated management of these resources. Typical modules in ERP systems include financials (accounts receivable, general ledger, accounts payable, etc.), distribution services, human resources, and product life cycle management (marketing). Some ERP systems also include other enterprise systems applications, such as customer relationship management (CRM), supply chain management (SCM), and warehouse management systems (WMS).

Like other enterprise systems, ERP applications are built on a centralized database, as shown in Figure 12.2, and offer system and business integration. The ERP system itself can reside on a centralized server, or it can be distributed across modular hardware and software units that are interconnected. Many vendors offer commercial ERP packages (such as SAP, Oracle PeopleSoft, and Microsoft Dynamics) that tend to include the collective experience of their consultants. As a result, such packages are said to include best practices from the industry. However, the biggest challenge with ERP systems is to customize them to the needs of the organization and to the computing platforms used within the organization.

TABLE 12.2 Examples of Enterprise Information Systems	
Tool	Description
Customer relationship management (CRM)	An organization-wide strategy for managing an organization's multiple interactions with customers. It allows organizations to attract, retain, and manage customer relationships—for example, by offering personalized services and self-servicing tools. It also allows an organization to analyze customer data to find its most profitable products, services, and customers. It involves both technology and a philosophy of customer service.
Enterprise knowledge management (also called business intelligence [BI])	A set of tools used to analyze business data to find trends, issues, and opportunities. Companies have accumulated large amounts of data, such as sales, revenues, or expenses, to name a few, and BI tools show various views of these data to help decision makers. BI tools include a wide variety of analysis approaches, such as online analytical processing, business analytics, or data and text mining.
Enterprise resource planning (ERP)	ERP software allows organizations to manage resources throughout the organization, independently of which business function controls the resource. Typical modules in ERP software include financial management, order processing, or human resources management.
Content management systems (CMSs)	CMS applications allow companies to manage the vast amounts of information of various types (content) they have stored. This includes the capture, storing, preservation, and delivery of content. CMS refers to technologies, strategies, and approaches to managing content.
Unified communications (UC)	UC is a framework for the integration of the various modes of communication used by an organization, including email, voice mail, videoconferencing, instant messaging, texting, and voice over IP. For example, a customer may leave a voice mail that is automatically sent as an email to the account representative (based on preferences). Similarly, instead of sending an email, the voice mail can be transferred immediately to an instant message if the representative is currently online.
Enterprise application integration (EAI)	EAI is a set of tools and services (a framework instead of a single technology) to allow various enterprise systems (such as ERP or CRM) to share the data. It makes use of middleware (see Chapter 7) to perform this communication.
Supply chain management (SCM) systems	SCM software helps organizations manage the movement of materials or products from provision to production to consumption. SCM involves collaboration between partners both upstream and downstream from the focal organization. (Upstream and downstream flows are shown in Figure 12.3.)
Sources: Adapted from Pearlson and Saunders (2010); Piccoli (2008); Wikipedia.com.	

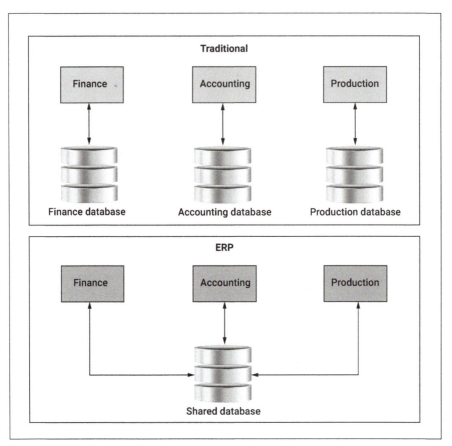

Figure 12.2 ERP Shared Database

Benefits and Disadvantages of ERP Systems

There are many benefits for organizations that successfully implement an ERP system. Efficiency is the most important of these, as it is quantifiable. Efficiency is achieved because ERP systems can reduce data redundancy (reduce direct costs of operation), improve communication (reduce indirect costs of operation through the streamlining of business processes), and in general allow complex systems to work together in an integrated fashion (also streamlining business operations and making sure tasks are performed in a standardized way). If properly implemented, ERP systems can also offer responsiveness to various stakeholders, including customers and suppliers, by providing more accurate information and more rapid responses to demands. ERP systems are also adaptable, since they can be configured to fit the organization's needs. However, when an ERP system becomes too customized, it can lose some of the advantages of integration and standardization offered by the standard packages. Finally, ERP systems benefit from the expertise of ERP software vendors and consultants who, through experience implementing many of these sys-

LEARNING ACTIVITY 12.2

The Online Beer Game

As seen in the focusing story, supply chain systems are very important enterprise systems for organizations that must receive raw materials or products needed for their manufacturing or resale operations. Before we discuss the concepts in greater depth, it is helpful to experience the issues surrounding proper management of supply chains. To do so, please connect to the Beer Game developed by MIT (Massachusetts Institute of Technology) professors, which can be found at https://beergameapp.com/. Use the Web-based version of the Beer Game for this exercise. Select a new game and set up an account. After setting up your account, you will create a free game as if you are the instructor. In doing so, you will need to select the default settings that are provided. When you play the game for the first time, choose one of the four positions and play the game.

After you have completed the game, be prepared to discuss the following questions:

1. How does information impact your supply chain decisions?
2. How could information systems help with the information needs of managing a supply chain?

tems over the years, have embedded into their software packages the state-of-the-art processes used in an industry.

ERP systems also suffer from a number of significant disadvantages. First, they represent an enormous amount of work and very high costs. Because most organizations do not use directly "off-the-shelf" applications but instead customize them substantially, ERP implementations can take anywhere from six months (for a few modules) to six years (for a full implementation). There are even stories of implementations lasting up to 12 years. One study reveals that in recent implementations of ERP, 72% of organizations said the implementation took longer than expected (Panorama Consulting Group 2017). Costs of ERP implementations vary depending on the size of the implementation, from a few hundred thousand dollars to many millions. One reason implementing an ERP system is not as simple as loading the software is that it involves changing the way business is done. Business processes that have been in place for a long time may need to be redesigned. Users accustomed to the old way of doing business may offer significant resistance to an ERP implementation. Departments that have to relinquish control of some data or applications may also be resistant to the implementation. The bottom line is that an ERP implementation is often risky and complex, but the benefits can be substantial and worth the effort.

A recent study confirms that the typical disadvantages of ERP implementations still exist. The study shows that ERP implementations take longer and cost more than expected and underdeliver business value. The study also reveals that organizations do not do a good job managing the organizational changes that result from an ERP implementation (Panorama Consulting Group 2017). Why is it that many ERP implementations are so difficult? Because, as we said, they often involve changes in processes and people's jobs; they therefore represent a third-order change (Chapter 11). In addition, they are often very large in scope and may require more customization than initially expected. Finally, ERP implementations require a strong champion, a person who believes enough in the implementation to see it through many obstacles and years of effort.

Supply Chain Management Systems

The focusing story at the beginning of the chapter introduced the concept of supply chain management (SCM) at Walmart. We discussed how managing the flow of goods in its supplier and distribution channels helps the company tightly control its inventory and reduce its costs. Supply chain management, however, can also apply to services, financial products, knowledge, and relationships, as well as goods.

The typical supply chain involves the flow of goods from the manufacturer to the wholesaler to the retailer and to the consumer. A typical supply chain is shown in Figure 12.3. Firms that provide or supply an organization's products or services are

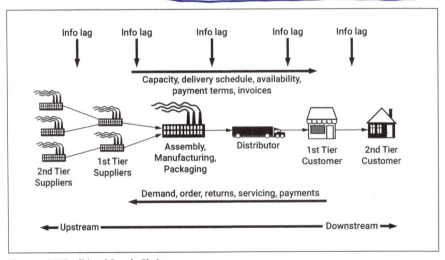

Figure 12.3 Traditional Supply Chain

BUSINESS EXAMPLE BOX 12.1

ERP Is Not Always a Horror Story: Cisco Implementation on Time and Budget

Enterprise resource planning (ERP) implementations make good horror stories for IT and business managers to talk about. As discussed in this chapter, they are often way over budget and delayed by months and sometimes years. However, there are also some good stories. For Cisco Systems, Inc., an ERP implementation went well and achieved its planned budget and time objectives.

Cisco, founded in 1984, produces telecommunications equipment. It quickly became a Fortune 500 company as a result of the growth of the Internet. To support its core business operations in the mid-1990s, Cisco used a UNIX-based software package that handled its core finance, manufacturing, and order entry functions. At the time, each functional area (business unit) used the system but made its own decisions regarding the future of its IT systems. As a result, the systems were often "upgraded" by each department for its own needs. When the CIO of Cisco at the time, Pete Solvik, decided to analyze the software applications, he found that the software package in use lacked the reliability, maintainability, and redundancy (backup) needed. A failure in the legacy system in 1994 that led to a two-day shutdown of operations resulted in the need for a new system.

As with many CIOs, Cisco's CIO was trying to avoid ERP because he was concerned that an ERP solution would involve implementation headaches and cost overruns. In particular, the independent department structure would not work well with an integrated ERP. However, the ERP did seem the best solution for the company's needs, so a consulting company was hired to help Cisco select a vendor and implement the ERP system. KPMG Consulting helped prepare the ERP request for proposal and then helped Cisco select Oracle as their ERP vendor. The agreement also involved Cisco helping Oracle market new releases to potential customers in exchange for a successful implementation within budget and on time.

The implementation team then prepared a budget and schedule, committing to top management that everything would be done within nine months for $15 million. Using rapid iterative prototyping to implement the ERP, 30 developers spent three months modifying the Oracle ERP system to support Cisco's business processes. After several phases of analysis, development, implementation, corrections, and training, the final implementation was done on time and below the $15 million budget. Meeting its goals, Cisco management offered the implementation team a bonus pool of more than $200,000.

Source: Datta, A. 2009. "Cisco Systems: Implementing 'Customized' ERP in Nine Months and within Budget." *Journal of Cases on Information Technology* 11(2): 56ff.

considered upstream, while firms that distribute an organization's products or services are considered downstream.

As you can tell from your experience with the Beer Game (Learning Activity 12.2), everyone in the supply chain benefits from collaborating with other members of the supply chain. You have also discovered (depending on how you did) that a poorly managed flow of goods can result in a disaster for a company. Some of the worst supply chain disasters are presented in Table 12.3.

TABLE 12.3 Supply Chain (and Other) Disaster Examples		
Year	Company	What Happened?
Halloween 1999	Hershey	Because of new order management and shipping systems (implementing both ERP and SCM simultaneously), Hershey could not fulfill critical Halloween orders. $150 million in revenues were lost as stock dropped 30%.
Christmas 1999	ToysRUs.com	The online retail division could not make its Christmas delivery commitments to thousands. Infamous "We're sorry" emails were sent December 23; eventually, Amazon took over fulfillment.
2001	Cisco	A lack of demand and inventory visibility as the market slowed led to a $2.2 billion inventory write-off and stock prices cut in half.
2001	Nike	A new planning system caused inventory and order woes. This was blamed for $100 million in lost sales and depressed the stock price by 20%. It also triggered a flurry of class-action lawsuits.
2005	Loblaws	Plans for a significant logistics network makeover ran into poor execution problems. The company blamed two quarters of poor results on high costs and lost sales from the effort. The CEO apologized to shareholders. Even though the logistics issues largely stabilized by early 2006, the stock price never recovered from the hit.
2007–9	Boeing	Plans to radically overhaul the supply chain for the new *Dreamliner* 787 aircraft sounded good, but massive problems with component deliveries led to a two-year delay in the aircraft's launch and some $2 billion in charges to fix supplier problems.
Christmas 2013	UPS	UPS did not have the capacity to supply deliveries for a spike in orders during Christmas 2013. In particular, on December 23, 2013, orders spiked by 63%. As a result, companies such as Toys R Us and Amazon.com had to inform customers that their products would not be delivered in time for Christmas.
April 2016	Toyota, Honda, and Sony	Two earthquakes hit Japan in April 2016. Because of suppliers being unable to deliver products and their own plants having sustained damage, all three companies slowed or temporarily stopped production.
2020	Amazon	When the global pandemic due to COVID-19 caused most people to stay at home, the number of orders for Amazon products significantly increased. As a result, Amazon did not have the capacity in their supply chain to meet the expectations of consumer demand. Customers complained about the delivery delays. However, because of the size of the organization and the limited competition, Amazon had the time to respond and saw their market share grow in the midst of the pandemic.

Sources: Gilmore (2009); Reuters (2016); SCDigest (2014).

STATS BOX 12.2

Supply Chain's Impact on Company Performance

Statistical evidence suggests that enhancements in their supply chain can help companies improve their performance. Here are some statistics that support this:

- The use of drones is making same-day delivery possible, especially in densely populated areas. Drones can travel at 50 miles per hour and carry up to five pounds; 86% of the items that Amazon.com sells are fewer than five pounds, which means there is a great potential for drone delivery.
- In 2014, 60% of organizations were automating their supply chains by pursuing increased mechanization. This is expected to increase to 83% by 2018.
- The Internet of Things has utilized RFID technologies in such a way that from 2012 to 2014, the value of the RFID market has increased from $6.96 billion to $8.89 billion. It is expected to reach $27.31 billion by 2024. The use of this technology enables smart devices in homes, cars, and elsewhere to seamlessly communicate with other devices over the Internet. This helps improve efficiencies in the supply chain.

However, unanticipated disruptions to the supply chain can have a negative impact on firms' performance. This was seen in 2020, when the COVID-19 pandemic caused significant numbers of manufacturing firms to close. This was most evident in the automobile industry, where the number of cars on dealerships' lots dwindled. People were still buying cars, but dealerships could not replace the vehicles quickly enough because there were not enough of them in the supply chain. It took several months for inventory to increase, and this cost dealerships several billion dollars in lost sales.

Source: Cerasis, https://cerasis.com/supply-chain-facts/.

Enterprise systems are key to the management of the supply chain. A well-managed supply chain can result in a reduction in inventory costs, since products can be ordered as needed as opposed to being stored for long amounts of time. It can also reduce the number of returned items through better inventory information. SCM can also improve a company's overall relationships with partners in the supply chain through more accurate and timely communication about the flow of products, information, and payments. These can be referred to, respectively, as the product flow (movement of goods), information flow (orders and delivery status), and financial flow (credit terms, payment schedules, and consignment and title ownership arrangements). A good SCM system also allows companies to offer just-in-time (JIT) inventory. This is when firms are able to maintain the lowest level of inventory while still being able to fulfill demand. Because SCM systems provide notices of inventory stock depletion in real time, personnel (or systems) are able to order new stock immediately.

The key to SCM's success rests in the sharing of information both upstream and downstream. Today, most companies use Web-based interfaces to access inventory, delivery, and other information from their SCM systems. The SCM systems also tend to be increasingly integrated with other enterprise systems, such as ERP and CRM systems.

LEARNING ACTIVITY 12.3

Blockchain and Supply Chain Management

Blockchain is a recent technological solution that is finding its way into many business processes, including supply chain management. Watch the video at the link https://www.ibm.com/blockchain to see how IBM is using blockchain to help manage the supply chain and then answer the questions that follow:

1. How is IBM using blockchain to help with supply chain management?
2. What is the advantage that blockchain provides?
3. Can you think of other implementations of how blockchain could be used in business?

Customer Relationship Management Systems

Customer relationship management (CRM) refers to an organization-wide strategy for managing an organization's multiple interactions with customers. For example, when dealing with a company to buy a product, you might talk to a person via a toll-free phone number, send an email, chat in an instant messaging box with a customer representative, or simply perform your transaction online without "talking" with anyone. These are all channels of interaction.

CRM involves a set of activities (including a philosophy of customer service) and technologies (the CRM system) meant to help companies understand the needs of current and potential customers. The CRM system allows organizations to attract and retain customers and to manage customer relationships—for example, by offering personalized services and self-servicing tools. It also allows a company to analyze customer data to find its most profitable products, services, and customers.

CRM includes three main goals (META Group 2001). First, CRM systems help organizations manage multiple channels of interaction with customers in ways that the customers prefer. The second goal of CRM systems is to provide an integrated picture of the customer across the various customer-facing parts of the organization. The third major goal of CRM systems is to enable the analysis of customer-related information. This information is gathered through interactions with customers, although external data (such as census data) may also be involved.

Customer Service Life Cycle

The **customer service life cycle (CSLC)** is a framework to help us understand the various tasks a CRM system can be useful for. The CSLC is a series of phases that customers pass through when interacting with an organization. There are a number of variations on the phases, but they are all similar. Our simplified CRM includes four phases: engage, transact, fulfill, and service. Table 12.4 summarizes the phases and shows some specific CRM-related applications that can be useful for each CSLC phase. The combined phases are referred to as a *cycle*, as shown in Figure 12.4, because completion of one round of the cycle often leads to another round. Note

TABLE 12.4 Phases in the Customer Service Life Cycle		
Phase	Description	Example of CRM System Use
Engage	Creating customer awareness of the product or service. The goal of the engage stage is to generate leads and then convert those leads into customers.	Campaign management Email marketing Lead processing Sales force automation Web-based catalogs
Transact	All activities associated with the purchase process. The goal of this phase is to efficiently and effectively complete the purchase process so that customers do not abandon the purchase process.	Order management Payment process and options Product configuration Product pricing
Fulfill	Delivery of the product or service to the customer.	Order tracking Supply chain integration
Service	Supporting the customer during the ownership of a product or service.	Call center automation Customer issues management Self-service
Source: Adapted from Van Slyke and Bélanger (2003).		

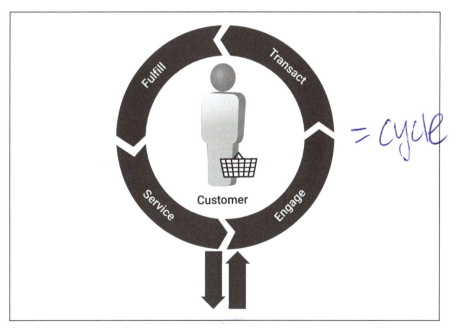

$= cycle$

Figure 12.4 Customer Service Life Cycle Source: Adapted from Van Slyke and Bélanger (2003).

that in the following discussion, we refer to products, but similar statements could be made about services.

Components of CRM Systems

CRM systems have three main components: operational, collaborative, and analytical (Van Slyke and Bélanger 2003). Figure 12.5 shows how the components work together.

The **operational components of a CRM** system help an organization improve day-to-day interactions with customers. They are involved with the operations of the company, such as sales or repairs.

The **collaborative components of a CRM** system help the organization interact and collaborate with its customers. They include older applications, such as email and automated response systems, but also newer ones, such as interactive chat facilities for customer service, **interactive voice response (IVR)** systems, and voice over IP (VoIP), which allows a customer to talk to a "live" customer service representative while online.

The **analytical components of CRM** consist of technologies and processes organizations can use to analyze customer data. A critical analytical component in most CRM systems is the data warehouse, which we discussed earlier. **Data mining technologies and methods are often used to discover patterns and groups in the data stored in a data warehouse.** Data mining involves using software that performs analyses of large data sets to find patterns that repeat themselves. For example, a grocery store might find that on college football weekends, certain items always seem to sell "together." The grocery store can then locate them together within the store. This is a very simple example, and data mining can find much more complex and detailed patterns. One analysis that many organizations find useful is customer segmentation (sometimes called *customer clustering*), which allows organizations to group customers that share certain characteristics. For example, Student Advantage, a company that provides college student-oriented marketing services, uses segmentation to place its student members into very small segments. This allows Student Advantage to provide extremely targeted marketing opportunities to a wide variety of clients. Data mining can also help in customer profiling. **Customer**

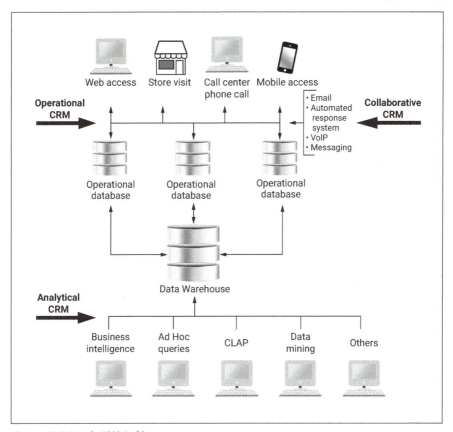

Figure 12.5 Sample CRM Architecture

S T A T S   B O X   1 2 . 3

CRM Vendors

The CRM market has been growing over the years, with revenues of $41.7 billion in 2017 and $48.2 billion in 2018. Research suggests that for 2018 the top vendors were Salesforce.com, with about 20% of the CRM market, SAP, with about 8%, and Oracle, with about 6% of the CRM market.

Source: https://softwarestrategiesblog.com/category/crm-market-share/.

LEARNING ACTIVITY 12.5

Self-Servicing

We discussed CRM applications as a way companies can gather information about their customers and as a way to manage their communications with them. An additional function of CRM applications is to permit **customer self-service**, allowing customers to take care of their service problems themselves. Before we discuss the pros and cons of doing this, you should try self-servicing.

1. *Scenario:* You ordered new noise-free headphones for taking on your multiple trips. You bought them on Amazon.com for $65. You opened the shipping package two months ago but never opened the actual package completely. (It has the original packaging.) You were busy with the start of classes, so you did not think about it again, but now your parents gave you the exact same item as a gift, so you want to return the package to Amazon.com.
2. Go to the Amazon website (http://www.amazon.com).
3. Figure out how much you are going to receive as a refund (not counting taxes, if any).
4. Estimate how long it is going to take for you to get your money once the headphones are shipped back to Amazon. Identify why this time is needed.
5. Figure out what you have to do to return the headphones.

Discussion Questions

1. How difficult was it to find the information needed?
2. How difficult do you think it would be to actually complete the return?
3. How likely are you to use this self-servicing application (Return Center) instead of calling someone at Amazon? Why?
4. What are the benefits to customers of self-servicing CRM applications?
5. What are the benefits to companies of self-servicing CRM applications?

profiling lets companies build a picture of a customer through data collected from their interactions with this customer. These profiles allow companies to better anticipate customer needs and better predict customer behavior.

CRM Implementation Options

Like most other enterprise systems, there are two main ways that CRM can be implemented: on premise or on demand. **On-premise CRM** implementation means that the CRM system is acquired, leased, or developed by the organization and installed on its local servers. This often has the advantage of providing a customized solution. However, the costs of development and maintenance could be substantial. **On-demand CRM**, also known as **software-as-a-service (SaaS) CRM**, refers to the remote use of a CRM provider's applications through the Internet. This offers advantages like up-to-date applications and no maintenance or implementation costs. Of course, the company has to use the standard functions, also known as *plain vanilla* software.

BUSINESS EXAMPLE BOX 12.2

DUFL Gains Competitive Advantage through Customer Service

DUFL is a company that strives to ease the burdens of traveling by shipping, storing, and cleaning its clients' business attire. Customers provide DUFL with their business attire, and DUFL stores the clothes. When it is time to travel, rather than pack their suitcases and take them to the airport with them, customers inform DUFL where they are staying, when they are arriving, and what clothes they want while they are there. DUFL then ensures that these clothes are at the hotel waiting for the clients. When the trip is over, the clothes are packed up and sent back to DUFL, where they are cleaned and stored for the next time they are needed.

A key aspect to DUFL's success is to provide a high level of customer service while keeping its costs low. To do so, DUFL relies on Salesforce to allow it to keep track of each customer's unique needs. Salesforce provides DUFL's customers and employees with a user-friendly interface that keeps track of every transaction and interaction. This results in consistent solutions when customers call back with similar requests. In addition to helping DUFL keep existing customers happy, the software also provides the ability to project future sales. All of this is done in a way that allows DUFL to focus on its core competencies of enhancing business travel and growing its business. The result has been 10% growth each month while maintaining a lean staff of 25 employees.

Sources: https://www.dufl.com/; https://www.salesforce.com/customer-success-stories/dufl/.

Benefits and Disadvantages of CRM

There are many expected benefits from the use of CRM systems for organizations. First, CRM systems are expected to provide greater customer satisfaction because customers can receive personalized services and products, faster response time to their questions, and more streamlined interactions with the company. For companies, this results in improved customer relations, which should lead to repeat business (called *customer retention*). In addition, CRM systems can help an organization maximize its profits by reducing the costs for acquiring and servicing customers, and increasing overall revenues. Revenues can increase because of customer loyalty but also because of what is called *cross-selling*, or the selling of products to customers based on other products they acquired. For example, Amazon.com often sends its customers book recommendations based on books they have previously bought.

There are also some potential disadvantages to the use of CRM systems. First, the implementation of a CRM system requires more than simply installing the software application. To be successful in implementing a CRM system, an organization must look beyond purely technical issues and ensure that all customer-facing employees are customer minded. The company must also make sure that it identifies all channels of interactions with customers. The ability to achieve buy-in from various areas of the organization is key to a successful implementation. Of course, there are complicated technological issues that also must be addressed. First, the organization must ensure that its CRM architecture is scalable so that growth in customer demands is met without problems. Just like ERP systems, CRM systems can be customized to meet the specific needs of the organization. Overall, the ability to deal with both nontechnical and technical issues is a requirement for successfully implementing almost any enterprise system.

Customer-Managed Interactions (CMI)

One issue with CRM systems is that they are very firm-centric in the sense that they focus on the existing transactional and behavioral data collected by the organization based on the prior and current interactions of the customer with the organization. While trends can be used to predict what customers might buy in the future, there are some acquisitions and customer activities that are unforeseeable unless the customer tells the company about it. For example, a customer might be interested in some products for a friend's upcoming wedding, but the company would not know of this need. To address this lack of information, some companies have turned their focus to customer-managed interactions (CMI).

Customer-managed interactions (CMI) involve letting customers store and manage data about themselves. Why would customers willingly tell companies of their buying intentions? For convenience, for money, or to obtain personalized service. Think of all your friends who create wish lists at some of the major online retailers like Amazon.com or Target.com. There is even a site dedicated to helping anyone create wish lists (WishList.com). Customers tell the company what they intend to buy; therefore, this information tends to be accurate. Once the company knows about intended purchases, it can perform targeted marketing, for example, by giving a specific discount, which may push the customer over into the buying phase of the customer service life cycle.

Warehouse Automation and Robotics

As organizations have focused on improving interactions with customers, they have also utilized advances in technologies to automate the operations of their warehouses. This has been made possible through the use of robotics. The way this often works is the robot completes jobs that are tedious and physically taxing, allowing the human workers to focus on less physically demanding aspects of warehouse operations. One downside to implementing the use of robotics is the loss of human jobs. However, at Amazon, the use of robotics has led to increased growth, resulting in increased jobs. The fact that these robots can perform physically taxing tasks makes the jobs being hired for safer.

Chapter Summary

In this chapter, we discussed enterprise systems. We first presented the main components, characteristics, and types of enterprise systems. We then explored in greater depth three enterprise systems: enterprise resource planning (ERP) systems, supply chain management (SCM) systems, and customer relationship management (CRM) systems. Here are the main points discussed in the chapter:

- Enterprise systems are information systems that serve the needs of the organization or parts of the organization.
- Enterprise systems are modular; they are integrated with one another (offering application integration) and use a central database (offering data integration).
- The purpose of enterprise resource planning (ERP) systems is to manage an organization's resources and enable information flow within and between processes (and departments) via a set of integrated information systems tools.
- Supply chain management (SCM) systems help organizations manage the flow of goods in their supplier and distribution channels to better control inventory, reduce costs, and improve communication with their business partners.
- CRM systems have three main components: the operational component that helps an organization improve day-to-day interactions with customers; the collaborative component that helps the organization interact and collaborate with its customers; and the analytical component, which consists of technologies and processes organizations can use to analyze customer data.
- The customer life cycle is a series of phases that customers pass through when interacting with an organization, and it can be used to identify important CRM applications.

Review Questions

1. What is the difference between a personal and an enterprise information system?
2. How are applications developed using the functional perspective? The hierarchical perspective?
3. Define business integration and systems integration.
4. What is an enterprise resource planning (ERP) system, and how does it support systems integration?
5. What is the purpose of a supply chain management (SCM) system?
6. What is customer relationship management (CRM)?
7. What are the main components of a CRM system? Briefly describe each one.
8. What are the main goals of a CRM system?
9. What is the difference between on-premise CRM systems and on-demand CRM systems?
10. What are customer-managed interactions? How do they differ from CRM?

Reflection Questions

1. What is the most important thing you learned in this chapter? Why is it important?
2. What topics are unclear? What about them is unclear?
3. How does the process perspective provide integration, and what types of integration are provided?
4. How does cloud computing relate to ERP and CRM applications?
5. How does the material in this chapter relate to the content of Chapter 4 on strategic information systems?
6. Explain how data integration is essential to the success of ERP, SCM, and CRM systems.
7. Which issues are similar for both the process and the hierarchical perspectives of developing enterprise systems? Why?
8. How can process modeling (Chapter 11) enable firms to successfully implement enterprise systems?
9. We discussed different types of integration in this chapter (system, business, and data). How do they relate to one another?
10. Can a very small business implement a CRM system? What would be the advantages and disadvantages of doing so?

Additional Learning Activities

12.A1. This activity requires you to complete a case analysis regarding an ERP acquisition. The case may be accessed through your university website, the Ivey Business School business case website (https://www.iveycases .com), or the Harvard Business School cases website (https://hbsp .harvard.edu/home/), or it may be provided to you by your instructor. Hajj, S. E. 2018. "The Diet Center: The SAP ERP Decision." Ivey School of Business, May 9. Product Number: 906E12-PDF-ENG.

12.A2. Watch the following short video: https://www.youtube.com/watch?v =XxJoPUVd3Yo. The video discusses how artificial intelligence is used to help with providing and analyzing data for a CRM targeted at realtors. Prepare a short report describing the main points of the video. Be sure to address the following questions:
   a. How do realtors benefit from using an artificial intelligence-based CRM?
   b. How do you think the CRM provider gets access to the different data that feeds into their system?
   c. How does a CRM make the users of the system more efficient?
   d. What would a realtor have to do if they did not use a CRM system such as this?

12.A3. Select a small organization with which you are familiar. This could be your employer, a company where you parents or friends work, or just a company you know well. First, evaluate the company's need for a CRM system based on your knowledge from the readings, discussions, and activities you have done in this chapter. Prepare a short report including your recommended features, applications, and strategies for this small business.

12.A4. After completing Learning Activity 12.A3, watch the "Salesforce Marketing Cloud: Overview Demo" from Salesforce.com and include any changes to your report based on the information in the demo: https://www.youtube.com/watch?v=s5n4CJbGfi4.

12.A5. Compare and contrast in a table the advantages and disadvantages of each enterprise system perspective (hierarchical, functional, and process).

12.A6. Imagine that you are a new student at a university that requires each student to have a computer but does not provide guidelines beyond the fact that they have to be laptops, use wireless networks, and have the capability to use basic office applications and statistical analyses. The laptops have to be useful for three years. Develop a list of activities from the customer service life cycle that apply to this situation. Identify the CRM applications that can be used for each activity.

12.A7. Self-driving vehicles are no longer a dream of science-fiction movies but are becoming a reality. As this has happened, people have discussed the idea that trucks that deliver products could eventually become driverless. This article, http://www.scdigest.com/ontarget/20-07-28_MIT_Autonomous_Trucking.php, suggests some of the challenges and issues facing this change. Read this article and then write a report that addresses the following three questions:

  1. What has enabled trucks to potentially be driverless?
  2. What are some of the challenges that need to be overcome before we can have driverless trucks?
  3. What are the benefits and costs to society as this possibly becomes a reality?

References

Gilmore, D. 2009. "The Top Supply Chain Disasters of All Time." *Supply Chain Digest,* May 7. http://www.scdigest.com/assets/FirstThoughts/09-05-07.php?cid=2451&ctype=content.

Panorama Consulting Group. 2017. *2017 ERP Report.* https://www.panorama-consulting.com/resource-center/erp-industry-reports/2017-report-on-erp-systems-and-enterprise-software/.

Pearlson, K., and C. Saunders. 2010. *Managing and Using Information Systems.* 4th ed. Hoboken, NJ: Wiley.

Piccoli, G. 2008. *Information Systems for Managers.* Hoboken, NJ: Wiley.

Reuters. 2016. "Toyota, Other Major Japanese Firms Hit by Quake Damage, Supply Disruptions." *Fortune,* April 17. http://fortune.com/2016/04/17/toyota-earthquake-disruptions/.

SCDigest Editorial Staff. 2014. "Supply Chain News: The Factors behind the UPS Failure to Deliver Christmas Goods." *SCDigest,* January 6. http://www.scdigest.com/ontarget/14-01-06-2.php?cid=7717

Van Slyke, C., and F. Bélanger, 2003. *Electronic Business Technologies.* New York: John Wiley & Sons.

Zornes, A. 2001. "META Group Report: Five CRM Trends for 2001–'02." *Datamation,* May 18. https://www.datamation.com/datbus/article.php/769281/META-Group-Report-Five-CRM-Trends-for-2001---02.htm.

Further Readings

SAP Information
- https://www.sap.com/products/enterprise-management-erp.html
- https://www.sap.com/products.html

Supply Chain Management Videos
- https://www.sap.com/products/supply-chain-management/supply-chain-planning.html
- http://www.youtube.com/watch?v=dDmGtPQAI24 (MS Dynamics)

CRM Video
- http://www.youtube.com/watch?v=-_6c018ZI2g

Glossary

Analytical components of CRM: Technologies and processes organizations can use to analyze customer data.

Application integration: Applications where modules are designed to be integrated with one another.

Business integration: The unification of business processes previously performed as separate activities to create cohesive and streamlined business processes.

Business intelligence (BI): A set of applications, technologies, and processes for gathering, storing, analyzing, and accessing data to help users make better business decisions.

Business process reengineering (BPR): The redesign of business processes to improve how work is done.

Call center automation: Applications that support call centers, which are often the main contact point between an organization and its customers.

Campaign management: Applications that help organizations plan, carry out, and analyze the results of marketing campaigns.

Collaborative components of CRM: Technologies and processes that help organizations interact and collaborate with their customers.

Content management systems (CMSs): Applications that allow companies to manage the vast amounts of information of various types (content) they have stored. This includes the capture, storing, preservation, and delivery of content.

Customer issues management: Applications that help organizations with tasks such as scheduling, dispatching, and communicating with field service personnel.

Customer-managed interactions (CMI): Applications that let customers store and manage data about themselves.

Customer profiling: Applications that help organizations group customers according to demographic characteristics or behaviors.

Customer relationship management (CRM): An organization-wide strategy for managing an organization's multiple interactions with customers.

Customer self-service: Software that allows customers to find solutions to problems without interacting with a customer service representative.

Customer service life cycle (CSLC): A series of phases that customers pass through when interacting with an organization.

Data integration: Applications where data are stored in one central database.

Data mining: The process of analyzing data to identify trends, patterns, and other useful information to make predictions.

Enterprise application integration (EAI): A set of tools and services (a framework instead of a single technology) to allow various enterprise systems (such as ERP or CRM) to share data.

Enterprise resource planning (ERP): Software that allows organizations to manage resources throughout the organization, independently of which business function controls the resource.

Enterprise systems: Information systems that serve the needs of the organization or parts of the organization.

Functional perspective: Software development approach where information systems applications are developed to meet the needs of individuals in a functional area, such as human resources or marketing.

Hierarchical perspective: Software development approach where software developers build information systems to mirror the organization so that applications are built to meet the needs of individuals at a given level in the organization.

Interactive voice response (IVR): Applications that allow customers to use a telephone to navigate through various types of systems, such as product request systems or customer service systems.

Lead processing: Applications used to qualify, assign, and track sales leads to maximize the probability that the leads eventually turn into sales.

On-demand CRM: Remote use of a CRM provider's applications through the Internet.

On-premise CRM: CRM specifically developed for the organization and installed on its local servers.

Operational components of CRM: Technologies and processes that help improve day-to-day interactions with customers, such as sales, repairs, and so on.

Order tracking: Applications that allow customers to determine the status of their order.

Personal information systems: Information systems applications for individual use.

Process perspective: Software development approach where information systems applications are developed to support the task being done (or process) regardless of the organizational level or functional area of the individual using the system.

Product configuration: Applications used to properly configure or choose complex products.

Radio-frequency identification (RFID) tags: Electronic tags that can be tracked remotely using wireless networks.

Sales force automation: Applications that provide functions directed at making sales representatives more efficient and effective.

Software-as-a-service (SaaS) CRM: Remote use of a CRM provider's applications through the Internet.

Supply chain: Entities involved in the flow of goods from the manufacturer to the wholesaler to the retailer and to the consumer.

Supply chain management (SCM) systems: Software that helps organizations manage the movement of materials or products from provision to production to consumption.

Systems integration: The unification of information systems and databases that previously operated as separate systems.

Unified communications (UC): A framework for the integration of the various modes of communication used by an organization, including email, voice mail, videoconferencing, instant messaging, texting, and voice over IP.

Chapter 13

Information for Electronic Business

Learning Objectives

By reading and completing the activities in this chapter, you will be able to:

- Describe the concept of e-business and various e-business models
- Identify e-business models, enablers, and impacts
- Explain search engine optimization
- Discuss the main business-to-business (B2B) e-business enablers
- Identify the major trends related to e-business

Chapter Outline

Introduction to E-Business
Focusing Story: The iPod and the Music Industry
    Learning Activity 13.1: Why Is E-Business Important?
E-Business Models
E-Business Enablers
E-Business Impacts
Design for E-Business
    Learning Activity 13.2: Why Would I Trust Them or Buy from Them?
Business-to-Business (B2B)
    Learning Activity 13.3: B2C versus B2B
Search Engine Optimization
    Learning Activity 13.4: Where Is My Web Page?
Trends in E-Business

The growth of the Internet has changed our everyday lives, just as it has changed how businesses and governments interact and conduct transactions with consumers. These digital interactions have led to the phenomenon of e-business, also known as digital markets. In this chapter, we explore various ways that organizations and individuals interact online and the implications that digital interactions have for us all, as you will see in the focusing story on digital music.

Introduction to E-Business

In Learning Activity 13.1, you will find several statistics about e-business. What you have already learned should have convinced you of the importance of e-business in

The iPod and the Music Industry

Who would have thought that a small device like an iPod could dramatically change a well-established major industry? Yet that is what happened when Apple Inc. launched the iPod and electronic music downloads. Because of this new way of selling songs and albums, the music industry has been one of the most affected by e-business.

In the "old" music industry, artists recorded songs and albums through record companies, which were the major players in the industry. Companies could sell CDs (and, earlier, LPs and cassettes) and make significant profits on each album customers bought, with some fees (rights) going to the artists and some commissions going to distributors and retailers. Eventually, the record companies decided not to release singles, only albums. That was not to the liking of music listeners. Then came the iPod, announced on October 23, 2001. It allowed users to download only the songs they wanted to a very portable device. The industry was never the same. The music industry faced two significant problems with the growing use of the Internet for access to digital music. The first is that digital music is easy to steal, and users started to download music illegally and share it with one another. In this scenario, everyone in the industry lost money. Even though illegal downloads had started years before, with Napster allowing illegal downloads for free between 1999 and 2001, the iPod made it easy to copy songs. At the same time, however, many users were happy to spend small amounts of money for legal downloads of the songs they really wanted. That is why Apple is said to have sold more than six billion songs on iTunes by the end of 2009.

A second phenomenon that occurred with the ability to download songs through the Internet was that some artists decided to sell their own music online. They were able to charge less to consumers because they only had to pay the e-business website for selling their music; they also made more profits this way. On iTunes, Apple keeps 30% of the profits from each song sold, with the rest going to the artist. So, on a song that sells for $1.29, Apple receives about $0.39, and the label and artist receive about $0.90. When artists produce their own music, they get the full $0.90 that normally goes to the artist and label. Record companies changed their licensing approaches to embrace new online services in response to this major threat to the industry. However, they did not envision online streaming, the new online threat to the music industry.

Now, many of you probably do not own songs, as you rely on online streaming music to get any song you want when you want. This is made possible because most people have a smartphone that is always connected to the Internet. This allows services like Pandora (http://www.pandora.com), Apple Music Plus, Amazon Music, Google Play Music, or Spotify (http://www.spotify.com) to provide access to almost all songs for free or relatively inexpensively. Some of these services allow you to download songs or playlists for offline listening. As users listen online instead of purchasing music, sales of music keep going down every year. The total music industry sales in 2000 were approximately $14 billion but were down to $10 billion by 2008. In 2014, revenue from streaming sources of music outpaced the revenue from CDs for the first time. By this time, sales of iPods had decreased dramatically because of the popularity of the iPhone and other smartphones. While today many of us listen to music on our phones rather than iPods, the impact of the iPod on the music industry remains.

Sources: Woollaston, V. 2013. "15,000 Songs a MINUTE Downloaded on iTunes as Apple's Music Service Celebrates Its Tenth Birthday." *Daily Mail,* April 29. http://www.dailymail.co.uk/sciencetech/article-2316473 /A-decade-iTunes-How-Apple-revolutionised-way-buy-music.html; MacMillan, D. 2009. "The Music Industry's New Internet Problem." *Bloomberg BusinessWeek,* March 6. http://www.bloomberg.com/bw/stories/2009-03

-06/the-music-industrys-new-internet-problembusinessweek-business-news-stock-market-and-financial
-advice; Mansfield, B. 2015. "Streaming Revenue Passes CDs for First Time." *USA Today,* March 19. http://www
.usatoday.com/story/life/music/2015/03/19/streaming-revenue-passes-cds-riaa-report/25023773/.

Focusing Questions

1. Will the streaming audio approach change the music industry once more? How?
2. How can the streaming audio business model be successful?
3. What other ways could companies use to leverage digital music?

LEARNING ACTIVITY 13.1

Why Is E-Business Important?

Most of you have already experienced **electronic business**, the topic of this chapter, when acquiring products electronically. E-business, however, is much broader than just buying something online. In this learning activity, we want you to think broadly about what e-business is and find two interesting, up-to-date statistics online about the size of electronic business. These can be for a subset (such as B2C or B2B), an industry, or a market segment. Make sure your statistics are current—say, within the last two years. Be careful about some of the older information available online. For example, you may find predictions from five years ago about expected e-business sales. It is much better to find out what current e-business sales are. Prepare a short report of your two statistics (make sure to include your citations!) and a brief discussion of why you think e-business is important today.

---

today's economy. However, the importance of e-business goes beyond any number you could find online. The reality is that e-business changed how most organizations interact with their customers and suppliers and even how they interact with the government. More important, complete industries have been transformed with the emergence of e-business applications. This is the case in the focusing story, which shows how e-business completely changed the music industry. What other industries have been changed by e-business? You may think of travel, publishing, and even education, just to name a few. Later in this chapter, we discuss specific impacts e-business has had on these industries.

What is e-business exactly? There are many definitions, but another way to look at it is as a digital market. Table 13.1 provides some terminology used when discussing e-business and digital markets. Note that **electronic funds transfer (EFT)** and electronic data interchange (EDI) provide tools for transferring documents or financial data between organizations. The other terms include variations on whether the Web, other Internet tools, or any other electronic means are used for business transactions, such as collaboration software, kiosk technologies, and so on. For our purposes, we will use the term *e-business,* which is often recognized as the broader term that includes various electronic means of conducting business.

TABLE 13.1 E-Business Terminology	
Term	**Definition**
Electronic funds transfer	Conducting financial transactions such as payments and remittances electronically
Electronic data interchange	Transferring documents between organizations electronically
Internet commerce	Conducting business using the Internet (for example, email, the Web, or file transfers)
Electronic commerce	Conducting business transactions with customers, suppliers, and external partners utilizing the Internet, electronic data interchange, or private networks
Electronic government	Interacting or conducting transactions with government agencies electronically
Electronic business	Conducting electronic commerce as well as executing internal business processes that facilitate conducting business, such as production and inventory management, electronically

Types of E-Businesses

In Table 13.1, we differentiate between commerce and government. There are other ways to categorize the various types of e-business. Some of the most common categorizations are as follows:

- **Business-to-consumer (B2C)**: This is when interactions are taking place between a consumer and a business. For example, when you buy books on Amazon.com, you are conducting a B2C e-commerce transaction.
- **Business-to-business (B2B)**: This is when interactions are taking place between two businesses. For example, when Walmart sends electronic requests for inventory to its suppliers, it is conducting a B2B e-commerce transaction.
- **Consumer-to-consumer (C2C)**: This is when interactions are taking place between two consumers. For example, when you buy goods from another individual on eBay.com, you are conducting a C2C e-commerce transaction.
- **Government-to-constituent (G2C)**: This is when interactions are taking place between a government agency and a constituent. Constituents could be citizens, businesses, or even other agencies. For example, when you pay your taxes online, you are having a G2C interaction with the government. This is most often referred to today as *electronic government*, or *e-government*.

STATS BOX 13.1

Retail E-Commerce in the United States

As you might have found in Learning Activity 13.1, there are many statistics on e-business that are often not consistent with one another. One good source of statistics for the United States is the U.S. Census Bureau, which provides quarterly statistics on retail e-commerce. The following table shows the statistics available as of August 2020. You can obtain the most recent statistics at the following site: http://www.census.gov/retail /#ecommerce/. Note that although retail sales numbers decreased in the second quarter of 2020, retail sales in e-commerce increased significantly. This was a result of people staying home because of the COVID-19 pandemic. It will be interesting to see how these numbers change when the pandemic is completely over.

Estimated Quarterly Total and E-Commerce U.S. Retail Sales, 2019–2020			
Quarter	Retail Sales Total	Retail Sales E-Commerce	E-Commerce as Percent of Total
2nd quarter 2020	1,310,973	211,505	16.1
1st quarter 2020	1,364,197	160,414	11.8
4th quarter 2019	1,381,250	156,581	11.3
3rd quarter 2019	1,374,212	153,274	11.2
2nd quarter 2019	1,359,250	146,394	10.8
1st quarter 2019	1,335,812	139,713	10.5

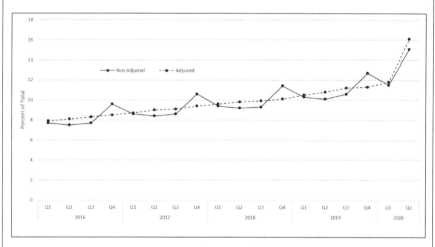

Estimated Quarterly U.S. Retail E-Commerce Sales as a Percent of Total Quarterly Retail Sales: 1st Quarter 2016–2nd Quarter 2020 Source: U.S. Census Bureau, http://www.census.gov/retail/# ecommerce.

TABLE 13.2 Types of E-Businesses			
E-Business Type	**Acronym**	**Definitions**	**Parties Involved**
Business-to-consumer	B2C	When interactions are taking place between a consumer and a business	• Consumers • Businesses
Business-to-business	B2B	When interactions are taking place between two businesses	• Businesses
Consumer-to-consumer	C2C	When interactions are taking place between two consumers	• Consumers
Government-to-constituent (e-government)	G2C	When interactions are taking place between a government agency and a constituent	• Citizens • Businesses • Government agencies

There are many more types of e-business, although they tend to be variations of the definitions in Table 13.2; for example, consumer-to-business (C2B), business-to-employee (B2E), or government-to-government (G2G). The main ones outlined above, however, seem to be the most popular categories.

E-Business Models

The various types of e-business previously discussed refer to who participates in the e-business interaction. Another important way to classify e-businesses is with respect to their business model.

What Is a Business Model?

A **business model** represents the way the organization functions and creates value. In other words, it is how an organization makes money. The business model often identifies the market that a business is in, the products or services that it offers, and the strategies and major activities it uses to seek competitive advantages in that market. Importantly, the business model should identify the organization's key business processes and organizational capabilities that allow the business to generate revenues and profits.

E-Business Models

There are many different business models in the world of e-business, some more popular than others, and many that did not really exist before the increased use of the Internet for business. Table 13.3 shows a variety of e-business models.

The e-business models presented in Table 13.3 continue to evolve as individuals find new ways to use the Internet and, increasingly, mobile technologies to create new approaches to generate revenue. In fact, the use of **location-based services (LBS)** on mobile devices (knowing where you are based on your smartphone's location) allows for a number of new business models. You are likely most familiar with

Adapting Its Business Model: Amazon.com

When the dot-com bust happened in the early 2000s, many e-businesses simply went out of business, unable to generate sufficient revenues to cover their costs. Others, however, were able to survive and thrive. One such company is Amazon.com. However, Amazon.com, which for most of its life span has existed only as a website, is now adapting to maintain its competitive advantage in the retail sector.

In 2015, Amazon.com opened its first retail store location on the campus of Purdue University. Students can place any orders that they want at the store but receive one-day shipping if they have it delivered to Amazon.com's on-campus location. Amazon.com plans to create more retail stores on other campuses in the future. Amazon expanded its physical stores to noncampus locations. Amazon has stores in several cities, including Chicago, New York, Seattle, Portland, Los Angeles, and San Diego, and provides the ability for customers to check out without using a cashier.

In 2017, Amazon.com surprised many by acquiring Whole Foods, a high-end chain of grocery stores. Almost immediately following the closing of the acquisition, Amazon lowered prices on many of Whole Food's best-selling items. Amazon Prime members also receive discounts and other in-store benefits. Amazon also utilizes the Whole Foods store as a warehouse to enable direct grocery delivery.

Amazon is also in the process of testing and obtaining permission to utilize drones, or pilotless aircraft, to deliver packages. Doing so would allow Amazon.com to deliver products without requiring a human delivery driver to bring products to the customer's location.

Sources: Smith, A. 2015. "Amazon Opens Its First Store." *CNN Money*, February 4. http://money.cnn.com/2015/02/04/technology/amazon-purdue/; Lapowsky, I. 2015. "Amazon Can Now Test Its Delivery Drones in the U.S." *Wired*, April 10. http://www.wired.com/2015/04/amazon-drone-faa-green-light/; "Whole Foods Market @ Amazon." https://www.amazon.com/b?ie=UTF8&node=17235386011; "Why Amazon Is the World's Most Innovative Company of 2017." *Fast Company*, February 13. https://www.fastcompany.com/3067455/why-amazon-is-the-worlds-most-innovative-company-of-2017.

the ride share business model, such as Uber and Lyft, which connects someone who is offering a ride with someone who needs a ride. The fact that almost everyone has a smartphone that knows their precise location and can facilitate a transaction makes such a business model easily possible. Other similar e-business models exist that facilitate the renting of a house (e.g., Airbnb, VRBO), the selling of handmade goods (e.g., Etsy) and finding a baby or dog sitter (e.g., https://dogvacay.com/). LBS also makes possible other e-business models that may not be as obvious, such as allowing marketing firms to offer targeted advertising and services that you may need at a specific moment based on where you are.

The Internet also offers the ability for other business models providing infrastructure services, such as PayPal.com, Venmo, and CashApp, as well as social media websites like Facebook.com or Delphiforums.com. There are even some fairly unique business models in existence today, like the name-your-own-price model of Priceline.com (http://www.priceline.com).

It is necessary for business models to evolve with today's rapid changes in technology

TABLE 13.3 E-Business Models

E-Business Model	Description
Online retailing	Offers products or services for sale to consumers online. Some businesses acquire the products for resale, while others act as electronic intermediaries for selling products.
Infomediary	Offers specialized information to consumers via the Internet that aggregates or analyzes products or services from several providers. The term comes from information intermediaries.
Content providers	Offers consumers content or relevant information and receives money from vendors for ads or downloads or from consumers for subscriptions.
Exchanges	Offers a location (marketplace) for buyers and sellers to transact online. Revenues come from fees for sellers or buyers or commission on sales.
Online community or social media	Offers individuals with similar interests and/or goals an online meeting place to interact with one another. Many start as not-for-profit but end up charging for select services or information or allow ads to be posted on the community.
E-business infrastructure provider	Offers infrastructure hardware, software, or services to other organizations for e-business. Revenues come from fees paid for the services.

and the changes in the ways individuals interact with one another. For example, as many students know, changes in your online profile on a social networking site result in changes in the types of ads you see when accessing the site. This change in the social media business model has occurred in the last few years. However, while keeping up with technological and social changes is important, the most important factor for success is to start with a proper business model. Back in the early days of e-business, many entrepreneurs jumped on the bandwagon of e-business without proper business models. Not surprisingly, many of them failed. This is one reason behind what was called the *dot-com bust,* when many individuals and companies lost money in their attempts to take advantage of e-business. Today, there are too many e-business successes to even attempt to summarize them. Some are very innovative, while others simply follow solid business models.

E-Business Enablers

The statistics previously presented and those that you found in Learning Activity 13.1 clearly show that B2C e-business continues to increase in value worldwide.

Important Facts	Examples
The ability to distribute products efficiently and effectively (fulfillment and logistics functions) is critical to success in this business model.	• LLBean.com • BestBuy.com • Staples.com
These companies do not sell products or services directly, and they do not have any inventory. The ability to maintain up-to-date information is crucial in this business model.	• Consumer goods: MySimon.com • Travel: Kayak.com • Automobiles: Edmunds.com
Content can be generated and published by the organization, or it can be user-generated content that the business publishes.	• News: Bloomberg (http://www.bloomberg.com) • Entertainment: Eonline.com • Historical and reference: Britannica.com • Travel: Tripadvisor.com
Generating a large number of buyers and sellers (a critical mass) is crucial for success in this business model.	• eBay.com • uBid.com
Online communities are virtual in the sense that the meeting place is online only.	• Facebook.com • Delphiforums.com
Online businesses can use these infrastructures through cloud computing (Chapter 7) or directly through the provider. This has substantially reduced the cost of starting online businesses.	• Online payment service: PayPal.com • Credit card processing service: Squareup.com • Web and database hosting services: Rackspace.com • Domain name registration and Web and database hosting: GoDaddy.com • Mobile Commerce Platform: Uber and Lyft

What has made this growth possible? A combination of technological and social factors can help explain why B2C e-business has grown so much over the years.

Technological Enablers

There are several technological requirements that must be met for consumers to successfully acquire goods or services online. First, there must be an easy-to-navigate website or smartphone application. The graphical user interfaces of today's browsers and apps as well as the search tools available are examples of technological improvements that have enabled e-business to grow. In addition, e-business requires that consumers have sufficient network bandwidth to access e-businesses' content. Think of how much time it takes sometimes to download images for goods you want to purchase. Most times, images and videos load instantaneously. However, when it takes too long for them to load, you might end up going to a different site. The bandwidth required depends on the images and other features of the site, and the overall bandwidth is a function of the vendor's network bandwidth (and server processing capability) and the consumer's Internet access bandwidth. In recent years, the use of broadband networks for Internet access (see Chapter 7) has significantly improved the overall bandwidth availability and therefore the online shopping experience of consumers. However, there are still many rural locations that lack reliable high-speed Internet service.

A less visible but just as critical technological requirement is interoperability. We discussed **interoperability** in Chapter 7, where we defined it as the ability of hetero-geneous systems to communicate with one another. Interoperability is key to e-business success because, by definition, e-business allows individuals using a wide variety of platforms (PCs, Macs, laptops, tablets, smartphones, etc.) to access busi-nesses that also use a wide variety of platforms (with different operating systems, databases, Web servers, etc.).

Critical Mass

Beyond technological factors, one key social factor that has had a huge impact on the growth of e-business is the attainment of a critical mass of users. The impact of crit-ical mass is best understood under the concept of network effects, which we define further below. In e-business, attaining **critical mass** means that there are sufficient buyers to sustain the business of vendors and sufficient vendors to attract buyers to the Internet.

Network effects or network economics can be used to discuss how value is cre-ated in a network. Recall from Chapter 4 that a firm can get competitive advantages when it owns a resource that is rare or unique. This is considered *value in scarcity*. In networks, however, there is also something completely opposite, which is *value in plentitude*. In a network (think of a group of people), the value of the network

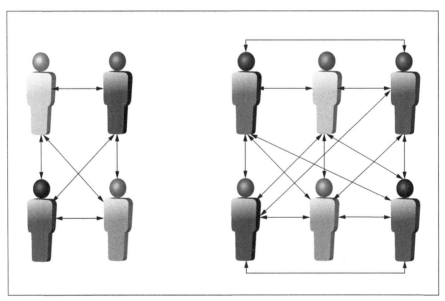

Figure 13.1 Networks of People

increases every time a new member is added to the network. Let us take the example of email. If only two of you use email, it makes one possible link. Eventually, you might find it limiting to only be able to send email to this one other person. Now if you add a third person to your group, you each have two potential people to send email to (or the existence of three links in the network). The addition of the third person adds value to your network. Now consider the networks in Figure 13.1. How many links are possible now?

Extend this concept to e-business, and you realize that the more vendors there are on the Internet, the more interesting it is for you to shop online. For vendors, the same is true: the more consumers there are shopping online, the more interesting it is to use the Internet to sell their goods or services. Today, there are more than 1 billion hosts (servers) on the Internet, and more than 4.6 billion people have Internet access (Clement 2020a). We could say that the Internet and the Web have reached critical mass.

E-Business Impacts

We discussed early in the chapter how e-business has changed industries. For some, these changes have been extremely positive (lower costs), while for others, the changes may be perceived as very negative (loss of profits or even disappearance of some businesses). Two specific terms used to refer to two of the impacts of e-business are *channel compression* and *channel expansion*.

Channel Compression (Disintermediation)

Channel compression is an impact on the downstream portion of the supply chain. Recall from Chapter 12 that the supply chain consists of several distributors and retailers upstream and downstream of the focal firm. In the distribution channel, the distributors and retailers are called *intermediaries*. **Channel compression** refers to situations when the distribution chain is shortened by eliminating some or all of these intermediaries between the focal firm (product producer) and the end consumer. For this reason, channel compression is also called *disintermediation*. As previously discussed, this change in the distribution structure has led to some major restructuring in several industries, including the music, publishing, and travel industries. When organizations have to decide whether to eliminate members of their distribution channel, they are often faced with channel conflict, or when to eliminate distributors and when to work with them. For example, if the Dell company decides to offer products through Dell retailers but also to sell directly online, how can it price products so that its resellers can make a profit but its consumers still feel they are getting the best prices when they buy directly online?

Channel Expansion

While channel compression has resulted in fewer organizations involved in the supply chain for some industries, channel expansion offers the reverse impact. **Channel**

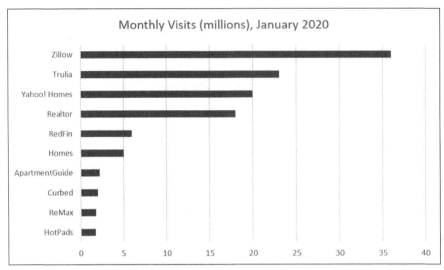

Figure 13.2 Top Real Estate Websites, January 2020 Source: https://www.statista.com/statistics/381468/most-popular-real-estate-websites-by-monthly-visits-usa/.

expansion is the addition of intermediaries in an industry whose purpose is to aggregate and provide information or brokering functionalities. In other words, intermediaries facilitate bringing buyers and sellers together or bringing relevant information to buyers and sellers. Think of the role of Kayak.com. It does not offer its own flights but pulls information from different vendors (airlines) as well as other search engines (like Expedia.com or Priceline.com). These intermediaries can make money by receiving fees from vendors or by allowing ads on their sites. The larger the volume of users, the better they can sell ad placement (selling ads on their websites based on the number of viewers). An overall important effect of these intermediaries is to reduce search costs and the information asymmetries that existed before the Internet and e-business. A good example is provided by real estate e-brokers. Figure 13.2 shows the most visited real estate websites in January 2020. How many of these companies existed before e-business?

Design for E-Business

There are many important design features that can make e-business websites successful. In Learning Activity 13.2, you identified several of them. For example, most of you will have identified the security of the website as a requirement for trusting the Web merchant and its privacy practices as a requirement for being willing to share information with the website. In Chapters 8 and 9, we discussed various features that increase (or reassure customers about) the security and privacy of information systems. In fact, many researchers have identified trust as one key determinant of individuals' intentions to buy from online merchants. Ease of use of the system in

LEARNING ACTIVITY 13.2

Why Would I Trust Them or Buy from Them?

In this activity, visit two retail websites and identify the factors that would make you more willing to trust the Web merchant and to buy and/or share information with the website. Select *two* of the following Web merchants:

- Dell: http://www.dell.com
- Amazon: http://www.amazon.com
- BestBuy: http://www.bestbuy.com
- Sierra: http://www.sierra.com
- Road Runner: http://www.roadrunnersports.com
- Murad: http://www.murad.com

Create a table, as shown below, to list the specific features that would make you trust, buy, or share information:

	Trust	Intentions to Buy	Willingness to Share Information
Site A	Feature 1 Feature 2 (etc.)	Feature 1 Feature 2 (etc.)	Feature 1 Feature 2 (etc.)
Site B	Feature 1 Feature 2 (etc.)	Feature 1 Feature 2 (etc.)	Feature 1 Feature 2 (etc.)

terms of navigating, checking out, and accessing information are usually important design features as well. The overall look of the website is also a factor in online buying decisions. Sites that are overloaded with information can be distracting for some users, while sites that have features and graphics that are too simplistic may look unprofessional. There are also nontechnological features for website success, such as return policies, shipping policies, and communication tools. Research also suggests that websites that offer customer reviews see increased site traffic and overall conversion rate (http://www.eMarketer.com). The conversion rate is the rate at which consumers who are browsing the website end up buying from the website. Other features that are found to annoy customers on websites include pop-up ads, the need for extra software to view a site, dead links, confusing navigation, the need to log in to view content, slow-loading pages, and out-of-date content, to name a few (eMarketer.com 2015).

Business-to-Business (B2B)

Most of this chapter so far has discussed e-business using B2C examples, mainly because you are more familiar with this type of e-business but also because many of the basic principles also apply to B2B e-business. However, there is a significant difference between B2C and B2B e-business beyond the type of players involved,

LEARNING ACTIVITY 13.3

B2C versus B2B

Every year, the U.S. Census Bureau collects data on e-commerce. Their studies reveal that the amount of money spent on e-commerce grows each year. For this learning activity, visit the website https://www.census.gov/programs-surveys/e-stats.html and collect the amount of e-commerce sales over the past three years. While you are doing this, differentiate how many sales occur through B2C and B2B channels. Compare these numbers and write a short response as to what they tell you about these two markets:

and that is the size of the market involved. B2B e-business represents a significantly larger market than B2C, both in the size of the individual transactions and the overall market share of interacting in this way.

There are several other differences between B2B and B2C. Of course, B2B involves two or more organizations as opposed to individual consumers. In B2B, these organizations typically know their trading partners fairly well, and, as a result, B2B is often relationship based, except for marketplaces that we will describe later.

B2B E-Business Enablers

B2B e-business involves the sharing of electronic documents, funds, and/or information between organizations for transactions to occur. There are many ways to implement this, which are summarized in Table 13.4. First, B2B e-business is enabled by **supply chain management (SCM)**, which we discussed in Chapter 12. In fact, when organizations share knowledge electronically across the supply chain to facilitate inventory control, reduce time delays in billing, and improve customer handling, they are conducting B2B e-business. Two other models of B2B e-business need further explanation: **electronic data interchange** and **e-marketplaces**.

Electronic data interchange (EDI) is the electronic exchange of information between two or more organizations using a standard format. The types of information exchanged are defined within the standards that are used and usually include business documents like bills, purchase orders, payment slips, invoices, and so on. EDI has been in existence for a very long time, allowing trading partners to lower transaction costs and improve profits through faster billing cycles, reduced errors, and improved customer responsiveness. Figure 13.3 shows a simplified sample set of EDI transactions.

An important characteristic of EDI is that organizations must use standardized formats for the documents they are exchanging. In fact, every organization has its own internal systems to handle purchase orders, invoices, and other business documents; therefore, each organization stores and handles data in different formats. An EDI transaction starts with the conversion of the documents and data from the format used by the internal system of the organization to an EDI format. This is called the **outbound transformation**. The EDI documents are then sent to the part-

TABLE 13.4 B2B E-Business Enablers	
Electronic data interchange (EDI)	Electronic exchange of information between two or more organizations using a standard format.
Supply chain management (SCM)	Allows organizations to manage the movement of materials or products from provision to production to consumption. SCM involves collaboration between partners both upstream and downstream from the focal organization.
E-marketplaces	Also known as an *exchange*. Allows vending companies and buying companies to "meet" electronically to conduct transactions.

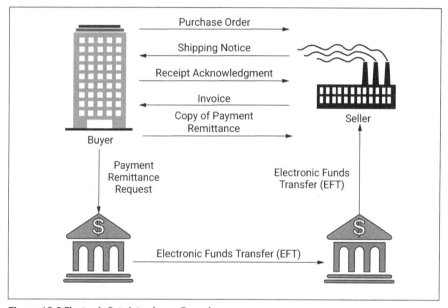

Figure 13.3 Electronic Data Interchange Example

ner organization, where the company's systems must perform an **inbound transformation**, which involves the system converting the EDI documents into the format used by the internal systems of that organization. The most-used standardized formats for EDI are **ANSI X.12** and **EDIFACT**.

EDI used to be limited to one-to-one communications between two trading partners. Today, it has evolved into a multipartner environment, allowing transmission of documents across a variety of information systems, even if they are incompatible in terms of hardware and software, since EDI documents themselves have to be in a very strict format. The use of a strict format is both an advantage and a disadvantage of EDI systems; on the one hand, it allows for fast processing and support for the various platforms, but on the other hand, it limits the flexibility of transmission that organizations want.

Today, organizations want to conduct more complex activities than simply trading documents with one another to support their business processes, and they want to use the Internet as opposed to proprietary networks for the transmissions. This is where **extensible markup language (XML)** becomes a popular way of conducting B2B transactions. XML is a markup language that allows organizations to give meaning to their data by inserting tags that describe the data within documents. A **markup language** (such as HTML or XML) is used to give "meaning" to information by placing a beginning and end tag around the information. For example, in HTML, we can use bold tags: this is bolded. This will appear as **this is bolded** on the Web page.

XML is very flexible, since organizations can define their own tags. Once documents are tagged with XML, a variety of systems can read and understand the documents, as long as the systems know the definitions of the tags. In conducting B2B, therefore, organizations have to agree on the meaning of the tags. For example, if one company decides to code its standardized documents in XML using the term *client* (<CLIENT> Mr. Jones </CLIENT>), but its trading partner uses the term *customer* (<CUSTOMER> Mr. Jones </CUSTOMER>), their respective systems will not be able to read each other's data. Some industries are tackling this problem by developing industry-wide XML definitions or standards, such as **electronic business extensible markup language (ebXML) for e-business**.

Finally, B2B can also be conducted in marketplaces, also known as exchanges. An **exchange** allows vendors or sellers to meet electronically. When a limited number of buyers or sellers are allowed to use a marketplace, we refer to this as a *consortia marketplace* or even a *private market*. When a large number of buyers and sellers exist, it is considered a **neutral auction**. Sometimes many buyers bid on a seller's products (a **seller-oriented or forward auction**), or sometimes many sellers offer their products to a single buyer (a **buyer-oriented or reverse auction**). Over the years, many B2B marketplaces have come and gone, although some have succeeded. For example, Buyerzone.com allows small businesses to buy and sell products to one another. However, marketplaces are difficult to sustain because they need to have a sufficient number of buyers and sellers to ensure the growth of the marketplace.

Search Engine Optimization

Anyone involved in e-business today would probably tell you that one of the most important factors for success online is being visible, which means showing up in searches. In fact, research has shown that if an organization's link is not in the top 30, it does not exist. Other research says that it is the first page of results that counts. In a study of which links individuals click on in the search results page, 42% of individuals clicked on the first organic link (explained below); 12% on the second; 8% on the third; and then 6%, 5%, and 4% on the following ones, with 3% for the remainder of the links on the first page of search results (http://www.redcardinal.ie/search -engine-optimisation/). Clearly, being at the top of the search results page is import-

LEARNING ACTIVITY 13.4

Where Is My Web Page?

In this activity, you will select two different Internet search engines to find some specific items. Record your results in a word processing file.

1. First, search for "intellectual property rights." Identify the top three sponsored (if any) and the top three organic links found.
2. Use the same two search engines to find the following products: a Trek mountain bike and an Apple AirPods Pro.

Indicate in your document the search engine used, the links you found, and your preferred search engine. Explain why you prefer a specific search engine.

ant for e-businesses. But how does an organization make sure it shows up toward the top of the search results? This is the role of **search engine optimization (SEO)**, which is the series of practices an organization can use to improve its visibility or optimize how its Web pages or website shows up on the search engine results pages. SEO uses a combination of how the Web page is designed, which keywords and language are used on the Web page, and which search engines are most used by specific audiences.

SEO is not a one-time effort, since search engines change their requirements often, people change their searching patterns, and competitors also try to optimize their own websites. In terms of design, one of the most important first steps for an organization is to clearly identify which keywords best represent what the organization's *intended audience* would use to find them. The key is to identify what the targeted audience thinks when they search online. It is also important to note that a keyword could actually be a keyword phrase, such as "mortgage lender" or "best place to eat." With this information, Web designers can do several things to help the ranking of the Web page. They can include the main keywords more often on the page and in more prominent places (headings, for example). They can include keywords in meta tags, which is information included with the page but not seen on the Web. There are many other factors that impact the ranking, such as including in-links, or inbound links, where other sites link to the organization; the age of the website domain (the older the better); and the overall size of the website. The exact details of how a website is more highly ranked vary over time and across search engines. This is why many consultants have started to offer SEO services to small, medium, and even large e-businesses.

Another way to have an organization's website show up higher in the rankings is to pay for placement. These are considered **sponsored links**, as opposed to **organic links** (ranked by design and SEO only). Some research suggests that most individuals prefer organic links, with 67.6% of all clicks being on the first five organic links (https://www.zerolimitweb.com/organic-vs-ppc-2020-ctr-results-best-practices/). Sponsored links show up separate from the search results and are based on the keywords entered in the search engines, as seen in Figure 13.4. The figure shows an example of Google AdWords, which is Google's sponsored link service.

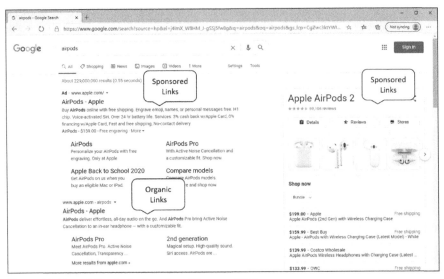

Figure 13.4 Search Results for Organic and Sponsored Links

Keyword bidding is another form of paying for a better ranking. It is also known as pay-per-click (PPC). The concept is that an organization pays the search engine owners only when someone clicks its link on the search results page. In most cases, companies bid for keywords, and the company that bids the highest will show up higher in the results when someone uses the keywords in a search. There is another type of PPC called *flat-rate PPC,* where the price of keywords is fixed. Nevertheless, you can easily see why PPC is an important source of revenues for companies like Google AdWords, Yahoo!, Search Marketing, and Bing Ads. Every time a user clicks on a link, the organization pays the search engine. In fact, many companies increase their paid search engine use for the holiday season. This results in significant revenue increases of around 50% during the holidays due to online sales directly attributable to paid searches (Enright 2011; Stambor 2019). What does it cost to "buy" a keyword from a search engine? The cost varies substantially depending on the month of the year and what is going on at that time. In an earlier edition of this book, we reported that in May 2014, the most expensive keywords were "insurance" at $54.91 per click and "loans" at $44.28 per click. In December 2017, the list was substantially different; the most expensive keywords were "business services" at $58.64 per click, with "bail bonds" close behind at $58.48 per click. In September 2020, this list included "insurance" as the most expensive at $54.91 per click and "gas/electricity" close behind at $54.62 per click. Table 13.5 shows some of the other expensive keywords as of September 2020. As you can see, keywords can be very expensive. Therefore, organizations are often told to find the best and most representative keywords for their businesses and to aim to be number two instead of number one in the search results.

TABLE 13.5 Most Expensive Keywords, September 2020	
1. Insurance	$54.91
2. Gas/Electricity	$54.62
3. Mortgage	$47.12
4. Attorney	$47.07
5. Claim	$45.51
6. Loans	$44.28
7. Lawyer	$42.51
8. Conference Call	$42.05
9. Recovery	$42.03
10. Donate	$42.02
Source: Based on data from WordStream, available at https://www.wordstream.com/articles/most-expensive-keywords.	

Paid search engines are a huge business, and the main players in the United States are Google, and Bing. However, when considering the global market, another player enters the scene: Baidu. As of mid-2020, Google owned about 70% of the market; Bing, 13%; and Baidu, 12%. In China, the search engine Baidu owns the majority of the market share, helping it reach third place in the global rankings (https://www.netmarketshare.com/search-engine-market-share.aspx).

Trends in E-Business

E-business has become mainstream, with individuals and organizations becoming increasingly comfortable transacting and interacting electronically. We discuss three major trends in e-business in the remainder of the chapter.

E-Government

A final topic we discuss is e-government, which was introduced at the beginning of the chapter. **E-government** involves using information technologies to enable and improve the efficiency with which government services are provided to citizens, employees, businesses, and agencies. It can occur at various levels of government, such as municipalities, counties, districts, states, provinces, or countries. It has become popular not only for obtaining information from government agencies but also for conducting transactions such as paying taxes, renewing licenses, downloading government forms, or paying government-provided utilities, such as water bills. In the United Nations, 57 countries have a very high E-Government Development Index, which means that the government has information available online and the infrastructure for people to access it, and citizens regularly use these services (Clement 2020b). E-government is provided through a number of platforms beyond websites, including Twitter, Facebook, YouTube, and Instagram. The future of e-government is

---

STATS BOX 13.2

The Global Nature of E-Business

E-business is a global phenomenon. eMarketer.com, a leading digital research firm, offers the following statistics:

- In 2012, B2C e-commerce sales topped $1 trillion for the first time.
- In 2017, B2C e-commerce reached $2.3 trillion worldwide. Estimates indicate that e-commerce sales will reach almost $4.5 trillion in 2021.
- B2B e-commerce is still much larger than B2C e-commerce. In 2017, B2B e-commerce reached $7.7 trillion, making B2B e-commerce more than three times the size of B2C e-commerce.
- In 2017, China was the largest e-commerce market in the world ($673 billion), followed by the United States ($340 billion), the United Kingdom ($99 billion), Japan ($79 billion), and Germany ($73 billion).
- Because of the COVID-19 pandemic in 2020, global retail sales growth slowed from 20.2% in 2019 to an anticipated 16.5% in 2020.

Sources: eMarketer. 2013. "Ecommerce Sales Topped $1 Trillion for First Time in 2012." *eMarketer.com*, February 5. http://www.emarketer.com/Article/Ecommerce-Sales-Topped-1-Trillion-First-Time-2012/1009649; Business.com Editorial Staff. 2020. "10 of the Largest Ecommerce Markets in the World by Country." *Business.com*, April 14. https://www.business.com/articles/10-of-the-largest-ecommerce-markets-in-the-world-b/; Orendorff, A. 2019. "Global Ecommerce Statistics and Trends to Launch Your Business Beyond Borders." *ShopifyPlus*, February 14. https://www.shopify.com/enterprise/global-ecommerce-statistics; Cramer-Flood, E. 2020. "Global Ecommerce 2020." *eMarketer.com*, June 22. https://www.emarketer.com/content/global-ecommerce-2020.

---

also likely to include some form of **Internet voting**, where citizens will be able to cast their votes electronically via the Internet. In fact, some countries and states have already started to allow citizens to vote via the Internet.

Chapter Summary

In this chapter, we discussed e-business. We first presented various ways to categorize e-business and then explored specific business models in greater depth. More specifically, we explained how e-business is conducted and what its enablers and impacts are. We also discussed how business-to-business (B2B) differs from B2C. The chapter concluded with search engine optimization and a discussion of important trends related to e-business.

Here are the main points discussed in the chapter:

- E-business involves electronic means to interact or conduct business with individuals, companies, or government agencies. E-business can be classified according to type: business-to-consumer (B2C; interactions are taking place between a consumer and a business), business-to-business (B2B; interactions are taking place between two businesses), consumer-to-consumer (C2C; interactions are taking place between two consumers), and government-to-constituent (G2C; interactions are taking place between a government agency and a constituent).

• E-business models include online retailing (offering products or services for sale to consumers online), infomediaries (offering specialized information to consumers via the Internet that aggregates or analyzes products or services from several providers), content providers (offering consumers content or relevant information), exchanges (offering a location [marketplace] for buyers and sellers to transact online), online communities or social media (offering individuals with similar interests and/or goals an online meeting place to interact with each other), and e-business infrastructure providers (offering infrastructure hardware, software, or services to other organizations for e-business). Enablers of e-business include technological enablers (ease of use, bandwidth, and interoperability) as well as a social enabler (critical mass). Finally, impacts of e-business include channel compression and channel expansion.

• Search engine optimization (SEO) is the series of practices an organization can use to improve its visibility, or optimize how its Web pages or website show up on the search engine results pages. SEO uses a combination of Web page design, keywords and languages used on the Web page, and listings on search engines most used by specific audiences. One way to optimize placement is through sponsored links, where companies pay for placement or use keyword bidding, which gives companies with the highest bids higher placement in search engine results.

• Business-to-business (B2B) e-business involves the sharing of electronic documents, funds, and/or information between organizations. B2B e-business models include supply chain management, electronic data interchange (EDI), and e-marketplaces. EDI is the electronic exchange of information between two or more organizations using a standard format. The most-used standardized formats for EDI are ANSI X.12 and EDIFACT, but organizations are increasingly using XML (extensible markup language) to conduct B2B transactions because it is much more flexible in terms of the types of information it allows organizations to exchange with one another. B2B marketplaces, or exchanges, allow vendors or sellers to "meet" electronically.

Review Questions

1. Define e-business and compare this to e-commerce, e-government, and Internet commerce. Is Internet commerce considered e-business?
2. Explain what an e-business model is and what it should include.
3. Describe the various ways to classify e-business models, including B2C and B2B models.

4. Identify the main technological factors that have enabled e-business growth and describe how the growth in mobile device use has provided for further growth.
5. Explain the concept of network effects or network economics and its importance to e-business.
6. Explain how e-business has led to disintermediation and provide examples.
7. Explain how e-business has led to channel expansion, discuss its impacts, and provide examples.
8. Describe the goals of search engine optimization, the various ways it is performed, and other options organizations have for marketing their websites.
9. Compare and contrast B2B e-business and B2C e-business.
10. Describe how electronic data interchange (EDI) functions and how it is evolving today.

Reflection Questions

1. What is the most important thing you learned in this chapter? Why is it important?
2. What topics are unclear? What about them is unclear?
3. Is it important to distinguish among e-business, e-commerce, and Internet commerce? Why or why not?
4. How is the material in this chapter related to the material on transmitting information in Chapter 7?
5. What is the relationship between network effects and the material on competitive advantages in Chapter 4?
6. What is the role of security in e-business success?
7. What is the role of privacy in e-business success?
8. If both channel compression and channel expansion occur in an industry because of e-business, does that mean that e-business has limited impacts on that industry? Why or why not?
9. A businessperson the authors know said that search engine optimization is as much art as science. What does he mean?
10. Why have new services like Uber and Lyft only been available since people have increasingly started using mobile devices? Why are Uber and Lyft likely more available in large cities than they are in small rural towns?

Additional Learning Activities

13.A1. This activity requires you to complete a case analysis regarding e-business. The case may be accessed through your university website or the Ivey Business School business case website (https://www.iveycases.com), or it may be provided to you by your instructor: Su, N., P. Bansai, and P. Laugland. 2017. *Alibaba Group: Technology, Strategy, and Sustainability.* Ivey Publishing. Product Number: 9B16E036 (revised).

13.A2. E-business has become a global phenomenon. However, not all countries are equal when it comes to having infrastructures, laws, and business practices that support e-business. The Economist Intelligence Unit (EIU; http://www.eiu.com) ranks each country on its digital economy readiness (see http://country.eiu.com/AllCountries.aspx). In this learning activity, you will be assigned a country and will need to explore its e-business readiness. Your instructor will assign you a country to research. Bring a short report to class that includes the following:
- The country you were assigned
- The EIU ranking for digital economy readiness
- Any e-business statistics you can find about your assigned country
- Your research of potential explanations for the ranking of your country (You will have to make assumptions.)

13.A3. Conduct research on the Web to give examples of each type of e-business defined in Table 13.2.

13.A4. Extend the existing business models presented in Table 13.3 with the use of location-based services (LBS). Describe how mobile users could make use of these models.

13.A5. Identify the business models of the following websites: http://www.pandora.com, http://www.consumerreports.org, http://www.yahoo.com, and http://www.weightwatchers.com.

References

Clement, J. 2020a. "Global Digital Population as of July 2020." *Statistica.com,* October 29. https://www.statista.com/statistics/617136/digital-population-worldwide/#:~:text=How%20many%20people%20use%20the,in%20terms%20of%20internet%20users.

———. 2020b. "World E-Government Leaders Based on E-Government Development Index (EGDI) in 2020." *Statistica.com,* August 26. https://www.statista.com/statistics/421580/egdi-e-government-development-index-ranking.

eMarketer. 2015. "Referrals Fuel Highest B2B Conversion Rates." *eMarketer.com,* February 10. http://www.emarketer.com/Article/Referrals-Fuel-Highest-B2B-Conversion-Rates/1012000.

Enright, A. 2011. "Paid Search Pays Off over the Holidays: Budgets Were Up 52% during November and December, and Related Sales Rose 69%." *Internet Retailer,* January 13. https://www.digitalcommerce360.com/2011/01/13/paid-search-paid-over-holidays/.

Smith, A. 2010. "Government Online: The Internet Gives Citizens New Paths to Government Services and Information." *Pew Internet and American Life Project Report*, April 27. https://www.pewinternet.org/2010/04/27/government-online/.

Stambor, Z. 2019. "Retailers Expect to Increase Ad Spending This Holiday Season." *DigitalCommerce360.com*, November 7. https://www.digitalcommerce360.com/2019/11/07/retailers-holiday-ad-spending.

Glossary

ANSI X.12: One of the most-used standardized formats for EDI.

Business model: The way the organization functions and creates value.

Business-to-business (B2B): A type of e-business where interactions are taking place between two businesses.

Business-to-consumer (B2C): A type of e-business where interactions are taking place between a consumer and a business.

Buyer-oriented or reverse auction: Many sellers offer their products to a single buyer.

Channel compression (disintermediation): A situation in which the distribution chain is shortened by eliminating some or all of the intermediaries between the focal firm (product producer) and the end consumer.

Channel expansion: The addition of intermediaries in an industry whose purpose is to aggregate and provide information or brokering functionalities.

Consumer-to-consumer (C2C): A type of e-business where interactions are taking place between two consumers.

Content providers: An e-business model where companies offer consumers content or relevant information and receive money from vendors for ads or downloads or from consumers for subscriptions.

Critical mass: A point when there are sufficient buyers to sustain the business of vendors and sufficient vendors to attract buyers to the Internet.

E-business infrastructure provider: A company that offers infrastructure hardware, software, or services to other organizations for e-business.

EDIFACT: One of the most-used standardized formats for EDI.

Electronic business (e-business): An electronic means to interact or conduct business with individuals, companies, or government agencies.

Electronic business extensible markup language (ebXML) for e-business: An industry-wide set of XML definitions or standards for e-business.

Electronic commerce (e-commerce): An electronic means to conduct business transactions, including the Internet, EDI, or private networks.

Electronic data interchange (EDI): A B2B e-business model focusing on the electronic exchange of information between two or more organizations using a standard format.

Electronic funds transfer (EFT): An electronic means to conduct financial transactions such as payments and remittances.

Electronic government (e-government): An electronic means to interact or conduct transactions with government agencies.

E-marketplaces (exchanges): Websites that allow vendors or sellers to "meet" electronically.

Exchange: An e-business model where companies offer a location (marketplace) for buyers and sellers to transact online.

Extensible markup language (XML): A markup language that allows organizations to give meaning to their data by inserting tags that describe the data within documents; a popular way of conducting B2B transactions.

Government-to-constituent (G2C): A type of e-business where interactions are taking place between a government agency and a constituent.

Inbound transformation: Converting the EDI documents into the format used by the internal systems of that organization.

Infomediaries: An e-business model where companies offer specialized information to consumers via the Internet that aggregates or analyzes products or services from several providers.

Internet commerce: An electronic means to conduct business using the Internet (for example, email, the Web, or file transfers).

Internet voting: When citizens are able to cast their votes electronically via the Internet.

Interoperability: The ability of heterogeneous systems to communicate with each other.

Keyword bidding (pay-per-click: PPC): When organizations pay the search engine owners when someone clicks their links on the search engine results page.

Location-based services (LBS): Services that make use of knowing where users are based on their cell phone location to offer targeted advertising or services at a specific moment.

Markup language (such as HTML or XML): A coding language used to give "meaning" to information by placing a beginning and end tag around the information.

Network effects or network economics: Explains how value is attained in a network when critical mass, or a plentitude of members, is achieved. The value of the network increases every time a new member is added to the network.

Neutral auction: A marketplace with a large number of buyers and sellers.

Online community or social media: An e-business model where organizations offer individuals with similar interests and/or goals an online meeting place to interact with one another.

Online retailing: An e-business model where companies offer products or services for sale to consumers online.

Organic links: Links on search engine results achieved by design and search engine optimization.

Outbound transformation: Conversion of the documents and data from the format used by the internal system of the organization to an EDI format.

Search engine optimization (SEO): A series of practices an organization can use to improve its visibility or optimize how its Web pages or website shows up on the search engine results pages.

Seller-oriented or forward auction: Many buyers bid on a seller's products.

Sponsored links: Links that show up higher in the search engine results because the organization has paid for placement.

Supply chain management (SCM): A B2B e-business model focusing on the electronic exchange of information along the supply chain.

Web commerce: An electronic means to conduct business transactions using the Web as an interface.

Chapter 14

Information and Knowledge for Business Decision-Making

Learning Objectives

By reading and completing the activities in this chapter, you will be able to:

- Explain why knowledge management is important to organizations
- Compare and contrast explicit and tacit knowledge
- Describe the main processes for knowledge management
- Compare and contrast major knowledge management technologies
- Describe the main categories of decision support systems
- Describe the main purposes of and techniques used in business analytics
- Explain how artificial intelligence and robots are affecting business

Chapter Outline

Knowledge Management
Focusing Story: Managing Knowledge by Texting
    Learning Activity 14.1: How You Manage Knowledge
Why Managing Knowledge Is Important
Main Processes for Knowledge Management
    Learning Activity 14.2: Wikis for Managing Knowledge
Knowledge Management Technologies
Decision Support Systems and Collaboration Systems
    Learning Activity 14.3: Using a Decision Support System
    Learning Activity 14.4: Slack for Student Team Projects
Business Analytics
    Learning Activity 14.5: Effective Visualizations
Artificial Intelligence
    Learning Activity 14.6: Are AI and Robots Good or Bad for Society?

Ultimately, a business's ability to compete depends heavily on how its employees apply knowledge to make decisions and solve problems. Information systems and technologies can help organizations better manage knowledge. In addition, information systems and technologies can facilitate better decision-making. In this chapter, we help you understand knowledge management and how technologies can help organizations manage and apply knowledge. Knowledge is often applied

Managing Knowledge by Texting

Several years ago, I was teaching an onsite class to information technology (IT) profession-als working at a major logistics company. While waiting for class to start one night, I over-heard one of the students asking another if he had received a text message regarding a problem he was having with a computer program. This surprised me. At the time, such mes-saging was generally done more for personal than business reasons. So, I asked the student who sent the text how frequently they exchanged such messages. The student replied, "All the time. Whenever one of us runs into a problem we haven't seen before, we send a mes-sage to the group. Usually someone has seen a similar problem, so they can help." Essential-ly, the group had informally created a "knowledge network" of shared expertise. By sending a simple text message, an individual worker can tap into the knowledge of the entire group. The result is hours of saved time and frustration.

More recently, people are using social networking sites such as LinkedIn and Facebook as a means of accessing the knowledge of others. For example, when I had to write a job description for an unusual faculty job, I turned to several LinkedIn groups for advice. Some individuals responded with helpful advice. When you are faced with a decision, it is often necessary to reach out to those who may have knowledge useful to the decision. This may be as simple as calling your car-loving brother when you are buying a new vehicle, or it may mean convening a meeting of experts to help a corporate board decide whether to pursue a merger. Regardless of the decision, the intent is the same. Additional knowledge can have great value for decision-making. Whether you use text messaging to get help solving a com-puter problem or you fly in experts from around the world, you are still trying to leverage the knowledge of others to help you make a decision. Fortunately, information technologies can aid in the quest to use the knowledge of others to help make better decisions.

Focusing Questions

1. When have you used the knowledge of others to help you make a decision? Briefly describe the decision. Who did you contact to help you? What knowl-edge were you trying to access?
2. How have you used information technologies to help you connect with oth-ers for the purpose of tapping their knowledge? Why did you use those tech-nologies?

---

to decision-making. Because of this, we also discuss decision support systems. As organizations gather an increasing amount of data, being able to leverage these data for decision-making and action is an increasingly important aspect of business. Business analytics systems, which we also discuss in this chapter, help organizations make sense of this sea of data. Artificial intelligence and robots are helping use data in interesting and important ways.

Knowledge Management

In an earlier chapter, we introduced you to the concepts of data, information, and knowledge. You may recall that data are unconnected facts, information is data that have been processed so that they are useful, and knowledge is information that is

LEARNING ACTIVITY 14.1

How You Manage Knowledge

This chapter helps you understand organizational knowledge management, but individuals also manage their knowledge. The goal of knowledge management is to make sure the right knowledge is available in a useful form to the right people. Personal knowledge management is the process of determining what knowledge you need and how to find, organize, and use that knowledge. Prepare a one- to two-page paper on how you manage your personal knowledge. What tools do you use? How effective are they? Compare your response with another student's response.

applied to a decision or action. If you think about these differences, you may see that as we progress from data to information to knowledge, there is a greater level of human contribution and greater value.

However, as we move up the hierarchy toward knowledge, things get harder to manage, in part because of the increased level of human contribution. It is relatively easy to manage data, but managing knowledge is quite difficult. Table 14.1 shows some characteristics of data, information, and knowledge. As you can see from the table, knowledge is hard to structure, transfer, and capture, which leads to knowledge being much harder to manage than data or information.

TABLE 14.1 Characteristics of Data, Information, and Knowledge		
Data	**Information**	**Knowledge**
Easily structured	Made up of processed data	Hard to structure
Easily captured	Requires unit of analysis	Difficult to capture
Easily transferred	Requires human involvement	Often tacit
Often quantified		Hard to transfer
Made up of mere facts		Information applied to some decision or action
Source: Adapted from Pearlson and Saunders (2006).		

Knowledge management is a process that allows organizations to generate value from their knowledge-based assets. Knowledge management involves capturing and documenting what employees and other stakeholders know and developing systems that make it easier to share and use that knowledge. Later in this chapter, we discuss the process of knowledge management. Before getting into that, however, we help you understand why knowledge management is important.

Why Managing Knowledge Is Important

Knowledge is among the most important resources of any organization. Recall that knowledge is information directed toward action and decision-making. To be successful, organizations not only must effectively manage information, but they must also manage knowledge. Unfortunately, knowledge management is often messy and complex.

Organizations have always managed knowledge. More recently, however, organizations have begun using information and communication technologies to help manage knowledge. Organizations that manage knowledge well make better decisions, are generally more innovative, and tend to be more successful overall than those that do not.

Effective knowledge management can bring about big benefits, including the following:

- Better problem-solving
- Improved customer service
- More effective product management
- Increased innovation
- More efficient and more effective processes
- Increased intellectual capital
- Better leveraging of intellectual assets

Types of Knowledge

Understanding more about knowledge will help you better understand important aspects of knowledge management systems. We can divide knowledge into two categories: explicit knowledge and tacit knowledge. **Explicit knowledge** can be expressed relatively easily. Examples include operations manuals, books and articles, scientific formulas, and statistical analysis. Because explicit knowledge can be easily expressed, it can be more easily shared, stored, and managed than our other category of knowledge, tacit knowledge. It may be helpful to think of explicit knowledge as "knowing that," while tacit knowledge is more about "knowing how."

Tacit knowledge is not easy to express or communicate. It is internalized and highly individualized. An individual's tacit knowledge is rooted in life experiences, values, and biases. Ask an artist how to create a beautiful piece of art. The chances are that he or she will have trouble expressing exactly how striking art is created. This is an example of tacit knowledge. The artist may be able to create such art but may not be able to easily communicate the process to you. In fact, sometimes the person who possesses the tacit knowledge may not even be aware that he or she has the knowledge.

Tacit knowledge also exists in more mundane contexts. A good salesperson knows when to close a sale and when to back off. A knowledgeable computer programmer can quickly diagnose an error message. There are examples of tacit knowledge in any organization and in any profession. One of the most interesting examples comes from the world of poultry. Determining the sex of a baby chick immediately after hatching is important for commercial egg producers. Egg producers prefer female chicks, since they will be able to lay eggs. Egg producers employ chicken sexers, who have the uncanny ability to discriminate between female and male chicks with a quick glance. What is interesting about this ability is that most expert chicken sexers cannot tell you how they make their decisions, yet they are very accurate. They have the tacit knowledge but are unable to easily convert this into explicit knowledge. Capturing tacit knowledge and codifying it so that it can be shared is one of the major challenges of knowledge management.

Now that you understand more about knowledge and its characteristics, we can turn our attention to how to manage knowledge.

Main Processes for Knowledge Management

A complete knowledge management system should consist of knowledge creation, capture, codification, storage, retrieval, transfer, and application, as shown in Figure 14.1. Knowledge management processes form a cycle. Applying knowledge often leads to the creation of new knowledge, which restarts the cycle. While information technology is not required for any of these processes, as you will see, IT can facilitate each process in a knowledge management system. In this section, we discuss each of these processes.

Information technology is critical to modern knowledge management, but technology alone cannot ensure effective knowledge management. Effective knowledge management also requires social and structural mechanisms that support knowledge management. For example, a culture of trust can greatly enhance knowledge sharing within an organization. Poor organizational processes and culture can bring about knowledge management failure.

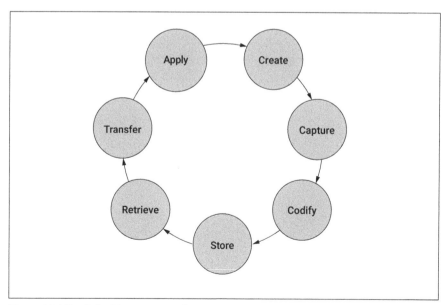

Figure 14.1 Knowledge Management Cycle

Creating Knowledge

While knowledge exists in individuals, organizational knowledge can also be created. Nonaka and Takeuchi (1995) proposed the SECI model of organizational knowledge creation. The model is shown in Figure 14.2. According to the model, knowledge creation is a continuous process consisting of interactions between implicit and explicit knowledge. There are four processes by which knowledge is created:

- *Socialization* involves sharing tacit knowledge through direct communication or shared experience. This is tacit-to-tacit communication.
- *Externalization* is tacit-to-explicit communication. Tacit knowledge is converted to explicit knowledge by developing specific concepts, models, and the like. This conversion allows the knowledge to be understood and interpreted by others. This also serves as a foundation for creating new knowledge.
- *Combination* is the process of combining the externalized explicit knowledge to form broader concepts, models, and theories.
- *Internalization* occurs when explicit knowledge transforms to tacit knowledge and becomes internalized by individuals within the organization. This can start a new cycle, beginning with the socialization of the new tacit knowledge.

These four processes form a cycle of increasing knowledge. Completing one cycle can start a new trip through the four processes, but with an ever-increasing amount of knowledge. This is represented by the spiral in Figure 14.2.

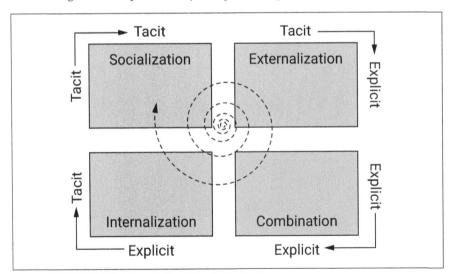

Figure 14.2 SECI Model of Organizational Knowledge Creation

Knowledge can also be "purchased" by hiring knowledgeable individuals as either consultants or employees. Knowledge can also be acquired by purchasing reports or even entire companies. An organization may acquire another company to gain access to the company's knowledge assets.

Capturing and Codifying Knowledge

Once knowledge is created or otherwise acquired, it needs to be captured and codified. There are many techniques for capturing knowledge, including interviewing experts, observing groups as they make decisions, forming focus groups, holding "lessons learned" debriefings, and performing task analysis.

Codifying knowledge means converting tacit and explicit knowledge into a form that organizational members can use. It is representing knowledge so that it can be reused. This also includes documenting knowledge that was previously undocumented. To codify knowledge, organizations should do the following (Davenport and Prusak 1998):

- Define the strategic intent. (How will the codified knowledge serve the organization?)
- Identify the knowledge necessary to achieve the intent.
- Evaluate the existing knowledge's usefulness and ability to be codified.
- Determine the best way to codify and distribute the knowledge.

Codifying knowledge is difficult, but there are tools that can help. Some of these tools are described below.

Cognitive maps represent the mental model of the expert's knowledge. Key concepts and the relationships among them are shown. Software such as Freemind, MindMeister, and Visio can help document the maps. Figure 14.3 illustrates a cognitive map of Web 2.0 concepts and tools.

Decision tables show a list of conditions and their values along with a list of conclusions or actions. The conditions necessary for each conclusion are indicated. Figure 14.4 shows a decision table for making MBA admission decisions. In each column, Y indicates that the condition was met, while N means that the condition was not met. The X entries indicate the appropriate action.

Decision trees show the alternate paths that impact decisions. The tree shows various paths that can lead to certain outcomes. Here the idea is to use decision trees as a way to document how a decision is made. Figure 14.5 shows a decision tree for making MBA admission decisions. Decision trees are often used in expert systems, which we discuss later.

The purpose of these and other codification tools is to systematically document knowledge that is "stored" in the minds of experts. By working with the experts to create these diagrams and tables, knowledge modelers can help capture and document experts' knowledge. Documenting the knowledge allows us to store, organize, and share the knowledge so that others can use it.

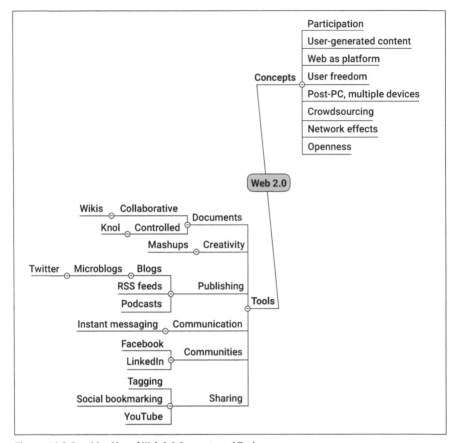

Figure 14.3 Cognitive Map of Web 2.0 Concepts and Tools

Condition Stub		Condition Entry						
		1	2	3	4	5	6	7
If (condition)	Applicant has undergraduate degree	Y	Y	Y	Y	Y	Y	N
	Applicant has > 2 years work experience	Y	Y	Y	Y	N	N	
	GMAT > 600	Y	Y	N	N	N	Y	
	Undergraduate GPA > 3.0	Y	N	Y	N	N	Y	
Then (action)	Admit	X						
	Conditionally admit		X		X			
	Advise to retake GMAT			X				
	Decline admission					X	X	X

Figure 14.4 Decision Table Example

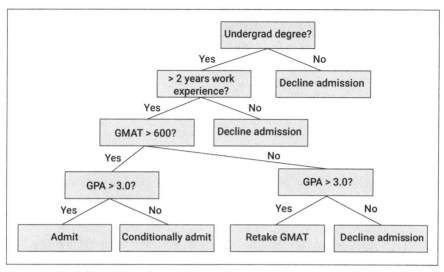

Figure 14.5 Decision Tree Example

BUSINESS EXAMPLE BOX 14.1

The Issue of Trust in Knowledge Management

Managers who want to have good knowledge management systems must work hard to create a culture that rewards rather than punishes knowledge sharing. Knowledge is a critical asset for any business professional. The more useful knowledge you possess, the more valuable you are to the organization. As is the case with any resource, the rarer the knowledge, the more valuable it is. In other words, if you have knowledge that few others have, you are more valuable to the business.

This simple fact can hinder knowledge management. Since sharing knowledge makes it more widespread (less rare), sharing decreases the value of possessing that knowledge. So, if you share your knowledge, you become less valuable to the organization. Because of this, trust is important to knowledge management. Experts who possess knowledge must trust that their organization will not unfairly take advantage of the individual who shares the knowledge. If this trust does not exist, the expert may hoard the knowledge to maintain its value.

Storing and Retrieving Knowledge

Organizing knowledge so that it can be easily accessed is challenging. One reason for this is that knowledge can exist and be documented in many different forms. As a result, there is no single knowledge base for an organization. Knowledge is stored in a variety of forms, including some that you might find surprising. For example, a well-organized corporate directory that lists areas of expertise can be thought of as a knowledge store. Such a directory effectively points to experts who are repositories of knowledge. Document storage systems are also examples of knowledge stores.

Knowledge storage and retrieval are closely interrelated. Even the most well-designed knowledge storage system is ineffective if knowledge consumers cannot retrieve the knowledge. Search technologies are important knowledge retrieval tools. Think about how many times you have used Google or some other search tool to locate knowledge. Accurate, fast search results are a boon to anyone seeking knowledge.

Transferring and Applying Knowledge

Knowledge needs to be applied to be useful, and application implies that knowledge is available to those who need it. Knowledge maps can facilitate knowledge transfer. Knowledge maps show the location of knowledge within the organization. A knowledge map can identify individual experts, networks of experts, or documents and databases.

To manage knowledge transfer, you must consider the sources of knowledge, the media used to transfer the knowledge, and the knowledge consumers. Sources of knowledge include experts, knowledge bases (databases for knowledge), and document repositories. Media may include computer networks and other communication media. Consumers are the individuals and groups that apply the knowledge.

You have engaged in knowledge transfer, even though you may not have realized it. Education is a form of knowledge transfer; your instructor is transferring some of her or his knowledge to you. Have you ever contacted a friend for help on an assignment or project? If so, you used your social network for transferring knowledge.

While humans have always used their social networks to access knowledge, the emergence of social networking technologies is increasing this sort of knowledge transfer. The use of instant messaging in the focusing story is an early example of using software to tap into others' knowledge. New social networking tools make this sort of knowledge transfer even more effective.

Knowledge Management Technologies

Information and communication technologies have helped with knowledge management for a long time. Even the telephone can be used to help individuals tap external knowledge. In this section, we discuss several knowledge management technologies.

Repositories

The goal of **knowledge repositories** is to make it easy to find and retrieve documents that contain knowledge. According to Pearlson and Saunders (2006), there are three main types of knowledge repositories:

- Externally focused knowledge
- Structured internal knowledge, such as research reports and marketing materials
- Informal internal knowledge, such as "lessons learned" reports and discussion databases and frequently asked questions collections

Communication-Based Tools

There are many communication-oriented tools for knowledge management. Email is an example. We often email colleagues when we need knowledge that we do not possess. For example, a sales representative may email another sales rep for advice on how to position a certain offering. While email can be effective, you need to know who has the knowledge you need. You also need access to that individual. Social networks can help overcome these limitations. **Communities of practice** that exist within networks are especially helpful. LinkedIn groups are an example. Some of these groups are effectively communities of practice. For example, the group "Technology-Using Professors" is a network of college faculty who are interested in how technology can enhance learning.

Collaboration tools combine elements of repositories and communication-based knowledge management tools. Systems such as Google Docs allow for document sharing and coediting. (You may recall that we mentioned Google Docs in Chapter 7.)

Some knowledge management experts see a trend from a content-focused view of knowledge management to one that is more connection focused. This means that rather than being centered around the ideas of collecting knowledge, the focus is shifting to tools that help you manage your connections to those who have helpful knowledge. The social networking tools we mentioned earlier are among the technologies that can help you with this form of knowledge management. For example, LinkedIn, Facebook, and Google+ allow groups and communities. These groups are good examples of how social networking tools can be used for connection-focused knowledge management.

Dashboards

Digital dashboards provide graphical views of key data along with graphical warnings when data indicate areas that need attention. Table 14.2 lists a number of companies that create dashboard products. Visit these to view the various ways visuals are used to present data using dashboard software. Later in this chapter, we discuss visualizations in some detail. Digital dashboards make extensive use of visualizations but also present data in tables, depending on which might be more easily and quickly understood.

TABLE 14.2 Dashboard Products	
Company	Link
BOARD	http://www.board.com/en/solutions-overview
ClicData	https://www.clicdata.com/
IBM	https://www.ibm.com/analytics/business-intelligence/
iDashboards	https://www.idashboards.com/solutions/
Klipfolio	http://www.klipfolio.com/guide-to-business-dashboards
SAP	https://www.sap.com/products/analytics/business-intelligence-bi.html
Tableau	http://www.tableau.com/solutions/business-dashboards

Expert Systems

Expert systems help users solve problems or answer questions in a way that mimics an expert's thought processes. An expert system typically has a narrow focus on a particular problem domain. For example, an insurance company can use an expert system to guide a salesperson through the process of selecting the most appropriate product for a customer. Many organizations use expert systems to guide nonexpert employees through complex decisions or problems, such as technical troubleshooting. There are also systems for medical self-diagnosis, although these systems should be used with caution, as they do not make a true diagnosis but rather provide probabilities of underlying causes for symptoms.

To close this section, we want to make sure that you understand that all organizations need to manage knowledge, even if they do not have formal knowledge management systems. Managing the knowledge resources of an organization is a critically important aspect of organizational success. While knowledge management is seen by some as an esoteric concept favored only by academics, in our opinion, organizational leaders are well advised to pay more attention to how they manage knowledge.

Knowledge is typically applied to some decisions; decision-making requires knowledge. While knowledge management technologies are relatively immature, decision support technologies are much more mature in comparison. Decision support systems provide technology tools for helping decision makers deal with the data and information that are necessary for applying knowledge. The next section discusses these tools.

Decision Support Systems and Collaboration Systems

Decision support systems (DSSs) are computer-based systems that help decision makers use data and models to solve semistructured or unstructured problems. There are many different types of DSSs, so it is convenient to divide them into categories. Our categories are based on Alter's (1977) and Power's (2001) DSS frameworks (http://www.dssresources.com).

LEARNING ACTIVITY 14.3

Using a Decision Support System

Pick one of the following decisions and use the associated decision support system (DSS) to help you analyze the decision. (Your instructor may give you a different decision and DSS.)

- What car can you afford? DSS: http://www.edmunds.com/calculators/affordability.html
- Should you trade your "gas guzzler" for a more fuel-efficient vehicle? DSS: http://www.edmunds.com/calculators/gas-guzzler.html
- Should you buy or rent a house? DSS: http://www.realtor.com/mortgage/tools/rent-or-buy-calculator/

Partner with another student and describe your experience using the DSS. Was it useful? Why or why not?

- A **data-driven DSS** focuses on the retrieval and manipulation of data that are stored in an organization's data stores. Data warehouses (which we discuss in the next section) are often important data-driven DSSs. While these DSSs are useful, they do little to guide the decision maker in making the decision. So if the decision maker does not use the data properly, a poor decision is likely. However, in the case of unstructured decisions, a data-driven DSS may be particularly useful because of the amount of data retrieval required.
- A **model-driven DSS** focuses on providing the decision maker with the ability to access and manipulate analytical models. These models are used to help the decision maker perform sophisticated analysis that would be difficult and time-consuming without the DSS. Model-driven DSSs are appropriate for semistructured decisions. Without some degree of structure, it is difficult to build an appropriate model. A model-driven DSS can be applied to decisions in a wide variety of domains, including logistics, portfolio management, demand forecasting, and optimization.
- A **document-driven DSS** focuses on managing and retrieving documents that may help with decision-making. The term *documents* should be taken broadly to include chat session transcripts, graphics, and even video and audio files. Document-driven DSSs are particularly useful for less structured decisions. Because there is no analytical aspect of document-driven DSSs, they draw heavily on the decision maker's analytical ability and cognitive models.
- A **communication-driven DSS** facilitates collaboration and group-based decision-making. According to Power (2001), communication-driven DSSs support one or more of the following: communication, information sharing, collaboration, and/or group-based decision-making.

Groupware, which we discuss next, is one type of communication-driven DSS. Email, document sharing, coediting systems (such as Google Docs), and remote meeting systems are other examples.

Collaboration systems (sometimes called **groupware** or team support systems) are network-based systems that help groups communicate and collaborate. Time and place are often used to categorize groupware. *Synchronous* and *asynchronous* describe whether the users collaborate at the same time (synchronous) or at different times (asynchronous). When group members are in the same physical location, they are *colocated*. Figure 14.6 uses the dimensions of time and place to categorize common collaboration system functions, which are described following the figure.

	Same time (synchronous)	Different time (asynchronous)
Same place (colocated)	Group decision support systems (decision room), presentation support	
Different place (distant)	Group decision support systems (remote), videoconferencing, chat, remote presentation, shared whiteboards	Discussion databases, email, email lists, workflow, document coediting

Figure 14.6 Collaboration System Examples

- *Group decision support systems:* Group decision support systems (GDSSs) facilitate group-based decision-making tasks. GDSSs often include applications for brainstorming, anonymous commenting, and voting. The goal of GDSSs is to make group decision-making more rational by supporting established group methods that encourage equal participation. Some organizations have "decision rooms," which consist of individual computers that access GDSS software, along with shared devices, such as projection systems. GDSSs can also be used remotely.
- *Shared whiteboards:* Whiteboards have long been used to help workgroups communicate. You have probably used one yourself. Shared whiteboard systems allow users who are not colocated to share drawings or notes in the same way. Often, shared whiteboard systems assign different colors to each user so that you can tell who is drawing each element.
- *Videoconferencing:* Videoconferencing allows individuals to communicate using both voice and video. You have probably used a videoconferencing system such as Zoom or Microsoft Teams. Some organizations have sophisticated videoconferencing rooms that have special equipment and software installed.
- *Chat systems:* Online chat systems allow users to write real-time messages in a public space. (Note that we are specifically referring to

LEARNING ACTIVITY 14.4

Slack for Student Team Projects

Slack is a popular collaboration system used by businesses to help their teams work together more effectively. Visit https://slack.com/ and familiarize yourself with its basic features. Write a one- to two-page report that addresses the following questions. Be sure to explain your answers.

1. What features would be the most useful for team projects?
2. How would you go about convincing your teammates to use Slack to help with team projects? What arguments would you use?

group-oriented chat systems.) Some systems allow for a moderator to lead the interaction. Many chat systems capture a transcript of a session, which can be useful for documentation purposes.

- *Remote presentation:* Remote presentation software allows noncolocated groups to view a presentation remotely. These systems often have a chat window that allows the presenter and remote audience to interact.

- *Email and email lists:* Email is one of the simplest groupware applications. Most of us have used email to distribute documents or ideas to a group. Email lists facilitate group communication by making broadcasting emails to the group easier. Some systems allow you to subscribe to an email list. These often let you choose whether you want to see each individual message or periodically receive digests of messages that show all messages sent to the group in a single email message.

- *Discussion databases:* Similar to discussion forums, discussion databases store threaded, online, asynchronous discussions in a way that allows users to search and retrieve relevant information. These are different from chat session transcripts in that discussion databases are oriented toward asynchronous discussions. Discussion databases are particularly useful for new group members who need to come up to speed on the group's discussions.

- *Workflow systems:* Workflow systems route documents through a group in a relatively predetermined manner. Workflow systems ensure that documents are routed properly so that the relevant group members are provided with documents in a timely manner. These systems also make sure that the proper authorities within the group review the documents.

- *Document coediting:* Document coediting systems, which we discussed earlier, help groups with their work by allowing group members to work on documents synchronously rather than having to email documents back and forth (or otherwise share them). You may have used a coediting system such as Google Docs for team projects. These systems also help with versioning by having all members work on the same version of a document. (If you have ever spent time editing the wrong version of a document, you will appreciate the importance of versioning.) While we list document coediting in the "asynchronous" column of Figure 14.6, most systems allow for simultaneous coediting.

Sometimes these functions are provided by standalone applications, and other times they are bundled into a single, integrated application. For example, Slack (a popular collaboration system) provides discussions, chat, and videoconferencing.

Business Analytics

You may have heard about **business analytics (BA)**; it is a very hot topic. Essentially, business analytics is the process of transforming data into insights to improve operations and decision-making. Business analytics is a very close cousin to **business intelligence (BI)**. In fact, the terms are often used interchangeably, although some view business intelligence as being primarily concerned with descriptive tasks, while business analytics is often thought to be more concerned with making predictions. So, business intelligence concerns making current decisions based on information about the past, while business analytics concerns making predictions to help organizations decide how to move forward. Since business analytics is the more current term, we will use that terminology throughout this section. However, much of what you read in this section also applies to business intelligence, especially if the topic is focused on past events and decisions about current operations. For most people, the terms can be used interchangeably.

The overall goal of BA is to help people make sense of data. The interest in BA is easy to understand—businesses are drowning in a sea of data. Managers know that information derived from these data potentially represents a gold mine of value but are often unsure how to transform the raw material into an information treasure. Business analytics systems may hold the key to unearthing the value of this information.

Data warehouses are critical elements of many BA systems. The data stored in data warehouses come from various internal and external sources, including internal transaction processing systems. One way to think about these data sources is to consider them the creators of the data. The data warehouse stores and organizes the data in a way that is better suited for supporting decision-making. Analysts and managers use data mining tools to make sense of the data stored in the data warehouse. The data warehouse stores and organizes the data in a way that is well-suited for supporting decision-making and analysis. Analysts use the BA tools to make sense of the data. This basic BA framework is shown in Figure 14.7. As you can see, data warehouses are an important part of a BA system. The next section provides an overview of data warehousing.

Business analytics tools help many different types of organizations reduce spending and increase sales. BA software helped eHarmony evaluate the effectiveness of their marketing program, causing them to save $5 million by eliminating an affiliate

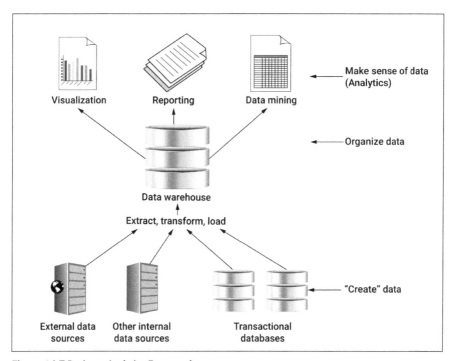

Figure 14.7 Business Analytics Framework

program that only provided low-quality traffic. The consulting firm Deloitte developed a pricing analytics system that helped a global food ingredients company increase their revenue by 2%. (This may not sound like much, but this is a multibillion-dollar company, so 2% is a lot of money!) UPS used BA to optimize their delivery routes, which resulted in reducing the miles driven by their trucks by more than 300 million miles, lowering fuel usage by more than 39 million gallons. Microsoft used BA to help relocate employees to improve collaboration and reduce the time employees spent traveling to meetings. The project reduced meeting travel time by 46%, which saved 100 hours per week across the affected employees. American Airlines used BA to save $5 million by detecting fraudulent ticket sales. There are many similar examples.

Sources: Moore, J. 2014. "Business Intelligence Meets Mobility at eHarmony." *CIO*, November 11. http://www.cio.com/article/2845557/mobile-application-management/business-intelligence-meets-mobility-at-eharmony.html; Petersen, R. 2016. "37 Big Data Case Studies with Big Results." *{grow}: Marketing, Strategy, Humanity*, December 6. https://businessesgrow.com/2016/12/06/big-data-case-studies/; Gavin, M. 2019. "3 Examples of Business Analytics in Action." *Business Insights*, January 15. https://online.hbs.edu/blog/post/business-analytics-examples; The Business Intelligence Guide. "American Airlines BI Systems." http://www.thebusinessintelligence-guide.com/industry_solutions/airline/Amercian_Airlines.php; Strain, M. "The Advan-

tages of Business Intelligence." *Chron.* http://smallbusiness.chron.com/advantages
-business-intelligence-24548.html.

Data Warehousing

Data warehouses support decision-making, which makes them important to business
analytics. Understanding this is a key to understanding the purpose of data warehousing. Supporting transactions requires that databases perform many data additions,
deletions, and updates. Decision support, in contrast, is more about data retrieval.
Unfortunately, a database that is very efficient for data updates usually does not perform as well for data retrieval. There is another problem: decision-making often
requires getting data from multiple transaction databases. Finally, transaction databases usually store data at a very detailed level. Aggregated data are often better for
decision-making. So gathering all the data necessary for decision-making can be time
consuming and inefficient. Data warehousing addresses these issues. Data warehousing is a process, the goal of which is to gain value from an organization's information
through the use of data warehouses. A **data warehouse** is a copy of transactional data
(and other data) that is formatted so that it is useful for decision support.

Data warehouses have several characteristics that set them apart from transactional databases:

- Data warehouses are subject oriented. They are organized around particular subjects such as marketing, human resources, sales, or production.
- Data in data warehouses are integrated from a variety of internal and
  external sources.
- Data in data warehouses are typically transformed from their original
  format. Detailed data are often aggregated. For example, a data warehouse for a restaurant chain would aggregate individual sales into total
  sales from some time period (such as per hour).
- Data in data warehouses typically are nonvolatile, which means that the
  data do not change. Once data are in the warehouse, they stay in the
  warehouse and are not changed. (Data that are no longer of value may
  be deleted from the warehouse.)

Data from various sources are gathered together, prepared, and loaded into the
data warehouse through the extract, transform, load (E/T/L) process, as mentioned
earlier. The E/T/L process is easy to understand conceptually but is often very complex and time consuming to perform in practice. Each element is described below:

- *Extract:* Data are pulled from the source systems (such as the transactional databases). Deciding exactly what data to extract is important.
- *Transform:* Data must be changed into a form that is suitable for decision
  support. Often this involves aggregating detailed data. Data cleansing is
  also an important part of the transform process. Data in transactional

databases are often messy. This problem gets worse when data are extracted from multiple systems. These messy data must be cleaned up before being loaded into the warehouse. Think about all the ways a company such as AT&T could be represented in different databases: AT&T, ATT, American Telephone and Telegraph, American Tel and Tel, AT and T, and so on. All these need to be recognized as the same company; otherwise, you could make decisions based on faulty data. Data cleansing is often complicated and expensive.

- *Load:* The cleaned data needs to be put into the data warehouse. This must be repeated periodically. Figuring out how often data should be loaded can be difficult. It primarily depends on how up to date the data need to be to support good decision-making. The more time-critical the data, the more frequent the loads.

Once you have the data in the data warehouse, you need to be able to retrieve and analyze it to gain value from the information. In the next section, we discuss some methods used to analyze data.

Business Analytics Methods

Business analytics goals can be classified into three categories: descriptive, predictive, and prescriptive analytics. Descriptive analytics provides insights into the past. Descriptive analytics methods include reporting, data mining, and visualization. Predictive analytics is directed at understanding the future by predicting likely outcomes based on analysis of historical data and other factors. Prescriptive analytics provides advice on preferred outcomes. The three categories are summarized in Table 14.3.

Table 14.3 Business Analytics Categories		
Category	Focus	Key Question
Descriptive	Insights into the past	What has happened?
Predictive	Insights for the future	What could happen?
Prescriptive	Advising on outcomes	What should we do?

The categories shown in Table 14.3 describe the purposes of BA. These categories are useful for understanding how BA can be applied. However, the specific approaches to analyzing data may cut across these categories. For example, data mining (discussed later in this chapter) can be applied to any of the categories. In this section, we describe three analytics approaches: visualization, reporting, and data mining. Later, we will discuss **artificial intelligence**, which is often used in business analytics.

Visualization

We discussed visualization in Chapter 6 but offer some additional insights in this section. Data visualization is the use of images to communicate information. You are

BUSINESS EXAMPLE BOX 14.2

Explaining Profitability Problems through Visualizations

Consider the visualizations shown below. KPI stands for key performance indicator. KPIs are important metrics that management uses to assess the performance of an organization. The three simple visualizations explain that one possible explanation for lower than desired profitability is sales with large discounts. The top visualization shows that profitability is not as strong as management would like, with several region-category combinations marked as "bad" and only one as "great." The bottom charts shown that if we only consider sales with no discounts, all region-category combinations show "great" profitability. In contrast, if we only include sales with large discounts, all region/category combinations show "bad" profitability. So, one possible explanation for the lower than desired profitability is the sales with high discounts. Of course, this is only a starting point. Management would need to analyze other data to decide on an appropriate course of action. This is typical with business analytics—an initial analysis often leads to additional analysis. Rarely do we get to the heart of an issue on the first try.

Profitability KPI
✓ Great
! OK
✗ Bad

Profitability is only 20% or more for one category/region combination

Profitability KPI By Category

Category		Region		
	Central	East	South	West
Furniture	✗	✗	✗	✗
Office Supplies	✗	!	!	✓
Technology	!	!	!	!

Sales with no discount

Category		Region		
	Central	East	South	West
Furniture	✓	✓	✓	✓
Office Supplies	✓	✓	✓	✓
Technology	✓	✓	✓	✓

If we consider only sales with no discounts, profitability is high across regions and categories

Sales with discounts >= 30%

Category		Region		
	Central	East	South	West
Furniture	✗	✗	✗	✗
Office Supplies	✗	✗	✗	✗
Technology	✗	✗	✗	✗

The picture changes when we look at sales with high discounts (30% or more)

probably familiar with charts and graphs, which are commonly used visualization methods. The executive dashboards mentioned earlier in this chapter often include both tables of data and visualizations. Visualizations use aesthetic design elements, including, position, shape, color, line characteristics (width and type), and size to express information. For example, a simple bar chart might show sales on the vertical axis and year on the horizontal axis, with the color of each bar indicating different product categories. Shapes, line types, and colors are often used to show different types of things (such as retail and wholesale customers). The size of a shape indicates different values, with smaller shapes indicating lower values.

LEARNING ACTIVITY 14.5

Effective Visualizations

Tableau is a widely used system for creating visualizations. Visit the Tableau Public Gallery of visualizations (https://public.tableau.com/en-us/gallery/?tab=featured&type =featured). Find one effective and one ineffective visualization. (Hint: The dropdown box may make it easier to find visualizations that are of particular interest to you.)

- Write a brief (one-page) paper that describes the key differences between the effective and ineffective visualizations. Focus your discussion on these questions. What makes the visualizations effective or ineffective? How could the ineffective visualization be improved?

Although software such as Excel and Tableau makes it relatively easy to create complex visualizations, understanding how to create an effective visualization is much more difficult. An effective visualization should strike a balance between providing enough information to be useful without using so many different visual elements that the visualization becomes confusing. A visualization crowded with many different visual elements is often the mark of an inexperienced analyst. More experienced analysts understand that a good visualization includes as many visual elements as necessary, but no more. Of course, understanding the line between enough and too much requires expertise and experience.

Experienced analysts know that before starting the design of a visualization they must understand its purpose. Generally, visualizations serve one of three purposes: exploration, explanation, or persuasion. Visualizations focused on data exploration tend to display more information that those intended to explain, although this is not always the case. Exploratory data visualization helps you get a handle on the data that you have. This often involves creating many visualizations to gain a better understanding of important relationships in the data. These visualizations often include a lot of detail, some of which turns out to be unimportant noise. Exploration is a process, though, so going down some unproductive paths is expected. One way to think about exploratory visualization is that its purpose is to find the story in the data.

Explanation typically follows exploration—once you know the story that lives in the data, you create visualizations that tell the story to your intended audience. Understanding your audience is very important. It determines the appropriate level of detail and complexity. Expert audiences can deal with detail and complexity better than audiences that have less expertise. For example, if you are creating visualizations for an audience of accountants, you would probably provide more detail than if you were presenting to an audience of consumers. Stories are effective ways to tie multiple visualizations together in a meaningful way. Stories string together visualizations in a narrative sequence. For example, a story might start with visualizations that illustrate some problem, then progress through visualizations that end in a suggested solution.

The final purpose we discuss is persuasion. Some experts might consider persuasion as a special case of explanation, but we like to treat it separately. The point of a persuasive visualization is to influence one or more people to accept a specific action or point of view. Persuasive visualizations use data, presented visually, as evidence of the suitability of a particular action or perspective. Often people are better able to interpret visualizations than statistics, especially when the visualizations are skillfully designed. However, it is important to accurately represent the data you are visualizing. Unethical designers can misrepresent data and trends by choosing misleading axis scales and endpoints, size, incomplete data selection, and other means. As intelligent consumers of data visualizations, you should always be on the lookout for such shenanigans.

Data Mining

Data mining is the process of analyzing data to identify trends, patterns, and other useful information. Data mining typically involves applying statistical techniques to identify trends and patterns. Some people consider querying and visualization to be related to data mining. Since we covered those topics in earlier chapters, the focus here is on statistical data mining.

There are wide applications of data mining in areas ranging from business to picking players in a pro sports draft. The basic process of data mining is shown in Figure 14.8. Existing data are analyzed using the appropriate data mining technique to build a model. The new model is then applied to new data to make a prediction, for example.

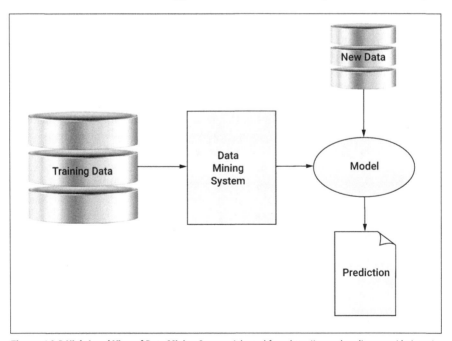

Figure 14.8 High-Level View of Data Mining Source: Adapted from http://www.thearling.com/dmintro/dmintro_2.htm/.

TABLE 14.4 Common Data Mining Techniques		
Technique	Description	Example
Association detection	Discovering patterns in which multiple events are connected	Customers tend to purchase eggs and butter together.
Sequence analysis	Discovering when one event leads to another event	A student graduates from college and then buys a new car.
Classification	Dividing data into mutually exclusive groups based on the variable you want to predict	Students are classified according to whether they are likely to drop out.
Clustering	Dividing data into mutually exclusive groups using all available data (not just the variable you want to predict)	Hotel guests are grouped according to various preferences.
Forecasting	Making predictions using discovered patterns	Past repayment and employment histories predict whether a mortgage applicant will default.

While data mining uses many statistical techniques, it is important to understand the difference between traditional statistical analysis and data mining. In most traditional statistical analyses, you create a model (such as a regression equation) and then test that model using data. Essentially, you have a hypothesis and then use data and statistical analysis to test that hypothesis. While this is often effective, it is limited by your knowledge and creativity. You can only test what you can hypothesize. Data mining can help overcome these limitations. In data mining, the emphasis is on discovering the model and then testing the validity of the model. Statistical techniques are used to uncover the relationships in the data that lead to the model. Table 14.4 shows common data mining techniques with descriptions and examples.

Data mining has many applications in business. You may have talked about customer segmentation in a marketing class. Customer groups (segments) can be built through cluster analysis. Cluster analysis divides information into mutually exclusive groups. (Each member can be in at most one group.) The goal is to build the groups so that each member in a group is as close as possible to every other member of the group and the different groups are as different as possible. For example, a hotel chain might use cluster analysis to segment customers into different groups, such as short-stay business customers, vacationers, and long-stay business customers. Advertisements could then be tailored to each of these customers, which should increase the ads' effectiveness. CRM systems can use cluster analysis to customize product or service offerings.

Market basket analysis identifies affinities among customers' product or service choices. Retailers often use checkout scanner data to identify products that customers tend to buy together. For example, a restaurant might determine that customers who order wine typically buy dessert, while beer drinkers usually buy appetizers. The restaurant could develop special offerings based on this information. Market basket

	A	B	C	D	E	F	G
1							
2							
3	Sum of Sales revenue	Column Labels					
4	Row Labels	Q1	Q2	Q3	Q4	Grand Total	
5	⊟California	1,899,680.30	1,760,147.80	1,930,516.50	1,889,224.70	7,479,569.30	
6	e-Fashion San Francisco	770,502.70	779,742.50	822,383.60	886,011.70	3,258,640.50	
7	e-Fashion San Diego	1,129,177.60	980,405.30	1,108,132.90	1,003,213.00	4,220,928.80	
8	⊟Colorado	525,682.20	500,076.00	510,776.90	523,740.10	2,060,275.20	
9	e-Fashion Dillion	525,682.20	500,076.00	510,776.90	523,740.10	2,060,275.20	
10	⊟DC	766,821.70	706,446.60	692,258.20	796,423.40	2,961,949.90	
11	e-Fashion Washington Midtown	766,821.70	706,446.60	692,258.20	796,423.40	2,961,949.90	
12	⊟Florida	515,687.70	489,997.80	387,809.70	485,663.30	1,879,158.50	
13	e-Fashion Miami Beach	515,687.70	489,997.80	387,809.70	485,663.30	1,879,158.50	
14	⊟Illinois	846,408.40	850,595.00	610,765.20	714,889.80	3,022,658.40	
15	e-Fashion Chicago Loop	846,408.40	850,595.00	610,765.20	714,889.80	3,022,658.40	
16	⊟Massachusetts	312,896.40	291,431.00	249,529.00	429,850.20	1,283,706.60	
17	e-Fashion Boston Newbury	312,896.40	291,431.00	249,529.00	429,850.20	1,283,706.60	
18	⊟New York	1,987,114.70	2,028,090.70	1,672,580.70	1,894,434.50	7,582,220.60	
19	e-Fashion New York 5th	804,084.40	795,193.60	637,277.10	723,811.40	2,960,366.50	
20	e-Fashion New York Midtown	1,183,030.30	1,232,897.10	1,035,303.60	1,170,623.10	4,621,854.10	
21	⊟Texas	2,875,569.20	2,499,276.80	2,146,302.80	2,596,515.50	10,117,664.30	
22	e-Fashion Dallas	555,458.70	450,588.30	467,925.90	496,061.30	1,970,034.20	
23	e-Fashion Houston	614,284.60	560,650.20	465,677.10	662,570.90	2,303,182.80	
24	e-Fashion DFW	775,482.70	667,850.30	581,470.40	674,869.80	2,699,673.20	
25	e-Fashion Houston Downtown	930,343.20	820,188.00	631,229.40	763,013.50	3,144,774.10	
26	Grand Total	9,729,860.60	9,126,061.70	8,200,539.00	9,330,741.50	36,387,202.80	
27							
28							
29							

Sheet1 / Report 1

Figure 14.9 Pivot Table Example

analysis is an example of association detection. Association examines the degree to which variables in a data set are related by considering the frequency with which the variables occur together.

Have you ever played the 20-questions game? If so, you have some familiarity with the concept of decision trees, which are a classification technique. A decision tree makes a prediction based on a series of questions. The next question in the series depends on the answer to the current question. You might have concluded that decision trees are useful devices, since we have discussed them twice in this chapter. Earlier, we discussed decision trees as a method for knowledge capture. Here the focus is on prediction rather than knowledge capture.

There are many other data mining techniques, including statistical methods such as regression analysis, time series analysis, neural networks, and text analysis. As we mentioned earlier, statistical analysis emphasizes testing preexisting models, while data mining tries to discover the models. Many of the same statistical techniques can be applied to both. One important point to keep in mind is that you should always assess whether your model and its predictions are reasonable. Many data mining techniques are based on mathematics; they are all about the numbers. The mathematics does not consider whether the underlying associations have any real meaning. That is where human interpretation and judgment come in.

We discussed information reporting tools in Chapter 6 in the section "Information Retrieval and Analysis Tools," so our remarks here will be brief. One of the most

	A	B	C	D	E	F
3	Sum of Sales revenue	Column Labels				
4	Row Labels	Q1	Q2	Q3	Q4	Grand Total
5	⊟California	1,899,680.30	1,760,147.80	1,930,516.50	1,889,224.70	7,479,569.30
6	⊟e-Fashion San Francisco	770,502.70	779,742.50	822,383.60	886,011.70	3,258,640.50
7	2 Pocket shirts	27,876.20	33,324.60	30,122.00	31,773.80	123,096.60
8	Belts,bags,wallets	49,517.90	40,896.40	17,621.40	29,908.20	137,943.90
9	Bermudas	994.90		340.30	4,883.40	6,218.60
10	Boatwear	11,091.60	2,065.60	2,040.30	8,949.70	24,147.20
11	Cardigan	36,084.10	22,446.50	14,003.60	1,176.40	73,710.60
12	Casual dresses	4,293.20	4,458.50	12,908.10	5,014.70	26,674.50
13	Day wear	4,344.90	28,676.30	29,980.00		63,001.20
14	Dry wear	1,396.10		6,236.20	10,935.60	18,567.90
15	Evening wear	26,296.00	4,699.80	32,976.70	101,981.50	165,954.00
16	Fancy fabric	4,758.00	3,607.60	3,646.70	4,964.70	16,977.00
17	Full length	1,935.50		4,459.90	15,833.50	22,228.90
18	Hair accessories	4,852.80		3,388.10	7,065.60	15,306.50
19	Hats,gloves,scarves	66,386.40	24,783.20	19,091.00	3,175.00	113,435.60
20	Jackets			6,570.00	381.30	6,951.30
21	Jeans	4,238.70	1,711.80	5,678.80	10,353.60	21,982.90
22	Jewelry	185,083.10	209,951.90	43,683.00	1,815.20	440,533.20
23	Long lounge pants	970.80	424.60	5,234.90	7,961.90	14,592.20
24	Long sleeve	17,300.00	10,490.60	40,848.40	61,425.20	130,064.20
25	Lounge wear	6,423.70	16,771.60	16,017.80	5,639.40	44,852.50
26	Mini city	818.00	3,445.20	5,697.80	1,098.40	11,059.40
27	Night wear			1,837.80	2,846.00	4,683.80
28	Outdoor	3,151.00	3,503.40	7,544.60	6,310.50	20,509.50
29	Pants	572.90		1,831.60	946.20	3,350.70
30	Party pants	8,416.00	14,143.80	59,198.40	4,208.90	85,967.10
31	Samples	1,022.80	10,618.70	12,960.80	7,878.90	32,481.20

Sheet1 Report 1

Figure 14.10 Pivot Table with Additional Dimension

useful (and easy to use) reporting tools is spreadsheet pivot tables. Although they may seem complicated at first, once you understand the basics, pivot tables allow you to quickly build multidimensional views of a data set. For example, Figure 14.9 shows sales data from a small retail clothing chain. The data in this figure show the total revenue for each store, broken down by state and quarter.

You can quickly modify the dimensions by which you view the data. Suppose you want to add product categories to the analysis. In a matter of seconds, you can modify the pivot table to include the new dimension, as shown in Figure 14.10. (Note that only a portion of the table is shown in the figure.) The flexibility of pivot tables makes them ideal reporting and analysis tools for many common business decisions. You can learn how to build pivot tables in Appendix G.

Artificial Intelligence

You may recall that artificial intelligence (AI) was one of three emerging technologies we discussed in Chapter 1 (so long ago!). AI refers to a family of technologies that approximate human cognitive abilities. In this chapter, we go into more details about specific AI applications and techniques. However, AI is a vast topic, so we only scratch the surface of this fascinating area. AI applications exist in virtually any industry you can imagine. (See Business Box 14.3 for some examples.) As we pointed out in Chapter 1, AI is often used in combination with other technologies, such as the Internet of Things, to create applications that have tremendous impacts. In this section, we provide a high-level view of the some of the AI techniques that enable these applications.

Currently available AI systems can be classified as narrow AI (sometimes called artificial narrow intelligence, or weak AI). This means that the AI system focuses on a single, relatively narrow task. For example, virtual assistants, such as Apple's Siri or Amazon's Alexa, help users perform a relatively narrow range of tasks, such as getting directions to a nearby Thai restaurant or checking the weather. Although many people find these virtual assistants quite useful, they struggle to perform tasks outside of a limited range. Even more sophisticated systems like IBM's Watson are still classified as narrow AI because they do not have any sentience (roughly defined as self-awareness). In the future, we may see deep AI (also known as artificial general intelligence, or strong AI). To be classified as deep AI, the system must be indistinguishable from a human in the way that it thinks and acts. If you are a science-fiction fan, you may be aware of artificial superintelligence, which is when an AI system is not only self-aware, but it also has intelligence far beyond human capabilities. While narrow AI is becoming common, most experts think that deep AI and artificial superintelligence are a long way from becoming reality.

There are many existing uses of AI. A few are listed below:

- Computer vision: These systems use AI techniques such as neural networks to interpret visual data. (Neural networks are AI systems that are loosely based on how animal brains work.) You probably use computer vision technology regularly if you have a phone or tablet that you can unlock using facial recognition. Virtually all of us have been the subject of facial recognition technology used by many law enforcement agencies and governments. Newer applications of computer vision include reading medical images, self-driving cars, and robotics.
- Natural language processing (NLP): The goal of NLP is for computers to understand language as humans can. Applications of NLP include advanced email spam detection, language translation, chatbots, and digital virtual assistants.
- Recommendation systems: Recommendation systems are widely used by companies such as Amazon to show consumers products that are likely to be of interest. Streaming services such as Netflix and Hulu use recommendation systems to suggest movies and shows based on prior viewing behaviors. These systems use machine learning algorithms, often proprietary, to make these recommendations.
- Expert systems: Expert systems use a knowledge base (an organized collection of facts and rules-of-thumb) and an inference engine (a set of rules of applying the knowledge) to simulate the judgment of subject-area experts. Expert systems have been applied to many knowledge

BUSINESS EXAMPLE BOX 14.3

Applications of AI

AI is transforming the way businesses operate. For example, AI is being used in human resources management to recruit and screen job candidates, personalize employee benefits, and automate some aspects of performance management. Many companies are using AI-based chatbots (discussed later in this chapter) to provide customer service. AI applications are helping energy companies reduce maintenance costs and reduce downtime. For example, companies such as General Electric, Siemens, and ABB Ltd. use AI to monitor industrial machines to detect potential problems and better schedule preventive maintenance. Many organizations are using advanced AI as part of their cybersecurity strategies. You may have received a fraud alert for your debit or credit card. If so, the potential fraud was probably detected through an AI machine learning algorithm. Spotify and Pandora use AI to personalize their song and artist recommendations. The impact of AI on business is clear and will grow substantially in the future.

Sources: HR Technologist Staff. 2019. "The Beginner's Guide to AI in HR." *HR Technologist*, February 5. https://www
.hrtechnologist.com/articles/digital-transformation/the-beginners-guide-to-ai-in-hr/; Phaneuf, A. 2020. "Artificial
Intelligence in Financial Services: Applications and Benefits of AI in Finance." *Business Insider*, September 9. https://
www.businessinsider.com/ai-in-finance; Change, E. 2019. "5 Energy Companies Using AI for Cost-Efficiency." *U.S.
News and World Report*, April 12. https://money.usnews.com/investing/stock-market-news/articles/2019-04-12/5
-energy-companies-using-ai-for-cost-efficiency; Faggella, D. 2020. "6 Examples of AI in Business Intelligence Applications." *Emerj.com*, March 14. https://emerj.com/ai-sector-overviews/ai-in-business-intelligence-applications/;
Sennaar, K. 2019. "Musical Artificial Intelligence—6 Applications of AI for Audio." *Emerj.com*, February 9. https://
emerj.com/ai-sector-overviews/musical-artificial-intelligence-6-applications-of-ai-for-audio/.

areas, including medical diagnosis, financial fraud detection, and manufacturing control, among others.

One of the most widely used AI-related technologies is machine learning. We provide an overview of machine learning in the next section.

Machine Learning

Machine learning is a branch of AI that involves developing applications that automatically get better through experience. Basically, the system learns through examples. A classic example is learning to recognize objects by being shown examples of the object and items that are not the object. For example, you might provide the system a set of images of dogs along with pictures that do not include dogs. Humans tell the system which pictures include dogs, and which do not. Over time, the system will learn to distinguish dogs from other animals by analyzing the images to identify characteristics of pictures that include dogs, then tweaking its analysis to improve its classifications. The more pictures of dogs you feed the system, the more accurate it becomes in distinguishing dogs from other images.

You may not know it, but you may have been helping train AI systems while using the Internet. You probably have encountered Google's Recaptcha system (that sometimes irritating "I'm not a robot" box) to prove that you are an actual human by identifying the pictures that include a bicycle, traffic signals, train, or some other common object. The photos used in this system come from Google Street View. When you identify the pictures of traffic signals, you are providing an AI system with training data that will help it more accurately identify traffic signals. Imagine the usefulness of an AI system that can recognize traffic signals with high accuracy—this is a big help for reliable and safe autonomous vehicles.

Machine learning involves four steps. There are other models of the machine learning process that include more steps, but these typically just break some of our steps into subsets.

1. **Choose and prepare a set of data called the training data set**. As the name implies, this data set is used to train the machine learning algorithm. Before it can be used, however, the data set needs to be put into a form that is usable to the algorithm. Sometimes you hold back some of the training set data to use to evaluate the model (see the third step).
2. **Choose the algorithm to use to analyze the training data set**. For example, you might choose regression algorithms, which predict the value of an outcome variable (the dependent variable) using one or more input variables (called the independent variables). For example, a company might want to predict the sales of hot dogs based on variables such as price, day of the week, and season of the year. The dependent variable is the number of hot dogs sold, and price, day of the week, and season of the year are the independent variables. The outcome of this step is a model, a regression equation in our example.
3. **Evaluate the model**. Evaluating the model involves using a new data set, or data held back from the original data set to determine its accuracy, precision, or other suitable criteria. Poorly performing models may be adjusted.
4. **Apply the model**. The improved model is put to use. In our regression example, the regression equation would be used to predict future sales of hot dogs.

The example above describes a category of machine learning methods called supervised learning. Supervised learning algorithms are trained by using a data set that contains correct answers. In our hot dog example, the training and evaluation data sets would be historical data, including the actual number of hot dogs sold. Unsupervised learning, in contrast, uses data that does not indicate the "right" answer. The algorithm tries to find some structure in the data. For example, a grocery store might analyze customer loyalty data to identify groups of customers who share common characteristics. Reinforcement learning algorithms use trial-and-error methods to determine actions that yield the highest rewards. The goal of the algorithm is to find the actions that provide the greatest rewards over a set period of time. The algorithm will try different sets of actions to determine which set is the best. Reinforcement learning is used for applications such as robotics, gaming, and autonomous vehicles.

One interesting application of AI is its use to make robots more autonomous. In the next section, we provide a brief introduction to robotics and its use in business. Business Box 14.4 discusses some interesting applications of robotics to agriculture.

Robots

Robots are computer-controlled machines that can carry out complex actions. Robotics is the study of robots—their design, construction, and operation. Although many movies often show robots in human-like forms (called humanoids), most robots are clearly machines. Industrial robots have been used for decades. For example, in the 1980s, automobile manufacturers used robots for jobs such as welding frames to bodies and painting. Today, robots are used for a wide variety of tasks, such as warehouse automation, crop harvesting, assisting with surgery, and even vacuuming your home. One of the authors has a pool-cleaning robot. (It is awesome to watch it in action!) Robots are even being used in arts and entertainment.

Even if you did not realize it, you have benefited from robotics. For example, you have probably ordered products online. If so, it is likely that some of the products moved through automated warehouses that made heavy use of robots. For example, Amazon uses small robots to move large stacks of products around its warehouses. Humans still do much of the work at these warehouses, but the level of automation is increasing.

BUSINESS EXAMPLE BOX 14.4
Agricultural Robots

Agricultural work is often dirty and dangerous. Finding farm labor is often difficult. Farmers often turn to chemicals to control weeds, which can be harmful to the environment. It is also challenging to monitor plants and livestock on large farms and ranches. Robots are helping to address some of these problems. Robots are helping harvest crops such as apples, strawberries, and lettuce, among others. Robots are also helping farmers reduce their use of weed control chemicals (called herbicides). Some of these weed control robots use AI to tell the difference between weeds and crops. Robots are also used in soil analysis and environmental monitoring, which can help increase crop yields. Farmers are using drones, which sometimes are considered robots, to monitor environmental conditions, and to spread seed and fertilizer. Ranchers are using drones to track their livestock and robots to herd cattle. Robots are even milking cows. Although the agricultural use of robots is in its early days, it seems clear that the impact of robots on farming and ranching will increase as robots improve and new applications are found. The increasing use of robots in agriculture has the potential to lower food costs and reduce agriculture's environmental impact.

Sources: Gossett, S. 2020. "Farming & Agricultural Robots." *Built In.com*, April 6. https://builtin.com/robotics/farming-agricultural-robots; NT, B. 2020. "Top 9 Best Robotic Milking Machines to Consider in 2020." *Robotics Biz*, May 10. https://roboticsbiz.com/top-9-best-robotic-milking-machines/; Cargill. 2018. "Meet the Robot That's Making Cattle Herding Safer." https://www.cargill.com/story/meet-the-cowboy-robot-thats-making-cattle-herding-safer

Are AI and Robots Good or Bad for Society?

Read the article "How Artificial Intelligence and Robotics Are Changing Our Lives" (https://www.sam-solutions.com/blog/ai-and-robotics-impact-on-our-lives/). Write a one- to two-page paper that provides your opinion on whether AI and robots are good or bad for society. Be sure to clearly provide your reasoning.

Robots are particularly useful when tasks are repetitive, require a high degree of precision, or are dangerous or beyond the capability of humans. Robots can work for long periods of time without rest, they follow instructions precisely, they can work in harsh environments (such as in space), and they can be built to exhibit incredible strength. For example, robots that handle fabric bales can lift almost 2,000 kilograms (around 4,400 pounds). Although some robots perform many functions, many do just one thing—for example, placing a particular engine part in a precise location. Some robots require human operators, while others work without direct human control. In this short section, we can do little more that introduce the topic of robotics and give you some sense of the impact they have on the world. Their impact will grow in the future. For example, before too long, swarms of tiny robots called nanobots will be commonplace for applications such as repairing human organs from inside the body and cleaning the environment. We are already becoming used to seeing robots at work. In the future, they will become so common that many will fade into the background, going about their work unnoticed.

Chapter Summary

In this chapter, we discussed knowledge and the processes and tools that help organizations manage knowledge. We also helped you understand technologies that support decision-making and that help organizations gain more value from their information resources. Here are the main points discussed in the chapter:

- Knowledge management is a process that allows organizations to generate value from their knowledge-based assets.
- Managing knowledge is important because knowledge is among the most important resources of any organization.
- Explicit knowledge can be easily expressed. Tacit knowledge is difficult to express and communicate.
- The main processes for knowledge management are create, capture, codify, store, retrieve, transfer, and apply.
- Information and communication technologies that facilitate knowledge management include knowledge repositories, communication and collaboration tools, executive information systems, dashboards, and expert systems.

- Decision support systems are computer-based systems that help decision makers use data and models to solve semistructured or unstructured problems.
- Business analytics is the process of transforming data into insights to improve operations and decision-making.
- Data warehouses and data mining systems are important business intelligence components.
- Data warehouses store and organize data for decision support.
- Visualizations uses aesthetic design elements to communicate and help make sense of data. Visualizations can also be used to persuade.
- Data mining is the process of analyzing data to identify trends, patterns, and other useful information.
- Artificial intelligence is having significant impacts on business and our daily lives through methods such as machine learning.

Review Questions

1. Briefly explain how knowledge and information differ in terms of ease of management and the level of human contribution.
2. Why is managing knowledge important to organizations?
3. State five benefits of effective knowledge management.
4. Contrast explicit and tacit knowledge.
5. Give an example of tacit knowledge.
6. Give an example of explicit knowledge.
7. Name the processes in the knowledge management cycle.
8. Briefly describe the four knowledge creation processes in the SECI model.
9. What is knowledge codification? Why is it important to knowledge management?
10. Compare and contrast decision tables and decision trees.
11. Explain why search technologies are important to knowledge management.
12. Give an example of each of the following: a knowledge repository and a communication-based knowledge management tool.
13. What is a digital dashboard?
14. Name and briefly describe the DSS categories described in the chapter.
15. Compare and contrast synchronous and asynchronous groupware.
16. Name and briefly describe the groupware examples discussed in the chapter.
17. What are the main components of a business analytics system? Briefly describe each component.
18. Describe the E/T/L process.
19. What are the three categories of business analytics?

20. Name and briefly describe the three purposes of visualization.
21. What is the purpose of data mining?
22. Name three data mining techniques.
23. Compare and contrast narrow AI and deep AI.
24. Name and briefly describe three uses of AI.
25. List the four steps in machine learning.

Reflection Questions

1. What is the most important thing you learned in this chapter? Why is it important?
2. What topics are unclear? What about them is unclear?
3. What relationships do you see between what you learned in this chapter and what you have learned in earlier chapters?
4. The chapter focused on organizational knowledge management, but knowledge management is also important to individuals. (This is called personal knowledge management.) Why is knowledge management important to your personal and professional development?
5. Have you ever tried to explain to someone how you are able to do something (hit a golf ball, draw, study, etc.)? Was it easy or hard to explain? What about the knowledge you were trying to transfer determined whether it was easy or hard to explain?
6. In Chapter 7, you learned about cloud computing. How can cloud computing facilitate knowledge management?
7. How does data warehousing facilitate data mining?
8. What is the relationship between business analytics and knowledge management?
9. How is the need for quick decisions related to the need to make ethical decisions?
10. Briefly describe three ways you might use visualizations in your classes.
11. What is the relationship between artificial intelligence and robots?

Additional Learning Activities

14.A1. Describe a situation in which you used technology to access knowledge. Address the following in your description: What knowledge were you seeking? How did you go about finding the knowledge? What technologies did you use? How effective were they?
14.A2. Form groups of two.
    a. One of you (the explainer) will answer the following question: Why is knowledge management important to organizational success?

b. When the first person finishes, the other person (the questioner) will seek clarification and provide her or his own ideas.

Record and be prepared to share your answers.

14.A3. Some music services, including Pandora and iTunes, can use the songs in your library (or playlists) to build smart playlists that you will like. Research how these services work and prepare a one-page paper on how these services relate to the data mining tools you read about in this chapter.

14.A4. Create a decision table or decision tree to document how you go about making the following decisions:

a. Deciding whether to take a class

b. Deciding which section of a class to take

c. Deciding on a major

d. Deciding whether to join a club

14.A5. Partner with two other students. Create a wiki on how to choose which section of a multisection class to take. You can use SlimWiki (https://slimwiki.com/), PB Works (http://pbworks.com), or any other wiki service to create your wiki. Be prepared to share your wiki with the class.

14.A6. Visit the Tableau Public Gallery (https://public.tableau.com/en-us/gallery/?tab=viz-of-the-day&type=viz-of-the-day). Examine at least three visualizations. Briefly describe your favorite visualization and discuss why you liked it.

14.A7. Research how companies are using data mining. Pick one example and prepare a one- to two-page report on how the company has used data mining.

14.A8. Find an example of AI related to your major. Briefly describe the application and discuss its impact on your discipline. Provide a link to the Web page where you found the example.

References

Alter, S. 1977. "Why Is Man-Computer Interaction Important for Decision Support Systems?" *Interfaces* 7(2): 109–15.

Davenport, T., and L. Prusak, 1998. *Working Knowledge: How Organizations Manage What They Know.* Boston, MA: Harvard Business School Press.

Iliinsky, N., and J. Steele, 2011. *Designing Data Visualizations.* Sebastopol, CA: O'Reilly Media, Inc.

Nonaka, I., and H. Takeuchi, 1995. *The Knowledge Creating Company.* New York: Oxford University Press.

Pearlson, K., and C. Saunders, 2006. *Managing and Using Information Systems.* 3rd ed. Hoboken, NJ: John Wiley & Sons.

Power, D. 2001. "Supporting Decision-Makers: An Expanded Framework." In

E-Proceedings of the Informing Science Conference, edited by A. Harriger, Krakow, Poland, June 19–22: 431–36.

Glossary

Artificial intelligence: A family of technologies that approximate human cognitive abilities.

Business analytics (BA): The process of transforming data into insights to improve operations and decision-making.

Business intelligence (BI): A set of applications, technologies, and processes for gathering, storing, analyzing, and accessing data to help users make better business decisions.

Codifying knowledge: Converting tacit and explicit knowledge into a form that organizational members can use.

Cognitive map: A graphical representation of a mental model of an expert's knowledge.

Communication-driven decision support system: A DSS that facilitates collaboration and group-based decision-making.

Community of practice: A group of people who share a common interest, usually in a craft or profession, and who interact to share knowledge.

Conceptual knowledge: Explicit knowledge that is embodied in language, symbols, and images.

Data-driven decision support system: A DSS that focuses on the retrieval and manipulation of data that is stored in an organization's data stores.

Data mining: The process of analyzing data to identify trends, patterns, and other useful information to make predictions.

Data warehouse: A copy of transactional data (and other data) that is formatted so that it is useful for decision support.

Decision support systems (DSS): Computer-based systems that help decision makers use data and models to solve semistructured or unstructured problems.

Decision table: A list of conditions and their values along with a list of conclusions or actions.

Decision tree: A diagram that shows alternate paths that impact decisions.

Digital dashboards: Computer-based systems that provide graphical views of key data along with graphical warnings when data indicate areas that need attention.

Document-driven decision support system: A DSS that facilitates the management and retrieval of documents that may help with decision-making.

Executive information systems (EISs): Information systems that help provide high-level managers with the information they need to monitor business activities and make decisions.

Experiential knowledge: Tacit knowledge that can be shared through common experiences and interpersonal communication.

Expert systems: Systems that help users solve problems or answer questions in a way that mimics an expert's thought processes.

Explicit knowledge: Knowledge that can be expressed relatively easily.

Groupware: A network-based system that helps workgroups communicate and collaborate.

Knowledge management: The process by which organizations create, capture, store, apply, and protect knowledge to achieve organizational objectives.

Knowledge repositories: Tools that make it easy to find and retrieve documents that contain knowledge.

Machine learning: A branch of artificial intelligence that involves developing applications that automatically improve through experience.

Model-driven decision support system: A DSS that provides the decision maker with the ability to access and manipulate analytical models.

Robots: Computer-controlled machines that can carry out complex actions.

Routine knowledge: Tacit knowledge that is embedded in the organization's practices.

Systemic knowledge: Explicit knowledge that has been systemized and packaged.

Tacit knowledge: Internalized and highly individualized knowledge that is difficult to express or communicate.

Index

Note: Page numbers in *italics* refer to figures; numbers in **bold** refer to tables. Page numbers preceded by "A" refer to material that is located in the Appendices, which is available online.

ABC (activity-based costing), A127, A128, A130
Access (advanced), A35–A49
  database queries, A35–A36
  Form Wizard, A43–A46, *A43–A47*
  forms, A43–A48
  Query by Example (QBE), A37, *A38,*
    A39–A42, *A39–A42*
  reports, A48–A49, *A48–A49*
Access (fundamentals), A23–A32
  creating a database, A24, *A25,* A32–A33
  creating relationships, A28–A29
  creating tables, A26–A28, *A26*
  data type error, *A31*
  data types, **A27**
  datasheet view of a table, *A30*
  edit relationship dialog box, *A29*
  entering table information, *A26*
  Getting Started screen, *A25*
  navigation pane, *A30*
  populating tables with data, A30–A31
  referential integrity violation error message,
    A31
  relationships button, *A28*
  relationships created, *A29*
  selecting primary key fields, A27–A28, *A28*
  show table dialog box, *A29*
Accounting services, 57
Ackoff, Russell, 9
Active content, 166
Activity-based costing (ABC), A127, A128, A130
Address resolution, A104
Agile development, 233, 244
Agility, 61
AgTech, 32
AI. *See* Artificial intelligence (AI)
All-or-nothing concept, 251
Alternatives, 108, 111

ALU (arithmetic-logic unit), A5
Anomalies, 87
Anonymity
  anonymous browsing, 203–206, 208
  perceptions of, 194
ANSI X.12, 305, 311
Antivirus software, 168, 172–173, A14
API (application program interface), A13–A14
App developers, *235*
App stores, 28, A19, A21
Application integration, 270
Application outsourcing, 240
Application program interface (API), A13–A14
Application software, A11–A13, A17–A20
  and app stores, A19, A21
  app usage, A20
  business or personal-use, A17
  classifying, A18
  finding an app, A19
  general productivity or special-purpose, A18
  horizontal- or vertical-market software, A18
  locally installed or cloud-based, A18–A19
Apps. *See* App stores; Application software
Architecture(s)
  client/server, 136, *137*
  and control, 135
  defined, 135
  and ease of implementation, 135
  and flexibility/interoperability, 135
  information systems, 62
  and maintainability, 136
  multitiered, 83
  network, 147
  peer-to-peer, 137
  principles of, 135–136, 141
  and scalability, 136
  security and reliability, 136

service-oriented, 138–139, *138*
systems network (SNA), A100
wireless, 137–138
See also Networking architectures
Arithmetic-logic unit (ALU), A5
Artificial intelligence (AI), 3, 5–6, 318, 341–346,
    347
    applications of, 343
    functions of, 6–7
    machine learning, 343–345
    narrow vs. deep, 342
    uses of, 342–343
    and virtual reality, 8
Asymmetric encryption, 178, *179*
Asymmetrical security warfare, 156
Attributes, A52
Audit logs, 168
Augmented reality, 8–9, *8*
Authentication, 158, 173
Automated driving systems. *See* Autonomous
    vehicles
Automation, 31; as first-order change, 260
Autonomous vehicles, 7, 125, 342
Availability, 158–159, 161, 164–165, 166, 168,
    183–184

B2B (Business-to-business), 294, 303–306,
    310–311
B2C (Business-to-consumer), 294, 298–299, 304,
    310–311
B2E (Business-to-employee), 296
BA. *See* Business analytics
Backbone networks (BBNs), **128**, 146
Backdoors, 163
Bandwagon, 105
Bar codes, 27
Beer Game, 273, 275
Behavioral biometrics, 173
Behavioral-based antivirus protection tools, 173
Benefits administration systems, 34
Berners-Lee, Tim, 131
BI (Business intelligence), 332
Biases, 51, 105
Big Data, 92–96
    3 V's (volume, velocity, variety), 92
    and decision-making, 116
    at Facebook, 94

integrating data, 94–95
    retrieving and disseminating data, 95–96
    storing data, 93–94
Binary relationship, A53
Biometrics, 169, 173–175
    behavioral, 173
    facial recognition, 175, 342
    physiological, 173
    sample solutions, *174*
    use for authentication, 173–174
    use for identification, 174–175, *175*
Bitcoins, 167, 258
Blockchain, 167, 258, 278
Bluetooth, 126
BPaaS (Business process as a service), 139,
    141–142
BPI (Business process improvement), 256–257,
    257, 261
BPR. *See* Business process reengineering (BPR)
Brainstorming, 330
Bring your own device (BYOD), 137–138
Broadband networks, 127, 129
Business analytics (BA), 332–341, 347
    categories, **335**
    and data warehousing, 332, 334–335
    framework for, *333*
    goals of, 332, 335
    methods, 335–341
    tools for, 332–333
    and visualization, 335–338
Business analytics systems, 318
Business information systems, 23
Business integration, 269
Business intelligence (BI), 332
Business models, 296
Business process as a service (BPaaS), 139,
    141–142
Business process improvement (BPI), 256–257,
    261
    stages of, *257*
Business process reengineering (BPR), 256–257,
    261
    goals for, **257**
    online ordering redesign, *259*
    stages of, *257*
Business processes, 12–13, 142, 249
    all-or-nothing concept, 251

for banking, 258
informal, 258
process modeling companies and software
    products, **255**
process modeling tools, **255**
redesigning, 258
subprocesses, 251–252, *252*, *254*
Business rules, 29–30, A63
Business-to-business (B2B), 294, 303–306,
    310–311
Business-to-consumer (B2C), 294, 298–299, 304,
    310–311
Business-to-employee (B2E), 296
Buyer-oriented auction, 306
BYOD (bring your own device), 137–138

C2B (Consumer-to-business), 296
C2C (Consumer-to-consumer), 294, 311
Cache, A7
California Consumer Privacy Act of 2018
    (CCPA), 208
Cardinality, A54, A56–A57
determining, *A55*
symbols, *A55*
Careers
in information analysis, 14–15
in information systems, 14–15
CASE (Computer-aided software engineering)
    tools, 231
CCPA (California Consumer Privacy Act of
    2018), 208
Cellular networks, 126
Cemex, 250
Central processing unit (CPU), 24, A5, A6–A7,
    A8
CERN (European Organization for Nuclear
    Research), 132
Channel compression, 301
Channel expansion, 301–302
Chat systems, 330–331
Chatbots, 342
Chen's notation, 89
Children's Online Privacy Protection Act of 1998
    (COPPA), 208
Christie, Chris, 40
CIA (confidentiality, integrity, availability),
    158–159, 166, 168, 183

Ciphers, 177
Ciphertext, 177
Cisco, 275
Clickstream data, 196
Clients, 136
Client/server architecture, 136, *137*
Cloud application, 26
Cloud computing, 26, 61, 139–142, *139*
business process as a service (BPaaS), 139,
    141–142
infrastructure as a service (IaaS), 139–140
infrastructure services, *140*
platform as a service (PaaS), 139–140
software as a service (SaaS), 139, 141, *141*
Cluster analysis, 339
CMI (Customer-managed interactions), 284
CMSs (Content management systems), 270
Cognitive maps, 323, *324*
Collaboration systems, 34, 330–332
chat systems, 330–331
colocated, 330
discussion databases, 331
document coediting, 331–332
email and email lists, 331
examples of, *330*
group decision support systems, 330
remote presentation, 331
shared whiteboards, 330
Slack for student team projects, 331
synchronous/asynchronous, 330
videoconferencing, 330
workflow systems, 331
Collaboration tools, 327
Commercial, off-the-shelf software (COTS),
    234–237
compared to custom development, *235*
Communication media, 25, 27
Communication-driven DSS, 329–330
Communication(s)
between computers (networking), A99–A106
improving, 63
of information in modern organizations,
    142–145
plan for, A141
support systems, 57
unified, 270
Communities of practice, 327

Competitive advantage, 57–60, 62, 67, 141, 234, 236, 242, 244, 259, 274, 283, 296, 300
  sustainability of, 72–74
Compiling, 237
Complete outsourcing, 239
Compliance and regulations, 61. *See also* Government regulations
Composite keys, A53
Composite primary key, 86
Computer hardware, A1–A8
  application software, A17–A20
  central processing unit (CPU), A5, A6–A7, A8
  components of, A4–A7
  generations, A1–A2, **A2**, A8
  in an information system, 25–26
  input and output devices, A6, A8
  microprocessors (chips), A5
  platforms, A2
  selection of options, A3
  solid-state drives (SSDs), A7
  storage technologies, A6–A7, A8
Computer software, *A12*, A20
  $6 billion bug, **222**
  antivirus, 168, 172–173, A14
  application software, 26, A11–A13
  backup, A19
  for business analytics, 332–333
  for business process modeling, **255**
  commercial, off-the-shelf software (COTS), 234–237, *235*
  computer-aided software engineering (CASE) tools, 231
  with data visualization capabilities, 115–116, *116*
  horizontal-market, A18
  in an information system, 25–26, 27
  infrastructure, 131
  installation of, A12
  methodologies for development, 223–224
  open source, 237–238
  purposes of, 337–338
  for remote presentations, 331
  statistical, 115
  system requirements for, 224
  system software, 26, A11–A17
  total cost of ownership (TCO), 236–237
  types of, A11–A13
  vertical-market, A18
  for visualizations, 337
  See also Application software; Utility software
Computer vision, 342
Computer-aided software engineering (CASE) tools, 231
Computers, level of protection, 157
Confidentiality, 158–159, 166, 168, 183–184, 193–216. *See also* Privacy issues
Confirmation bias, 105
Consistency, 86, 87
Consortia marketplace, 306
Consumer-to-business (C2B), 296
Consumer-to-consumer (C2C), 294, 311
Content management systems (CMSs), 270
Control, 23, 24, 31
Convergence, A1, A3
Cookie managers, 201–203, 215
Cookies, 196, 201–203
Coordination, improving, 63
COPPA (Children's Online Privacy Protection Act of 1998), 208
Corrective controls, 168
Cost reduction/controls, 61
COTS (Commercial, off-the-shelf software), 234–237, *235*
COVID-19 pandemic, 30, 58
CPU (Central processing unit), 24, A5, A6–A7, A8
Credit cards
  with chip technology, 180–181
  with contactless technology (tap-and-go), 181
  security, 179–181
Critical mass, 300–301
Critical success factors (CSFs), 70
  for improving business in Australia, 72
  CRM. *See* Customer relationship management (CRM)
Cross-selling, 283
Crowdsourcing, 143
Crow's foot notation, 89
Cryptography, 177–179
  types of, 178–179
CSLC (Customer service life cycle), 278, **279**, *279*
Customer activity database design, *99*
Customer clustering, 280

Customer profiling, 281
Customer relationship management (CRM), 60,
    240, 269–270, 278–284
  analytical components, 280–281
  benefits and disadvantages of, 282–283
  collaborative components, 280
  on-demand, 282
  goals of, 278
  implementation options, 282
  operational components, 280
  on-premise, 282
  sample architecture, *281*
  self-servicing, 282
  vendors, 281
Customer retention, 283
Customer segmentation, 280
Customer service life cycle (CSLC), 278, **279**,
    279
Customer-managed interactions (CMI), 284
Cybersecurity, 61
Cycles, 278

DARPA (Defense Advanced Research Projects
    Agency), 130–131
DAS (Direct-attached storage) model, 93
Data
  anonymized, 194
  characteristics of, **319**
  clickstream, 196
  connecting elements, 85
  defined, 9–10
  digital, 82
  errors in, 195
  hierarchy of, *10*
  improper access to, 195
  in an information system, 25–26
  management of, 61
  manipulation of, 28
  organization of, 27–28
  for a PivotTable, *A82*
  processing of, 24
  resources, 60
  retrieval of, 28
  social networking, 194
  storage of, 24, 27–28
  transformation of, 24
  See also Big Data; Data mining; Databases

Data analytics, 61
Data collection, 24
  on Facebook, 207
  threats, 196
Data integration, 269
Data lakes, 94
Data mining, 280, 338–341, 347
  cluster analysis, 339
  common techniques for, **339**
  decision trees, 323, *325*, 340
  high-level view of, *338*
  market basket analysis, 339–340
Data quality, 46. *See also* Information quality
Data visualization, 115–116, *116. See also*
    Visualization
Data warehouses, 329, 332, 334–335, 347
  characteristics of, 334
  E/T/L process (extract, transform, and load),
    94–95, *95*, 334–335
Database management systems (DBMS), 83
  when to use, 84–88
Database schemas, 88–90
Databases
  advanced concepts, A51–A65
  anomalies in, 87
  behind Facebook, 82
  creating in Access, A24, A32–A33
  defined, 83
  design of, *99*
  diagrams, 88–91
  discussion, 331
  interaction with applications, *84*
  for knowledge, 326
  online, 91–92
  redundancy vs. consistency in, 86–87, **88**
  shared, *272*
  vs. spreadsheets, 84–85
  unsecured, 199
  wish-list database schema, *A24*
  See also Database management systems;
      Relational databases
Data-less keys, A59
D'Aveni, Richard, 73
D'Aveni's 7 Ss, 73, **73**
DBMS (Database management systems), 83
Decision criteria, 108
Decision goals, 108

Decision rooms, 330
Decision support systems (DSSs), 328–330
   communication-driven, 329–330
   data-driven, 329
   document-driven, 329
   model-driven, 329
   using, 329
Decision tables, 323, *324*
Decision trees, 323, *325*, 340
Decision-making, 13, 41–42
   analyzing information for, 101–118
   check that the solution solves the problem, 111
   collaboration systems, 330–332
   decision matrix, 110–111, **110**
   defining the criteria, 108
   determining requirements and goals, 108
   evaluating the alternatives, 111
   goal-seek analysis, 113–114, *115*
   identifying alternatives, 108
   identifying and defining the problem, 107
   importance of good skills, 102–103
   improving, 63
   information and, 103–104
   information and knowledge for, 317–346
   nominal group technique, 108
   paired comparisons, 109–110
   processes for, 106–111
   pros/cons analysis, 109
   selection of techniques/tools, 109–111
   semi-structured, 104
   strategies for, **107**
   "what-if" analysis, 113–114
Decisions
   defined, 102
   structured (programmed), 104
   unstructured (nonprogrammed), 104
   See also Decision-making
Decryption, 177
Decryption key, 177–178
Defense Advanced Research Projects Agency
   (DARPA), 130–131
Defense in depth, 156
Denial-of-service (DOS) threats, 159
   distributed attacks, 165
   threat vectors for, 165–166
Denial-of-service attacks, 165
Design triangle, 223

Detective controls, 168
DevOps (Development-Operations), 232–234,
   233
Dictionary attack, 170
Digital dashboards, 327, **328**
Digital data, 82. *See also* Data
Digital music, 292–293
Digital transformation, 61
Direct changeover, 228
Direct-attached storage (DAS) model, 93
Disaster recovery, 165
Discussion databases, 331
Disintermediation, 301
Distance learning, 133
Distributed denial-of-service attacks, 165
Document coediting, 331–332
Document management systems, 113
Document-driven DSS, 329
Domain name server (DNS), A104
DOS. *See* Denial-of-service (DOS) threats
Dot-com bust, 298
Drawing programs, 33
Drive shredders, 169
DSSs. *See* Decision support systems (DSSs)
DUFL, 283
Dynamic addressing, A106

E-business
   attaining critical mass, 300–301
   bandwidth requirements, 299
   channel compression, 301
   channel expansion, 301–302
   defined, 293
   design for, 302–303
   enablers, 298–301
   global nature of, 310
   impacts of, 301–302
   importance of good skills, 293
   introduction to, 291, 293
   models, 296–298, **298–299**, 311
   real estate websites, *302*
   retail e-commerce, 295
   role of trust, 302–303
   and search engine optimization, 306–309
   terminology, **294**
   trends in, 309–310
   types of, 294, 296, **296**

EbXML (electronic business extensible markup language), 306

Economic feasibility, 225

EDI (Electronic data interchange), 34, 293, 304, 305–306, 311

EDIFACT, 305, 311

EFT (Electronic funds transfer), 293

E-Government Development Index, 309

Electronic business extensible markup language (ebXML), 306

Electronic data interchange (EDI), 34, 293, 304, 305–306, 311

Electronic funds transfer (EFT), 293

Electronic government (e-government), 294, 309–310

Elliot, T.S., 9–10

Email
  anonymity and, 194
  and CRM, 278, 280
  encryption of, 178
  and group collaboration, 331
  as information system, 31, 32
  networks, 301
  phishing, 160–161
  spam, 168, 342
  spoofing, 162
  as tool for knowledge management, 327
  viruses in, 172

Email lists, 331

E-marketplaces, 304

Encryption, 177–179
  asymmetric, 178, *179*
  breaking, 180
  for credit card security, 179–181
  symmetric, 178–179, *179*
  for wireless network security, 182

Encryption key, 177–178, 182

Energy management, 6

Engineering triangle, 223

Enterprise application integration, 270

Enterprise information systems, 265–285
  characteristics of, 270
  components of, 268
  examples of, **271**
  functional perspective, 268–269
  hierarchical perspective, 268, **268**, *269*
  and integration, 269–270

  vs. personal information systems, 267–269
  process perspective, 269
  and supply chain management, 277

Enterprise resource planning (ERP), 267, 270–274, 284
  benefits and disadvantages of ERP systems, 272–274
  shared database, *272*
  top ERP vendors, 274

Enterprise risk management (ERM), A112

Entities
  defined, A52
  relationships among, A53

Entity instance, A52

Entity occurrence, A52

Entity type, A52

Entity-relationship (E-R) modeling, A51

Environmental monitoring, 6

Equifinality, 22

E-R (Entity-relationship) diagrams (ERDs), 88–89, A52–A54, A57
  checking the model, A64–A65
  choosing primary keys, A59–A61
  complete order entry model, *A65*
  creating, A57–A65
  creating a database, A66
  determining cardinalities, A62–A63
  entity list for order entry form, **A59**
  excerpt from user interview, *A62*
  interpreting, A53
  modeling example (order entry form), *A57*
  modeling relationships, A61–A62
  modeling the entities, A57–A59
  order entry diagram, *A64*
  order entry form, *A64*
  order entry form entities and attributes, *A60*
  reading relationships in, A56
  steps in building, *A57*

E-R (entity-relationship) modeling, A51

ERDs. *See* E-R *(*Entity-relationship) diagrams (ERDs)

ERP. *See* Enterprise resource planning (ERP)

ERM (Enterprise risk management), A112

Errors in data, 195

Ethernet cables, 126

Ethernet local area network, *128*

Ethical feasibility, 225

Ethical issues
   information quality, 48
   PAPA framework (privacy, accuracy, property,
      accessibility), 211–212, **212**
   PLUS framework, 213–214
   privacy issues, 212
E/T/L process (extract, transform, and load),
   94–95, *95*, 334–335
European Commission, Data Protection Working
   Group, 206
Evaluating strategic initiatives
   critical success factors (CSFs), 70
   priority matrix, 71–72, *71*
Evolutionary prototyping, 229
Excel (advanced), A81–A97
   Goal-Seek analysis, A94–A95, *A95*
   PivotTables, A81–A86, *A82–A89*, A88–A89
   Scenario Manager, A89–A93, *A90–A94*
Excel (fundamentals), A69–A77
   charts, A74–A75, *A75*
   creating a new spreadsheet, A70, *A70*
   creating a new worksheet, A70, *A71*
   editing a worksheet, A71–A72, *A71*, A74, *A74*,
      A78–A80
   editing and formatting worksheets, A70–A71
   functions and formulas, A72–A74
   printing, A76–A77, *A76*, *A77*
   updating a worksheet, A72, *A72*, A73, *A73*
Expert systems, 5, 328, 342–343. *See also* Artificial
   intelligence (AI)
Explicit knowledge, 320
Extensible markup language (XML), 306, 311
Extreme programming (XP), 233

Facebook, 82
   Big Data at, 94
   privacy issues, 207
Facial recognition, 175, 342
Fair information practices (FIP) principles, 205
Fake news, 40–41
Family Educational Rights and Privacy Act
   (FERPA), 208, 210, 211
FCC (Federal Communications Commission),
   207
Feasibility, 225
Federal Communications Commission (FCC),
   207

Federal Trade Commission (FTC), 197, 198, 206
Feedback, 22, 23, 225, 231, 233
FERPA (Family Educational Rights and Privacy
   Act), 208, 210, 211
Fifth-generation cellular networking (5G), 3–4,
   127
   functions of, 6–7
   and virtual reality, 8
File transfer protocol (FTP), 146
Financial management systems, 34
Fingerprinting scanners, 173
FIP (Fair information practices) principles, 205
Firewalls, 175–178, *175*
   architecture for defense in depth, *177*
   types and terminology, **176**
Fitness trackers, 93
5G. *See* Fifth-generation cellular networking
   (5G)
Flat-rate PPC, 308
Flexibility, 61, 86, 106, 135, 147, 226, 238, 240,
   305, 341
Foreign keys, 86–87, 89, 96
Formjacking, 163
Forward auction, 306
4G. *See* Fourth-generation cellular networking
   (4G)
Fourth-generation cellular networking (4G), 3,
   127
Fraud, 164–165, 201
Fraud detection, 343
FTC (Federal Trade Commission), 197, 198, 206
FTP (File transfer protocol), 146
Full outsourcing, 239
Functional information systems, 34
Functional requirements, 226

Gambler's fallacy, 105
Gannt Chart, A139, *A139*
Gatekeepers, 41
Gateway, A105–A106
GDPR (General Data Protection Regulation),
   208, 209
GDSSs (Group decision support systems), 330
General Data Protection Regulation (GDPR),
   208, 209
Geosynchronous satellites, 126

GLBA (Gramm-Leach-Bliley Financial Services Modernization Act), 208
Global systems, 34
Goal-seek analysis, 113–114, *115*
Government regulations, 201, 206, **209**, 215
Government-to-constituent (G2C), 294, 311
Government-to-government (G2G), 296
Gramm-Leach-Bliley Financial Services Modernization Act of 1999 (GLBA), 208
Group decision support systems (GDSSs), 330
Groupware, 330–332

Hacking, 154, 159
    of big technology companies, 163
    formjacking, 163
    keystroke loggers, 159, 164
    spoofing, 162
    targets of, 162
    web-based attacks, 163
Hacktivism, 164
Handshakes, 179
HANs (Home area networks), **128**, 146
Hardware. *See* Computer hardware
Health care
    and the Internet of Things, 134
    mobile services, 58–59
Health Insurance Portability and Accountability Act (HIPAA), 208
Hickox, Kaci, 40
HIPAA (Health Insurance Portability and Accountability Act), 208
Home area networks (HANs), **128**, 146
Horizontal-market software, A18
Hotspots, 124, 126–127
HTTP (Hyper-text transfer protocol), 132
Human resources, 60, 69
    uses of AI in, 343
Hypercompetition, 72–74
    D'Aveni's 7 Ss, 73, **73**
Hyper-text transfer protocol (HTTP), 132

IaaS (Infrastructure as a Service), 140
IANA (Internet Assigned Numbers Authority), 130
IDEF1X, 89
Identity theft, 198–199
    black market for, 202
    and fraud type, 201

Immediate changeover, 228
Imperatives, 71
Improper access to data, 195
Inbound transformation, 305
Information
    for business decision-making, 101–118, 317–346
    characteristics of, **319**
    and decision-making, 103–104
    defined, 9–10
    for electronic business, 291–311
    and emerging technologies, 2–6
    evaluating sources of, 49
    filtering, 44
    gaining strategic value from, 57–75
    hierarchy of, *10*
    how businesses use, 12, 13–14
    how information systems help us deal with, 27–30
    how to use, 11
    as a product, 13
    promising careers, 14–15
    rapid growth of, 3
    resources, 60
    as second-order change, 260
    storing and organizing, 81–96
    transmitting, 123–147
    unauthorized secondary use of, 197–198
    uses of, 12–14
    value of, 1–2
    withdrawal from, 44
Information analysis, careers in, 14–15
Information evaluation, 39–52
    being a smart information consumer, 39, 41–42
    believability, 50–51
    comprehensiveness, 51
    currency, 50
    defined, 42
    evaluating sources, 49, 51
    framework for, *49*
    and information overload, 43–45
    objectivity, 51
    relevance, 50
    usefulness, 50
    See also Information quality
Information flow, 31
Information literacy, 11–14
Information overload, 43–45, 47, 49, 52

Information privacy
  concern for, 197
  government regulations, 208–209, **209**
  technologies and solutions for, 200–203
  threats to, 195
  See also Privacy issues
Information privacy organizations, A118, **A119**,
    A120
Information processing cycle, *24*
Information quality, 45–48
  accessibility, 46
  business impacts of, 43
  contextual, 46
  cost of, 46–47
  dimensions of, 45–47
  as ethical issue, 48
  intrinsic, 46
  poor-quality information, 42, 47
  representational, 46
  See also Information evaluation
Information retrieval and analysis
  reporting tools, 111
  SAP business objectives screen, *112*
  tools for, 111–116
Information security, 153–154
  access levels, *157*
  concepts, 156
  costs of data breaches, 167
  technologies and solutions, 168–183
  See also Information systems security; Security
Information systems (IS), 10, 13
  activity-based costing (ABC), A130
  careers in, 14–15
  and change, 30
  and data manipulation, 28
  and data organization, 27–28
  and data retrieval, 28
  and data storage, 27–28
  dealing with information, 27–30
  and on-demand service, 29
  development of, 221–244
  elements of, *25*
  foundations of, 23–27
  functional and management information
    systems, 34
  funding methods (allocation), A129
  funding methods (chargeback), A129

funding methods (overhead), A129
global systems, 34
Gorry and Scott-Morton's framework for, **106**
identifying, 21
integrated enterprise systems, 34
interorganizational systems, 34
levels of, *33*
managers, 19–21, 61
manual, 23
and organizational change, 30–33
outsourcing, 239–243
personal applications, 33
point-of-sale, 27
resources, 60
stand-alone systems, 34–35
strategic planning process for, 59–63, *59*
time, cost, and quality, 223
total cost of ownership (TCO), 236–237,
    A127, A128, A130
transaction processing systems (TPSs), 34
types of costs, A129–A130
vision, 61–62
See also Enterprise information systems;
    Information systems security
Information systems architecture, 62
Information systems assessment, 60
Information systems management, 19–21
Information systems projects, A133–A142
  classification categories, **A135**
  costs of changes and risks over time, *A138*
  IS project management life cycle, A136, A142
  network diagram, *A140*
  portfolio management, A134–A136
  project closing, A142
  project definition, A136–A137
  project execution, A141
  project planning, A137–A139, A141
  projects, programs, and portfolios, A133–A134
  responsibility matrix, *A137*
  scope statement, A137
  trade-offs in project management, A135–
    A136, *A136*
Information systems security
  jobs and certifications, A116, A118, **A118**,
    A121
  organizations, A116, **A117**, A121
  plan, A111–A116, *A112*, A120–A121

protocols for, A121
standards, **A116**
See also Information security; Security
Information technology (IT), 2, 10
activity-based costing (ABC), A127, A128
alignment with business, 61
failures, 230
funding methods (allocation), **A126**, A127
funding methods (chargeback), A126–A127, **A126**
funding methods (overhead), A126, **A126**
identifying system costs, A127–A129
infrastructure, 240, 266
levels of change, 259–260
managers, 47, 275
total cost of ownership (TCO), A127, A128
types of costs, A127–A128
See also Technologies
Information-based business, 68–69
Infrastructure
for BPaaS, 141
and e-business, 311
and e-government, 309
of Facebook, 82
health care, 58
of the Internet, 131
IT, 240, 266
networking, 135
open source software, 131
services, 297
technical, 225
Infrastructure-as-a-Service (IaaS), 140
Innovation(s), 61, 73, 145, 266–267, 320
Input, 24
Integrated enterprise systems, 34
Integrated services digital network (ISDN), 127
Integration, and enterprise systems, 269–270
Integrity, 158–159, 161, 166, 168, 183
Intelligent agents, 5. *See also* artificial intelligence (AI)
Intelligent connectivity, 6
Intelligent transportation systems, 6. *See also* Autonomous vehicles
Intended audience, 307
Interactive voice response (IVR), 280
Intermediaries, 301–302
Internet, 129–135

applications, 131–132, 132, 133
and e-business, 297
vs. intranet, 134
number of hosts connected to, *131*
users of, 130
and the Web, *132*
See also Internet of Things (IoT)
Internet Assigned Numbers Authority (IANA), 130
Internet backbones, 130
Internet of Things (IoT), 3–5, 134–135, 211
and artificial intelligence, 342
and cloud computing, 139
and decision-making, 116
functions of, 6–7
privacy issues, 197, 211
top application areas, 5
and virtual reality, 8
vs. Web 3.0, 146
See also Internet
Internet Protocol (IP), 130, 134, A103
IP address, A104
Internet Protocol Security (IPSec), A115, A121
Internet telephony, 146
Internet voting, 310
Internet2 network, 134
Interoperability, 62, 131, 135, 136, 138, 47, 300, 311
Interorganizational systems, 34
Intranet, 134
IoT. *See* Internet of Things (IoT)
IP address, A104
IPods, 292–293
IPSec (Internet Protocol Security), A115, A121
IPv4, 130, A106, A107
IPv6, 130, A105, A106, A107
Iris scanning, 173
ISDN (Integrated services digital network), 127
ITunes Match, 90
IVR (Interactive voice response), 280

Kernels, A13
Key pairs, 178
Keys
composite, 86, A53
data-less, A59
decryption, 177–179

encryption, 177–179, 182
foreign, 86–87, 89, 96
primary, 86, 88–89, 96
public, 178, 179
session, 180
Keystroke logger, 159, 164
Keyword bidding, 308, **309**, 311
keywords, 307–308, 311
Knowledge
for business decision-making, 317–346
capturing and codifying, 323
characteristics of, **319**
combination of, 322
creation of, 322–323, *322*
defined, 9–10
explicit, 320, 346
externalization of, 322
externally focused, 327
hierarchy of, *10*
informal internal, 327
internalization of, 322
issue of trust in, 325
purchasing, 323
SECI model of knowledge creation, *322*
socialization of, 322
storing and retrieving, 325–326
structured internal, 327
tacit, 320–321, 346
transferring and applying, 326
types of, 320–321
See also Knowledge management
Knowledge management, 318–319, 346
benefits of, 320
content-focused vs. connection-focused, 327
cycle, *321*
importance of, 319–321
main processes for, 321–326
technologies, 326–328
by texting, 318
wikis for, 326
See also Knowledge
Knowledge repositories, 327

Language translation, 342
LANs (Local area networks), 127–128, **128**, *128*, 146
LBS (Location-based services), 296
Legacy systems, 256

Legal feasibility, 225
LePage, Paul, 40
Local area networks (LANs), 127–128, **128**, *128*, 146
Location-based services (LBS), 296
Logical access controls, 168–169
Logical system design, 227
Loss aversion, 105
Lyft, 297

Machine learning, 5, 343–345, 347
four steps of, 344
See also Artificial intelligence (AI)
Malware, 166
Management
content, 270
database, 83–88
document, 113
energy, 6
enterprise risk (ERM), A112
financial, 34
information systems, 19–21
memory, A13
risk, 155, 185
security, A114
supply chain, 278
top concerns of, 61
See also Customer relationship management (CRM); Knowledge management
Management information systems, 34
Managerial control, 104
Managers, 30, 42, 59–60, 62, 65, 66–72, 74–75, 111, 113, 135, 168, 195, 234, 256, 259, 325, 332
business, 43–44, 275
cookie, 201–203, 215
front line, 268
functional, 61
information systems, 19–21, 61
information technology, 47, 275
making bad decisions, 105
middle, 32, 268
password, 171–172
personal information, 33
reports for, 50
senior, 61, 103
and strategic planning, 100
Manufacturing control, 343

Market basket analysis, 339–340
Markup language, 306
Mashups, 143, 147
Mason, Richard, 48
Mass collaboration, 143
Materials requirements planning (MRP) systems, 34
Maximum cardinality, A54, A56
Medical diagnosis, 343
Memory management, A13
Mendix, 231, 261
Mentoring database design, *99*
MHealth, 58–59
Microprocessors (chips), A5
Middle managers, 32, 268. *See also* Managers
Minicomputers, A4
Minimum cardinality, A54, A56
Misinformation, 40–41
Mixed reality, 8, *8*
Mobile devices
    and app usage, A20
    and cellular networks, 126
    location-based services (LBS), 296
    and mobile wi-fi hotspots, 124
    Open Android vs. Closed IOS, 238
    privacy issues, 196, 210–211
    use by managers, 59
    used for health care services, 58–59
    and wireless architecture, 137–138
Mobile wi-fi, 124
Model-driven DSS, 329
Moore's law, A4
Mortgage process, 260
MRP (Materials requirements planning) systems, 34
Multitiered architecture, 83
Music industry, 292–293

N-ary relationships, A53
NAS (Network-attached storage) model, 93
Natural language processing (NLP), 5, 342. *See also* Artificial intelligence (AI)
Near Field Communication (NFC), 181
Nearshore outsourcing, 243
Negativity bias, 105
Network adapter, 125
Network architectures, 147

Network economics, 300
Network effects, 300
Network interface card (NIC), 125, 146
Network-attached storage (NAS) model, 93
Networking
    domain name server (DNS), A104
    dynamic addressing, A106
    gateway, A105–A106
    IP address, A104
    IPv4 vs. IPv6, A106
    models and protocols, A99–A106
    network diagram, *A140*
    Open Systems Interconnection (OSI) model, A100–A102, *A101*, A106–A107
    social, 145
    subnet mask, A105
    Transmission Control Protocol/Internet Protocol (TCP/IP), A107
Networking architectures, 135–142
    architectural principles, 135–136
    business-process-as-a-service (BPaaS), 141–142
    client/server architecture, 136
    cloud computing, 139–142
    infrastructure as a service (IaaS), 140
    integrated services digital (ISDN), 127
    IPv4 vs. IPv6, A107
    peer-to-peer architecture, 137
    platform-as-a-service (PaaS), 140
    service-oriented architecture, 138–139
    software-as-a-service (SaaS), 141
    virtualization, 142
    wireless architecture, 137–138
Networking infrastructures, 135
Networks
    backbone (BBNs), **128**, 146
    broadband, 127, 129
    cellular, 126
    components and characteristics of, 123, 125
    defined, 123
    home area networks (HANs), **128**, 146
    hubs, 128
    local area networks (LANs), 127–128, *128*, **128**, 146
    network coverage, 126–129
    neural, 340
    personal area networks (PANs), **128**, 146

Transmission Control Protocol/Internet Protocol (TCP/IP), A102–A104, *A103*
types of, 126–129, **128**
types of cables, 126, *126*
virtual private networks (VPNs), 146, 179
wide area networks (WANs), 127–129, **128**, 146
Wi-Fi, 126–127
wired vs. wireless, 126–127
wireless, 126–127, *127*
Neural networks, 340
Neutral auction, 306
NFC (Near Field Communication), 181
NIC (Network interface card), 125, 146
NLP (Natural language processing), 342
Nominal group technique, 108
Nonfunctional requirements, 226
Nonrepudiation, 158
Normal forms, 87
NoSQL, 95
Note-taking systems, 33

Office automation systems, 33
Offshore outsourcing, 243
OLTP (Online transaction processing) systems, 34
On-demand CRM, 282. *See also* Software-as-a-Service (SaaS)
On-demand service, 29
One-to-many matching, 174
One-to-one matching, 173–174
Online databases, 91–92
Online mortgage process, 260
Online piracy, A120
Online transaction processing (OLTP) systems, 34
On-premise CRM, 282
Onshore outsourcing, 243
Open ports, 162
Open Source Initiative, 237
Open systems, 22
Open Systems Interconnection (OSI) model, A100–102, *A101*, A106–A107
application layer, A101
data link layer, A102
network layer, A102
physical layer, A102

presentation layer, A101–A102
session layer, A102
transport layer, A102
Operating systems, 26, A13–A14, A20
selection of, A14
Operational control, 104
Opting in/opting out, 197–198
Order process, 13, *13*
Organic links, 307, *308*
Organizational change, 30–33
Organizational culture, 321
Organizational feasibility, 225
Organizational processes, 249
Organizational structure, flattening, 32
Outbound transformation, 304
Output, 24
Outsourcing, 239–243
application, 240
benefits and risks of, 240–242, **242**
full or complete, 239
geographic considerations, 243
making the decision, 242
nearshore, 243
offshore, 243
onshore, 243
personnel, 239–240
process-based, 239
project-based, 240
reversing the decision, 241
selective or partial, 239
what to outsource, 239

Packets, 162–163
Paging, A13
PANs (Personal area networks), **128**, 146
PAPA framework (privacy, accuracy, property, accessibility), 211, 212, **212**
Parallel operation, 228
Partial least squares, 115
Partial outsourcing, 239
Password crackers, 170
Password managers, 170–172, **171**
Passwords
strength of, 170
tools for managing, 170–172
worst passwords (2020), **171**
Pay-per-click (PPC), 308

Payroll systems, 57
Peer-to-peer architecture, 137
People, and information systems, 25, 27
Personal area networks (PANs), **128**, 146
Personal calendars, 33
Personal information managers, 33
Personal information systems, 267–269
Personnel outsourcing, 239–240
Phased implementation, 228
Phishing, 159–163
  detecting, 160
  example, *160*
Physical access controls, 168–169
Physiological biometrics, 173
Pilot operation, 228
Piracy, online, A120
PivotTables, 341, *341*, A81–A86, *A82–A89*,
  A88–A89
Plaintext, 177
Platform-as-a-service (PaaS), 139–140
PLUS framework, 213–214
PMI (Project Management Institute), A134
Polymorphic viruses, 166
Porter, Michael, 65
Porter's Five Competitive Forces Model, 65–66,
  65, **66–67**
Porter's Value Chain Analysis, 66–68
Portfolios
  defined, A134
  management of, A134–A136
PPC (Pay-per-click), 308
Primary keys, **A60**, 86, 88–89, 96, A52–A53,
  A59–A61
Principle of least privilege, 169
Priority matrix, 71–72, *71*
  evaluate quadrant, 72
  imperatives quadrant, 71
  quick wins quadrant, 71–72
  stay away quadrant, 72
Privacy issues, 61, 193–216, A118–A121
  breaches and company reputations, 200
  consequences of privacy violations, 198–200
  data collection, 195–198
  and ethics, 212–215
  and Facebook, 207
  government regulations, 208–209
  information privacy organizations, A118,
    A119, A120

  information privacy threats, 195
  and the Internet of Things, 197, 211
  mobile information privacy, 210–211
  online piracy, A120
  perceptions of monitoring, 199
  privacy defined, 195
  and security, 215
  smartphones, 196
  technologies and solutions for information
    privacy, 200–203
Privacy Pizza video, 198
Privacy policies, 204–205, 211
  creation of, 205
  and fair information practices (FIP) principles,
    205
  and privacy seals, 206, 208
  self-regulation, 205–206
Privacy seals, 206, 208
Private market, 306
Procedures
 and information systems, 25, 27
 security, 182–183
Process modeling, 252–254
  companies and software products, **255**
  of a process to acquire materials, *253*
Process modeling tools, 254
  for business processes, **255**
Process-based outsourcing, 239
Processes
  defined, 249
  example (to acquire materials), *251*
  improving, 30, 250
  organizational, 321
  steps involved in, 251
  and technology, 254, 256
  See also Process modeling; Process modeling
    tools
Processing, 24
Profitability, 267, 336
Programs, A134
Project Management Institute (PMI), A134
Project Management Professional certification,
  A134
Project triangle, 223
Project-based outsourcing, 240
Projection, A36
Prototyping, 229, 231
Public keys, 178, 179

Public spaces, management of, 6
Purchase order form, *99*

QBE. *See* Query by Example (QBE)
Query by Example (QBE), A37, A39–A42
    completed query, *A40*
    create ribbon and show table dialog box, *A38*
    grid for multitable query, A41
    grid with fields and sorting, *A38*
    query results, *A39*, A41
    query with missing table, *A42*
Quick wins, 71
Quicken Loans, 260

RAID (Redundant array of independent disks),
    A7
Random-access memory (RAM), A7, A13
Ransomware, 154, 166–167
Rapid application development (RAD), 231–232,
    232
Read-only memory (ROM), A7
Recaptcha system, 343–344
Recommendation systems, 342
Recursive (unary) relationship, A53
Redundancy, A7, 86, 87, **88**, 272
Redundant array of independent disks (RAID),
    A7
Registration process, 20
Regression analysis, 340
Reinforcement learning, 344–345
Relational databases, 81, 85–88
    cross-referencing (relationships), 86
    diagrams, 88–90
    fields, 85
    instructors table, 85–86, **85**
    linking or intersection tables, 88
    many-to-many relationships, 88, *89*
    one-to-many relationships, 87
    one-to-one relationships, 88
    overview, 83
    records, 85
    small (example), *87*
    See also Databases
Relationships, A53
    and cardinality, A54
    degrees of, A53
    in an ERD, A56

many-to-many, A54
modeling example (order entry form),
    A61–A62
one-to-many, A54
one-to-one, A54
Remote presentation software, 331
Repeat business, 283
Repeaters, 125
Reputation seals, 206
Requirements, 108
Requirements elicitation, 226
Reverse auction, 306
RFID (radio-frequency identification) tags, 266
Risk management, 155, 185
Robotics, 284, 345–346
Robots, 318, 345–346
    advantages of, 346
    agricultural, 345
    industrial, 345
ROM (Read-only memory), A7
Rootkits, 163
Routers, 125, **125**

SaaS (Software-as-a-Service), 139, 141, *141*, 240,
    282
Sales force automation systems, 34
Scalability, 136
Scenario Manager, *114*, A89–A93, *A90–A94*
Scope statement, A137
Seal programs, 206
Search engine optimization (SEO), 306–309, 311
SECI model of knowledge creation, *322*
Secure HTTP (S-HTTP), A115, A121
Security, 61, A111–A118, A120–A121
    antivirus software, 168
    audit logs, 168
    credit cards, 179–181
    drive shredders, 169
    encryption, 177–179
    firewalls, 175–178
    goals, 158
    Internet of Things (IoT), 182
    levels of, 157–158
    outsourcing, A114, A121
    physical vs. logical, 168–169
    and privacy, 215
    protecting home wireless networks, 181

protocols for, A114–A115, **A115**
and risk management, 155
user profiles, 169–172
See also Information security; Security threats
Security certificate, *180*
Security controls, 184
  corrective, 168
  detective, 168
  preventive, 168
Security management, A114
Security policies, 182–185
  violation of, 178
Security procedures, 182–183
Security seals, 206
Security solutions, 184
Security threats, 153, 155, 158–168, 172, 176, 182, 183–184
  consequences of unauthorized access, 163–164
  denial-of-service (DOS), 159, 165–166
  hacking, 159
  malware, 166–167
  passive vs. active unauthorized access, 161–162, *161*
  phishing, 159–163
  social engineering, 159–163
  spamming, 168
  theft and fraud, 164–165
  two-factor authentication (2FA), 172
  unauthorized access, 159–160
  virus hoaxes, 168
  virus protection, 172–173
  viruses, 166, 167
Selection, A36
Selective outsourcing, 239
Self-driving vehicles, 7, 125, 342
Self-regulation, 205–206
Seller-oriented auction, 306
Sensor technologies, 7
SEO (Search engine optimization), 306–309, 311
Servers, 136
Service-oriented architecture (SOA), 138–139, *138*
Session key, 180
Shared whiteboards, 330
S-HTTP (Secure HTTP), A115, A121
SIM (Society for Information Management), 61
Slack, 331, 332

Smart cities, 6
Smart homes, 4–5, 134
Smartphones. *See* Mobile devices
SNA (Systems Network Architecture), A100
Sniffers, 162
SOA (Service-oriented architecture), 138–139, *138*
Social engineering, 159, 161
Social networking, 145
Society for Information Management (SIM), 61
Software. *See* Computer software
Software development methodologies, 223–224
Software security hole, 162, 165
Software-as-a-Service (SaaS), 139, 141, *141*, 240, 282
Solid-state drives (SSDs), A7
Source code, 237
Spam detection, 342
Spamming, 168
Spiral Model, 233
Sponsored links, 307, *308*
Spoofing, 162
Spreadsheets, 84–85
  for calculating grades, 113
  cost of errors, 117
  example, *114*
  pivot tables, 341, *341*
  showing formulas, *115*
  used for decision-making, 102, 113–115
  See also Excel (advanced); Excel (fundamentals)
Spyware, 166–167
SQL (Structured Query Language), 95, A37
SSDs (Solid-state drives), A7
SSL (Secure Socket Layer), A115, A121
Stand-alone systems, 34–35
Statistical analysis, 115, 340
Stealth virus, 166
Storage of data, 24, 27–28
Strategic business planning, 60
Strategic information systems, 57, 59
  evaluating strategic initiatives, 69–72
  frameworks for, 63–69
  identification of initiatives, 62–63
Strategic planning process, 59–63
  advantages of, 63
  and decision-making, 104
  for information systems, *59*

information systems architecture, 62
information systems assessment, 60
information systems vision, 61–62
strategic business planning, 60
strategic initiatives identification, 62–63
Strong, Diane, 46
Structured Query Language (SQL), 95, A37
Subnet mask, A105
Subprocesses, 251–252, *252*, *254*
Subsystems, 22
Supercomputers, A4
Supervised learning, 344
Supply chain, 274–275, *274*
    disaster examples, **276**
    impact on company performance, 277
Supply chain management (SCM), 265–267, 270,
    274–278, 284, 304
    and blockchain, 278
    and enterprise information systems, 277
Surveillance societies, 204
Sustainability, 32, 72–73, 267, 313
SWOT (strengths, weaknesses, opportunities, and
    threats) analysis, 63–65, *64*, 74
Symmetric encryption, 178–179, *179*
SYN flood attack, 165
System migration, 228
System shall statements, 226
System software, A11–A17, A20
    operating systems, A13–A14
    utility software, A14–A17
    See also Computer software
Systems
    business information, 23
    components of, 22
    defined, 21–23
    open, 22
    and their environments, *22*
    See also Information systems
Systems development
    alternative methodologies, 229, 231–234
    build or buy decision, 234–235
    commercial, off-the-shelf software (COTS),
        234–237
    design phase, 227
    developing by modeling, 231
    development phase, 227

DevOps (Development-Operations), 232–234,
    233
    functional requirements, 226
    hybrid solutions, 235
    implementation phase, 227–228
    life cycle, 224–225
    maintenance phase, 228
    nonfunctional requirements, 226
    planning phase, 225
    prototyping, 229, 231
    rapid application development (RAD),
        231–232
    requirements elicitation, 226
    requirements phase, 225–226
    Spiral Model, 233
    use case diagram example, *226*
    waterfall method, 224–225
Systems development life cycle (SDLC),
    224–225, *224*
    advantages and disadvantages of, 229
Systems integration, 270
Systems Network Architecture (SNA), A100

Tacit knowledge, 320–321
Targets of opportunity, 166
TCP/IP (Transmission Control Protocol/Internet
    Protocol), A100, A107
Technical feasibility, 225
Technical resources, 60
Technologies
    3D devices, A6
    diffusion of innovations timeline, 145
    emerging, 2–6
    enabling e-business, 299–300
    for knowledge management, 326–328
    and processes, 254, 256
    for secondary storage, A7
    sensor, 7
    solutions for information privacy, 200–203
    technology-free life, 4
    touch-screen, A6
    Web 2.0, 74
    See also Computer hardware; Information
        technology
Ternary relationship, A53
Text analysis, 340
Theft, 164–165

Threats
  data collection, 196
  denial-of-service (DOS), 159, 165–166
  external, 155
  information privacy, 195–197, 200, 210, 215
  internal, 155
  in SWOT analysis, 63–65, *64*, 74
  types and causes, *156*
  vectors of, 165–166
  See also Security threats
3D devices, A6
Time series analysis, 340
T-lines, 127
TLS (Transport Layer Security), A115, A121
To-do list systems, 33
Total cost of ownership (TCO), 236–237, A127, A128, A130
Traffic control systems, 4, 6
Transaction processing systems (TPSs), 34, 105, 224
Transformation
  inbound, 305
  outbound, 304
  as third-order change, 260, 273
Transmission Control Protocol/Internet Protocol (TCP/IP), A100, A107
Transport Layer Security (TLS), A115, A121
Trojan horses, 166
Trust, 51, 167, 206, 208
  culture of, 321
  in e-business, 302–303
  in knowledge management, 325
  and leadership, 214
Turing, Alan, 5
Two-factor authentication (2FA), 172

Uber Technologies, Inc., 29, 297
Unary (recursive) relationship, A53
Unauthorized access, *161*
Unauthorized secondary use of information, 197–198
Unified Modeling Language, 226
University Corporation for Advanced Internet Development (UCAID), 134
Unsupervised learning, 344
UPC (universal produce code), 27
Use case, 226, *226*

User profiles, 169–172
  on Facebook, 207
  possession, knowledge, and traits, 169
Utility software, A14–A17, A20
  antimalware, A16
  data backup, A15
  file compression and archiving, A15
  file synchronization, A16
  finding, A17
  system migration, A16
  system repair, A16

Value
  in plenitude, 300
  in scarcity, 300
Value chain analysis, 66–68
Value chains
  sample in manufacturing firm, *68*
  virtual, 68–69, *69*
Vertical-market software, A18
Videoconferencing systems, 34, 330
Virtual assistants, 342
Virtual machines, 142
Virtual private networks (VPNs), 146, 179
Virtual reality (VR), 8–9, *8*
Virtual teams, 32
Virtual value chains, 68–69
  handling information, *69*
Virtualization, 142, *142*
Virus hoaxes, 168
Virus signature, 166
Viruses, 166
  protection against, 172–173
Visualizations, 327, 335–338, 347
  and business analytics, 335–338
  data, 115–116, *116*
  effective, 337
  explaining profitability problems through, 336
  software for, 337
Voice over IP (VOIP), 146
VPNs (Virtual private networks), 146, 179
VR (Virtual reality), 8–9, *8*

Walmart, 266–267
Wang, Richard, 46
WANs (Wide area networks), 127–129, **128**, 146
WAPs (Wireless access points), 126–127

Warehouse automation, 284
Warehouse management systems (WMS), 270
Waterfall method, 224–225
Web 2.0, 143, 145, 147
    popular technologies, *144*
Web 3.0, 145–146
Web design, 307
WEP (Wired Equivalent Privacy), A115, A121,
    182
"What-if" analysis, 113–114
Whiteboards, 330
Wide area networks (WANs), 127–129, **128**, 146
Wi-Fi, 126–127
 mobile, 124
Wi-Fi Protected Access (WPA), A115, A121,
    182
Wikinomics, 143
Wikipedia, 49

Wired Equivalent Privacy (WEP), A115, A121,
    182
Wireless access points (WAPs), 126–127
Wireless architecture, 137–138
Wireless networks, 126–127, *127*
Wisdom
    defined, 9–10
    hierarchy of, *10*
WMS (Warehouse management systems), 270
Work breakdown structure (WBS), A138, *A139*
Workflow systems, 331
World Wide Web (the Web), 131
Worms, 166
WPA (Wi-Fi Protected Access), A115, A121,
    182
WPA2, 181–182, 184, A115, A121
XML (extensible markup language), 306, 311
XP (Extreme programming), 233